Lecture Notes in Statistics

Springer
New York
Berlin
Heidelberg
Barcelona
Hong Kong
London
Milan
Paris
Singapore
Tokyo

Anne Boomsma
Marijtje A.J. van Duijn
Tom A.B. Snijders
Editors

Essays on Item Response Theory

Springer

Anne Boomsma[1]
Marijtje A.J. van Duijn[2]
Tom A.B. Snijders[3]
Department of Statistics and
Measurement Theory
University of Groningen
Grote Kruisstraat 2/1
9712 TS Groningen
The Netherlands

[1]E-mail: a.boomsma@ppsw.rug.nl
[2]E-mail: m.a.j.van.duijn@ppsw.rug.nl
[3]E-mail: t.a.b.snijders@ppsw.rug.nl

Library of Congress Cataloging-in-Publication Data

Essays on item response theory / editors Anne Boomsma, Marijtje A.J. van Duijn,
Tom A.B. Snijders.
p. cm.— (Lecture notes in statistics; 157)
Includes bibliographical references and index.
ISBN 0-387-95147-4 (softcover: alk. paper)
1. Item response theory. 2. Psychometrics. I. Boomsma, Anne.
II. Van Duijn, Marijtje A.J. III. Snijders, T.A.B. IV. Lecture notes in
statistics (Springer-Verlag); v. 157.

BF39 .E717 2000
150'.28'7—dc21 00-047081

Printed on acid-free paper.

Camera-ready copy provided by the editors.
Printed and bound by Sheridan Books, Ann Arbor, MI.
Printed in the United States of America.

9 8 7 6 5 4 3 2 1

ISBN 0-387-95147-4 SPIN 10781894

Springer-Verlag New York Berlin Heidelberg
A member of BertelsmannSpringer Science+Business Media GmbH

To Ivo W. Molenaar

Preface

Over the past three decades, since the publishing of Lord and Novick's *Statistical Theories of Mental Test Scores* in 1968 and Fischer's *Einführung in die Theorie psychologischer Tests* in 1974, item response theory (IRT) has developed rapidly. This is demonstrated in the *Handbook of Modern Item Response Theory* (Van der Linden & Hambleton, 1997) with chapters on a wide range of topics in IRT. The study of individual responses to behavioral stimuli has clearly evolved into a major discipline of psychometric theory.

The Dutch statistician and psychometrician Ivo Molenaar has played an important role in the growth of IRT, not only in the Netherlands but worldwide. His contributions to item response modeling cover a broad area: item and person fit in both parametric and nonparametric models, for example, and polytomous nonparametric item response models including the development of the MSP program. With Gerhard Fischer he edited a book on Rasch models that was published in 1995, and in cooperation with Klaas Sijtsma he is now preparing a textbook on nonparametric item response theory. In the Netherlands in particular, he has encouraged researchers and doctoral students to advance into new areas of IRT. To honor such achievements we dedicate this volume on item response modeling to Ivo W. Molenaar.

Ivo's general attitude towards psychometrics is perhaps best characterized by the definition he gave at the European Meeting of the Psychometric Society in Lüneburg in 1999, when introducing his successor as president of that society: "Psychometrics is mathematical statistics in the service of substantive psychology." This statement reflects Ivo Molenaar's scientific efforts: after the completion of a Ph.D. in mathematical statistics and obtaining a chair in statistical analysis and measurement theory at the University of Groningen, The Netherlands, in 1971, he showed an increasing professional interest in psychometrics. In that sense Ivo's career resembles that of Georg Rasch (1901–1980) described by Andersen and Olsen in Chapter 1 of this volume. In his inaugural speech as a full professor (Molenaar, 1972), he explicitly referred to the work of Georg Rasch and Robert Mokken, exponents of parametric and nonparametric IRT, respectively. From then on he gradually used his skills in mathematics and statistics in the service of item response theory. For Ivo it was a challenge to study and further develop both classes of models with their intriguing statistical and practical problems. And in the end he became an aficionado of modern item response theory, one of his greatest scientific endeavors.

This book deals with several aspects of item response modeling. We arranged the contributions in two parts: parametric and nonparametric IRT. Within each part, the chapters are roughly ordered as (1) historic accounts and overviews, (2) new models, (3) new methods, and (4) applications and miscellaneous topics. It should be noted that some chapters, such as those of Junker and of Mellenbergh, cover both parametric and nonparametric topics.

This book honors Ivo Molenaar on the occasion of his retirement from the Department of Statistics and Measurement Theory at the University of Groningen as of September 2000. In that small department, which he guided, extended, and defended for almost 30 years, many students and colleagues were fortunate to benefit from his outstanding didactic qualities in teaching and consultation. His work is perhaps best characterized by his practical approach to the use of statistics in social science applications. The invited contributors to this volume are researchers who have cooperated more or less closely with Ivo in the area of IRT, including former doctoral students. During the past 30 years Ivo Molenaar has been a memorable actor on the psychometric stage.

This is a book that should be of special interest to psychometricians, empirical researchers, and graduate students. It gives an impression of the present state of the art of IRT models and methods. It may be seen as a complementary successor to the *Handbook of Modern Item Response Theory* (Van der Linden & Hambleton, 1997), and we hope its contents will set a direction for future developments in IRT.

We thank the reviewers for their expertise and more than useful comments on the chapters of this book. Finally, we appreciate the patience of the authors while trying to cope with our frequently demanding requests for major and minor revisions.

Groningen, September 2000

Anne Boomsma
Marijtje A.J. van Duijn
Tom A.B. Snijders

References

Fischer, G. (1974). *Einführung in die Theorie psychologischer Tests: Grundlagen und Anwendungen.* Bern: Huber.

Fischer, G.H., & Molenaar, I.W. (Eds.). (1995). *Rasch models: Foundations, recent developments, and applications.* New York: Springer-Verlag.

Lord, F.M., & Novick, M.R. (1968). *Statistical theories of mental test scores.* Reading, MA: Addison-Wesley.

Molenaar, W. (1972). *Dit is een uitdaging* [This is a challenge]. Oratie [Inaugural lecture], Rijksuniversiteit Groningen.

Van der Linden, W.J., & Hambleton, R.K. (Eds.). (1997). *Handbook of modern item response theory.* New York: Springer-Verlag.

Contents

II Nonparametric Item Response Theory

Part I

Parametric Item Response Theory

1 The Life of Georg Rasch as a Mathematician and as a Statistician

Erling B. Andersen and Lina Wøhlk Olsen[1]

ABSTRACT In this chapter an account of the life of Georg Rasch as a mathematician and as a statistician is given, with emphasis on the years 1952 to 1960 when the Rasch model was developed and justified by requirements of proper measurement. But Rasch's life before 1952 is also described, especially how he, forced by the employment situation in Denmark, had to turn to statistics after having graduated and defended his doctoral dissertation in pure mathematics. Finally, it is discussed at some length how Rasch, especially after 1960, formulated his requirements of proper measurement into the formalized concept of specific objectivity.

1 Introduction

This is an account of the life of Georg Rasch as a mathematician and statistician. Our main interest is what happened between 1952 and 1960, a short span of years relatively late in Rasch's life, where he managed to formulate four new models and justify them by requirements of proper measurement.

There is an earlier account, similar to ours, namely, the foreword and afterword by Benjamin Wright in the republication of Georg Rasch's 1960 book, *Probabilistic Models for Some Intelligence and Attainment Tests*, by the University of Chicago Press in 1980. The main source for Wright was a series of interviews with Rasch carried out by David Andrich in the summer of 1979 at the Island of Læsø, where Rasch lived. These interviews, which we call the Andrich-interviews, are also a main source for our account. They have been compared with interviews of people who knew Rasch personally (carried out by L.W.O.) and with several written sources including Rasch's publications about his models, some reports preliminary to Rasch's discov-

[1]Københavns Universitet, Økonomisk Institut, Studiestræde 6, 1455 København K., Danmark; e-mail: *erling.b.andersen@econ.ku.dk, lina@econ.ku.dk*

References in the text to the senior author, Erling B. Andersen, are (E.B.A.) while references to the junior author, Lina Wøhlk Olsen, are (L.W.O.).

FIGURE 1.1. Georg Rasch. [Pencil drawing by E.B. Andersen.]

ery of the Rasch model, and some accounts of seminars, notes, letters, and so on. Fortunately, Rasch (1977) contains an account of some of the main events that led him to derive the Rasch model. Not all questions can be answered completely from the available sources. So, although we have tried to be as faithful to the sources as possible, we occasionally make interpretations or guesses. We believe them to be close to the truth, but we can of course never be sure.

2 Early Life (1901–1945)

Georg Rasch was born in 1901 as the youngest of three brothers. His father ran a missionary school for sailors in the city of Svendborg on Funen, the middle island in Denmark. At an early age, Rasch showed an obvious talent for mathematics. This was greatly encouraged by his father, who was educated as a ships' officer and well acquainted with the art of navigation. In high school Rasch continued to show extraordinary abilities in mathematics, and in 1919 he decided to study mathematics at the University of Copenhagen. To support his studies he worked as an assistant to one of the mathematics professors, Niels Nielsen. After only three years of study, he published a scientific paper jointly with Professor Nielsen, a remarkable feat he later was proud to relate. In 1923 he was admitted to the oldest and most prestigious dormitory in Copenhagen, founded in the early 17th century by the Danish King Christian IV.

In 1922 a third chair in mathematics was created at the University of Copenhagen for N.E. Nørlund. Rasch was fascinated by Professor Nørlund. He followed his lectures and soon started to work for him.

After Rasch graduated with a masters degree in 1925, he continued to work for Professor Nørlund. His most important task was to find out whether the Danish mathematician J.L.W.V. Jensen (famous for Jensen's inequality in probability theory) had actually proved the Riemann Hypothesis. Jensen's library was preserved, so Nørlund asked Rasch to go through Jensen's papers to find the missing proof. Rasch never found the proof, but regarded this work as his introduction to mathematical research. In fact, some of Rasch's papers from the period 1925 to 1934, concerned with Lagrange equations, can be traced directly to his work on Jensen's library.

Another remarkable fact is that Rasch managed within four years after his graduation to write a doctoral dissertation *Matrix Calculus and Its Applications to Difference Equations and Differential Equations.* He defended his dissertation in June 1930 and received the title of Doctor of Science. At that time the title was regarded as qualifying for becoming a full professor. In 1930 Rasch applied for a chair in mathematics but did not get it. The next chair, which became vacant soon after, was also given to somebody else. In all likelihood the problem of not getting a chair had to do with Rasch's (and Professor Nørlund's) relationship to Harald Bohr, a younger brother of the famous nuclear physicist Niels Bohr. Harald Bohr was the dominant figure in Danish mathematics at the time and he and Rasch were never on good terms. Rasch confirms this fact in the Andrich-interviews. The people who were appointed to the two chairs were both students of Bohr, so there is probably some truth to the rumors, persistent over the years, that Bohr did not want Rasch to become a full professor, regardless of his obvious qualifications.

Even though Rasch never came to hold a position in pure mathematics throughout his entire life, he thought of himself as a mathematician and in 1979 he said in the Andrich-interviews:

> *So this is one point I want to stress: that, although, I have been most known as a statistician, my original training and my original gift is in mathematics.*

The years without a permanent position were hard for Rasch and his wife Elna (always called Nille). Fortunately, friends got Rasch various jobs, some of them connected with the analysis of data. It was of special importance that two medical scientists, Fog and Stürup, who knew Rasch from their student days, asked Rasch to look at some data in a paper on which they were working. This was in the beginning of the 1930s. In the Andrich-interviews Rasch says about the work for Fog and Stürup:

> *This is what started me in a different direction. ... Well I had no experience in dealing with that sort of thing. But they encouraged it as friends, and then I did it for friendships sake. ... I had no statistics in my mind by then, I didn't even know the least squares method, but did it by some simple calculations. ... That was my first experimental paper.*

In 1934 and 1935, he began working regularly for the University Department for Hygiene and the Danish Serum Institute. The work at the Serum Institute became especially important for Rasch in the coming years. Its director, Th. Madsen, was a member of the Danish Royal Academy, as was Professor Nørlund. It seems that Madsen and Nørlund discussed Rasch's future at a meeting of the Academy. To benefit more fully from the valuable work on data analysis Rasch had already provided for many scientists, they thought it important to give Rasch an opportunity to study mathematical statistics abroad. They managed to get a Carlsberg scholarship for him to study with the famous econometrician Ragnar Frisch in Oslo and a Rockefeller scholarship to study with R.A. Fisher in London. Frisch came to play a crucial role later, but it was primarily the stay in London from September 1935 to the summer of 1936, which had immediate importance for Rasch's future. In the Andrich-interviews, he says about the stay in London:

> *That was in 1935. I went over there in September and stayed there a full year. I learnt a lot there. ... I got hold of his paper from 1922 where he developed the theory about maximum likelihood. I was interested in that matter. ...*
>
> *That, I think, is really the high mark of what he ever did. That was the realization of the concept of sufficiency. Many may consider it just a mathematical trick, but I think it's much more than that. Anyhow, I feel, that this is the most important thing I got from Fisher. I have played quite a lot around with the concept of sufficiency. ...*
>
> *Well, Nille and I stayed for one year in London. I followed the course of the lectures given by R.A. Fisher. But in the same building, also were located Egon Pearson and Jerzy Neyman. I also followed their presentations of the same theories as I heard*

> *about from Fisher, but in a quite different version with their own fancy ideas. When I left London University then I was familiar with both traditions of modern statistical methods.*

In 1936 mathematical statistics was still a young science in Denmark, although there had been influential contributions from Danish mathematicians such as Thiele, Gram, and Westergaard, early in the century. But on his return from London in 1936, Rasch was the person in Denmark with the most recent and best education in the principles of statistical analysis as developed by R.A. Fisher. Rasch was also familiar with the fundamental work of J. Neyman and E.S. Pearson, but they never influenced him in the same way that Fisher did. It is thus fair to say that Rasch was the prime conduit of Fisher's ideas on mathematical statistics in Denmark.

From 1936 until 1962 when he finally became a professor at the University of Copenhagen, Rasch continued to make a living as a statistical consultant, in the late 1930s and the early 1940s primarily at the Serum Institute and as a coworker on many medical projects. However, in the early 1950s, he started to work on a regular basis for two research institutes, which became even more important for him later: the Danish National Institute for Educational Research (DPI) and the Military Psychology Group (MPA). The feeling for data and his extraordinary ability to "read" the structure of a data set, is undoubtedly due to the many data sets he analyzed as a consultant. As he expresses it in the Andrich-interviews:

> *I have somehow always felt that the methods I should use I would have to produce myself. Of course, I used sufficient statistics when I could get them. I also sometimes used the maximum likelihood estimation, but mostly I tried to make an intensive analysis of the data and tried to make the data tell me what they were about rather than my telling the data how they should behave.*

It is a curious fact that Rasch read few statistical papers or books during these years, possibly because he was not associated with a University Department. It seems as if Rasch as a scientist took pride in solving the theoretical problems himself as opposed to searching the literature for a solution. He had an easy way with mathematics and was always able to derive the necessary statistical tools, often using characteristic functions in his calculations. Another work method was his use of graphs. He strongly believed that the data structure would reveal itself if relevant graphs were produced. Hence the students working for him on projects had to spend endless hours calculating numbers and plotting points on paper for Rasch to inspect. He was familiar with all the main distributions and was in general able to find a distribution and a parametric structure to match his graphs. But it was equally important to him that the points on the graphs should really match the model apart from random errors. He was a pioneer regarding the idea of checking the fit of a model.

Especially at the Educational Institute, DPI, and at the Military Psychology Group, as well as at the Serum Institute, the directors gave Rasch great freedom to work with the statistical problems raised by the data, on many occasions even after the original project was terminated. This freedom became of immense importance for the development of his models, as we show.

His interest in psychometrics seems to have started in 1939, when two of his friends, Rubin and Tranekjær Rasmussen, both professors of psychology at the University of Copenhagen, asked him to teach their students statistics. This teaching continued for many years.

3 Rasch's First Analysis of an Intelligence Test (1945–1948)

Shortly after World War II the Military Psychology Group asked Professors Rubin and Tranekjær Rasmussen to develop an intelligence test to be used in connection with the selection of recruits for the Armed Forces. This test was named the IGP (Intelligence Group Test) and was developed in cooperation with Rasch who made the necessary statistical computations. This was Rasch's first analysis of an intelligence test (Rasch, 1947).

For the analysis of the IGP Rasch used only empirical descriptive methods. There are no traces of the *Rasch model* to be found. However, some of the methods he used can be found in his later articles on the Rasch model.

The IGP test had 46 dichotomous items and was tested on approximately 1000 new recruits. The recruits were divided into three groups according to

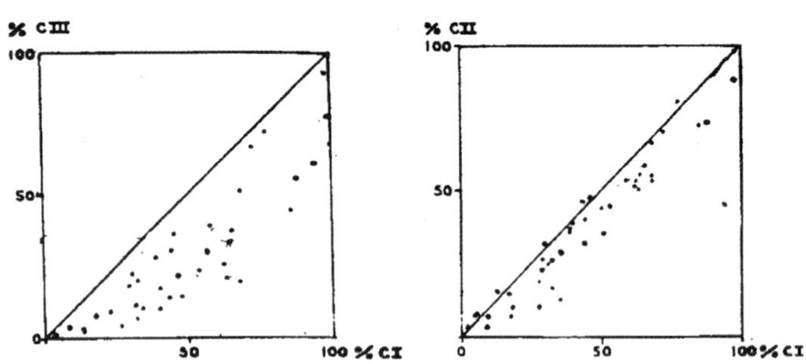

FIGURE 1.2. Error frequencies for the items in different educational groups plotted against each other. CIII against CI at left and CII against CI at right. *Source:* Rasch (1947).

level of education called CI, CII, and CIII. Rasch started the analysis by asking himself whether it was possible to define the degree of difficulty of a test in a unique way or if this was only possible within a certain subpopulation. He examined this question by plotting the error frequencies of the 46 items of the IGP two by two in the different educational groups. This is illustrated in Figure 1.2. Apart from the fact that the better educated seemed to make fewer mistakes on all items than the less educated, not much could be seen from these graphs.

He then transformed the error frequencies with a probit transformation. As Figure 1.3 shows, he now obtained straight lines with slope 1. This meant that the relationship between the probit of the error frequencies was the same within the different educational groups. Rasch interpreted the probit transformation of the error frequencies as local measures of item difficulty within a given educational group. The local measure in, for example, CI could then be transformed into a local measure in CII simply by adding to all the item difficulties the value obtained as the line's intersection with the y-axis. Weighting together local measures, Rasch thus obtained a global measure of the degree of difficulty of an item. It is not evident from the source Rasch (1947) how this weighting was done. However, when E.B.A. was a student of Rasch's in the early 1960s, Rasch always weighted estimates together the correct way, that is, with weights inversely proportional to the variance.

The estimates of the item difficulties were then used to evaluate the intelligence of each individual recruit. For each recruit a graph like Figure 1.4 was made. All the items were ordered according to difficulty on a horizontal axis. When a recruit was confronted with a set of items (not necessarily the same set for each recruit), those items the recruit was able to solve were noted by + and −. The typical picture was that a recruit was able to solve

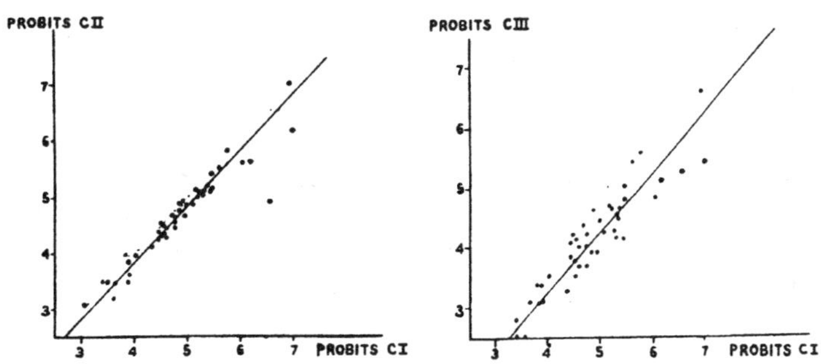

FIGURE 1.3. Probit of the error frequencies for the items in different educational groups plotted against each other. CIII against CI at left and CII against CI at right. *Source:* Rasch (1947).

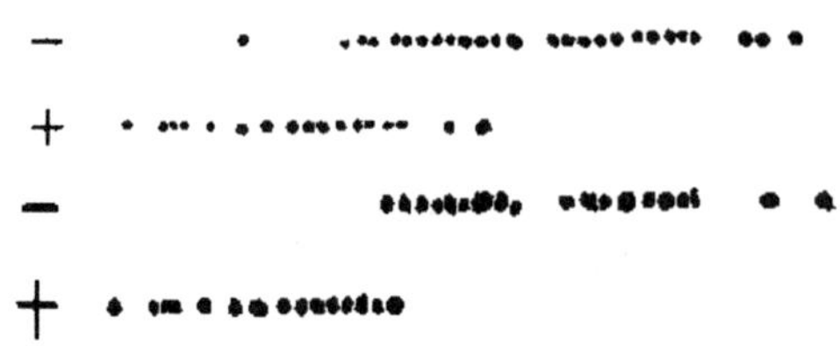

FIGURE 1.4. Registration of two recruits responses to the items as a function of item difficulty. A "+" indicates a correct answer and a "−" an incorrect. *Source:* Rasch (1947).

all items at the beginning of the test but at the end of the test the recruit was not able to solve any of the items. The intelligence of the recruit was estimated by an empirical evaluation of the difficulties of the items he was able to solve and of those he was unable to solve.

4 The Analysis of Slow Readers (1952)

Later in his career Rasch claimed that the development of the Rasch model had its beginning with the analysis of a data set in 1952. The data in question were from children who had trouble reading during their time in school and for that reason were given supplementary education. The children, whom we call slow readers, had been tested during their school years and again as adults in 1951. The aim was to find out whether the supplementary education had an effect on individual ability. There were several problems in the analysis of the slow readers. One was that the data had not been systematically collected. The children had, for example, not been tested with the same reading tests, and no effort had been made to standardize the levels of difficulty of the tests. Another problem was that World War II had taken place between the two testings. This made it almost impossible to reconstruct the circumstances of the test. It was therefore not possible to evaluate the slow readers by standardization as was the usual method at the time. Accordingly, it was necessary for Rasch to develop a new method where the *individual* could be measured independent of which particular reading test had been used for testing the child.

It has not been possible to find the data and the analysis from 1952. Therefore we do not know how he solved the problem or which models were used. Fortunately, Rasch has written about the 1952 analysis (Rasch, 1965, 1977). He writes that his aim was to give an estimate of the individual that was as well defined as the measurement of a length. Whether this was possible could not be known in advance. But by collecting supplemen-

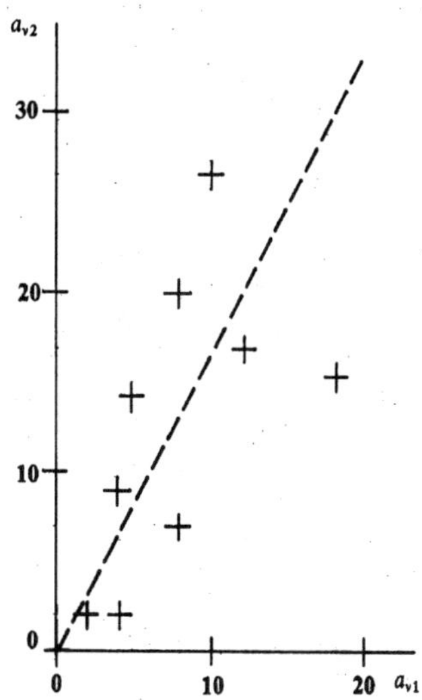

FIGURE 1.5. Number of misreadings of nine subjects in two tests. *Source:* Rasch (1977).

tary data he examined the question graphically, almost in the same way as in the analysis of the IGP. The method was as follows: two of the tests which had been used to test the slow readers were given to a sample of schoolchildren in January 1952. Rasch graphically compared the number of misreadings in the two tests by plotting the number of misreadings in test 1 against the number of misreadings in test 2 for all persons. This is illustrated in Figure 1.5. The graphical analysis showed that apart from random variations the number of misreadings in the two tests were proportional for all persons. Furthermore, this relationship held no matter which pair of reading tests he considered.

To describe the random variation Rasch chose a Poisson model. The probability that person number ν had misread $a_{\nu i}$ words in test number i he accordingly modeled as

$$p(a_{\nu i}) = e^{-\lambda_{\nu i}} \frac{\lambda_{\nu i}^{a_{\nu i}}}{a_{\nu i}!} , \tag{1.1}$$

where $\lambda_{\nu i}$ is the expected number of misread words. Rasch then interpreted the proportional relationship between the number of misreadings in the two tests as a corresponding relationship between the parameters of the model:

$$\frac{\lambda_{\nu 1}}{\lambda_{\nu i}} = \frac{\lambda_{01}}{\lambda_{0i}} \quad \Leftrightarrow \quad \lambda_{\nu i} = \frac{\lambda_{\nu 1}}{\lambda_{01}} \lambda_{0i} = \theta_\nu \delta_i \ . \tag{1.2}$$

Thus the parameter of the model factorized into a product of two parameters, a *person parameter* θ_ν and an *item parameter* δ_i. Inserting factorization (1.2) in model (1.1), Rasch obtained the *multiplicative Poisson model*

$$p(a_{\nu i}) = e^{-\theta_\nu \delta_i} \frac{(\theta_\nu \delta_i)^{a_{\nu i}}}{a_{\nu i}!} . \tag{1.3}$$

The way Rasch arrived at the multiplicative Poisson model was characteristic of his methods. He used graphical methods to understand the nature of a data set and then transferred his findings to a mathematical and a statistical formulation of the model.

But the graphical analysis was not Rasch's only reason for choosing the multiplicative Poisson model. In Rasch (1977) he writes on page 63:

> *Obviously it is not a small step from figure 1* [our Figure 1.5] *to the Poisson distribution (1.1) with the parameter decomposition (1.2). I readily admit that I introduced this model with some mathematical hindsight: I realized that if the model thus defined was proven adequate, the statistical analysis of the experimental data and thus the assessment of the reading progress of the weak readers, would rest on a solid—and furthermore mathematically rather elegant—foundation.*
> *Fortunately the experimental result turned out to correspond satisfactorily to the model which became known as The Multiplicative Poisson model. ...*

As we show, Rasch later developed the "elegant foundation" of the multiplicative Poisson model into a concept, although in the beginning of the 1950s Rash merely used it as a tool to estimate the ability of the slow readers by a method he called *bridge-building.* The point in using bridge-building is that one can estimate the attainment of the individual regardless of the particular item on which the individual has been tested. Bridge-building can be exemplified by the multiplicative Poisson model as follows: Rasch (1958, pp. 20–22) writes that the main point of bridge-building is that it should be possible to assign to each item a degree of difficulty *that is independent of the persons to whom the item has been applied.* This is possible in the multiplicative Poisson model, because the distribution of a person's responses to two different items *conditioned* on the sum of his responses only depends on the item parameters: $P(a_{\nu i}, a_{\nu j} \mid a_{\nu i} + a_{\nu j}; \theta_\nu, \delta_i, \delta_j) = g(\delta_i, \delta_j)$. The person parameter θ_ν is thus eliminated. Having estimated the item parameters in a distribution only depending on the item parameters, this estimate $\hat{\delta}_i$ may be inserted in the distribution (1.3) giving

$$p(a_{\nu i}) = e^{-\theta_\nu \hat{\delta}_i} \frac{(\theta_\nu \hat{\delta}_i)^{a_{\nu i}}}{a_{\nu i}!} , \tag{1.4}$$

which only depends on the person parameter. Hence it is possible to estimate the parameter θ_ν of the individual person even if there is a response to only one item. This is done by using a person's frequency of misreadings as an estimate of $p(a_{\nu i})$ and solving the equation (1.4) with regard to θ_ν.

The way Rasch solved the problem of parameter separation for the slow readers was not the method he used later. But it represents the first trace of the idea to separate the estimation of item parameters from the estimation of person parameters.

There are clear indications that Rasch used at least two models in connection with the analysis of the slow readers. The other model is a model for reading speed which we denote the *gamma model.* In order to analyze the time $t_{\nu i}$ student number ν has spent reading text number i, Rasch chose a gamma distribution with scale parameter $\theta_\nu \delta_i$ and shape parameter N_i, where N_i is the known length of text number i. The distribution of the variable $t_{\nu i}$ is thus given by the density

$$f(t_{\nu i}) = \exp(-\theta_\nu \delta_i t_{\nu i}) \frac{t_{\nu i}^{N_i - 1}}{(N_i - 1)!} (\theta_\nu \delta_i)^{N_i} \,. \tag{1.5}$$

As the multiplicative Poisson model this model has the property that one can deduce a distribution that depends only on the item parameters, but not on the person parameters. Bridge-building is thus possible.

5 Measuring Intelligence (1952–1953)

The Military Psychology Group was founded in 1952 with Major Poul Borking as head of the group. Major Borking decided that the intelligence test used by the military, the IGP, should be revised. Rasch was attached to the group as a statistical consultant because Borking knew Rasch had worked on intelligence tests.

According to the Andrich-interview Rasch tried to describe data from the IGP with a new model. This model was later to become known as the *Rasch model.* Using Rasch's notation the model states that the distribution of the dichotomous variable $a_{\nu i}$ for correct or incorrect answers to an item in the test is (Rasch 1960, p. 171)

$$p(a_{\nu i}) = \frac{(\theta_\nu \delta_i)^{a_{\nu i}}}{1 + \theta_\nu \delta_i} \,. \tag{1.6}$$

Normally, the model (1.6) now is written with $\alpha_\nu = \ln \theta_\nu$ and $\beta_i = \ln \delta_i$.

It is not clear from our sources that Rasch's statements regarding model (1.6) means he had actually written down the equation in 1952. Notes from 1952 found at The Education Institute strongly indicate, however, that the model is implicit in the way Rasch analyzed the 1952 recruit data. In the Andrich-interviews Rasch says about his discovery of the dichotomous Rasch model:

> *The discovery of that model actually was an achievement in connection with my work in 1952 in the analysis of the reading test and the study of the multiplicative Poisson model. ...*
>
> *Earlier, of course, I had taken great interest formally in the intelligence test while working with Professors Rubin and Rasmussen. Then it struck me that I might analyze this test, that is the test we had constructed and which had been taken over by the military intelligence group. I tried to see whether the model there worked and also whether it worked in some other test. The first thing I did was, in fact, to analyze the Raven tests. They worked almost perfectly according to the multiplicative model for dichotomous items. That was my first really nice example using the newly discovered model. Now I compared the results of the Raven's test and the result of an analysis of the Military tests which we had been taking over as a routine intelligence test for recruits. The intelligence test did not conform and I showed it to the Head of the military psychologist group. ...*
>
> *There were some 72 questions altogether, but they fell into roughly something like 7 different groups of* quite *different* kinds *of questions. That I pointed out to the head of the military psychology group, Major Poul Borking. He immediately saw the point. I had earlier talked to him about my attempts to make sense of intelligence tests by means of the model I had discovered in connection with the multiplicative Poisson model.*

Thus, instead of just fitting a new model to data from the IGP, Rasch undertook an investigation of *why* data from the IGP test did not conform to his new model. His analysis showed that the IGP test was measuring so many different abilities that no conclusion could be reached as to which item was the most difficult, nor as to which recruit was the most intelligent. When Rasch told Major Borking about his result, Borking's reaction was to decide to have a new test constructed, and a young psychologist, Børge Prien, was employed for the project. The resulting test was named BPP (in Danish "Børge Prien Prøven"), or the Børge Prien Test. Børge Prien has told one of the authors (L.W.O.) that in 1953 when he was going to construct the test, Rasch pointed out that it was very important that the test should make it possible to say whether one person had twice the ability of another person and whether one item had twice the difficulty of another item.

We find that Rasch in the Andrich-interviews tries to justify early developments of the models by the concept of specific objectivity, a concept he first developed much later. But in this case we believe that Rasch discovered this measuring property of his models in 1952, that is, prior to his talks with Børge Prien in 1953. In Rasch (1958) he explains the measuring trait as follows. In the multiplicative Poisson model the parameter $\lambda_{\nu i}$ is equal to the expected number of misreadings. The fact that $\lambda_{\nu i} = \theta_\nu \delta_i$ means that given $\theta_\nu = 2\theta_\tau$, twice the number of misreadings is expected from a person with parameter θ_ν than from a person with parameter θ_τ. Symmetrically, if $\delta_i = 2\delta_j$, then twice the number of misreadings is expected when responding to item number i than when responding to item number j, independent of the person. Rasch interpreted this as statements such as "one person has twice the degree of ability of another" or "one test has twice the degree of difficulty of another." For a model with dichotomous items, Rasch then asked whether it was possible to use a similar terminology. The answer he found was that if one chose a decomposition of the parameter $\lambda_{\nu i}$ into a product $\lambda_{\nu i} = \theta_\nu \delta_i$, these statements were valid, and a proper measurement was possible.

6 Discovery of the Dichotomous Rasch Model (1952–1958)

It is an intriguing question how Rasch discovered the dichotomous Rasch model, that is, model (1.6). According to the Andrich-interview though, Rasch's statements in Rasch (1958) and (1960, Part A) by and large are consistent with his original thoughts from 1952 to 1953. Rasch's statements (1960, part A) can be summarized as follows.

In order to obtain a proper measurement, the parameterization of the model must be multiplicative; that is, the model should depend on the item parameter δ_i and the person parameter θ_ν as the product $\theta_\nu \delta_i$. With $\zeta = \theta\delta$ Rasch (1960, p. 74) writes:

> *... the problem is to choose a function of* ζ *which only takes values between* 0 *and* 1. *And as we may find both very easy and very difficult problems and both very able and very unable persons,* ζ *must cover all values from zero to infinity.*
>
> *The simplest function I know of, which increases from* 0 *to* 1 *as* ζ *goes from* 0 *to* ∞, *is* $\zeta/(1+\zeta)$.

If one inserts $\zeta = \theta\delta$ above this is exactly the dichotomous Rasch model (1.6).

We are of the opinion that this statement covered Rasch's thinking prior to 1958. From his work at the Serum Institute, Rasch was well acquainted with the probit and logit approaches to bioassay data. But the quotation above does not give any hint as to why Rasch found the logit function simpler than, for example, the probit function that was commonly used as a statistical model in psychology in the 1950s. Our hypothesis is that Rasch chose the logit transformation because it gave sufficient estimates of the item parameters. From a historical point of view this hypothesis seems reasonable. We know that Rasch in the early 1950s had discovered the method of bridge-building, and we have seen that it is necessary for bridge-building to have estimates of the item parameters that are independent of the individuals. Since Rasch was fascinated by R.A. Fisher's concept of sufficiency it seems natural for Rasch to have chosen the logit function, because the item marginals would then be sufficient statistics for the item parameters. In Rasch's interpretation this would mean that only the item marginals would be relevant for the estimation of the item parameters and the item parameters could be estimated independently of the individuals. Many talks between Rasch and one of the authors (E.B.A.) in the early 1960s support this interpretation.

It will probably never be known when exactly between 1952 and 1960 Rasch discovered that the logit transformation had much more attractive properties than the probit transformation both distributionally and in particular as regards parameter interpretation and model check. As the discovery happened in an interplay between data analysis and formal thinking, it is most likely that he discovered the dichotomous Rasch model, including its logistic form, early in the 1950s, but that it was not until 1958, when he presented the results of the analysis of the BPP test, that he became fully aware of the importance and implications of his new model for measurement in psychology.

7 Work on the Models (1953–1958)

After his work on the BPP test in 1953, Rasch continued as a statistical consultant for the next five years working intensively with his models. In 1956 the MPA was made a permanent institution, Military Psychology Service (MPT), with Lieutenant Colonel F. Agersted as head. The change of head and organization did not put any restrictions on Rasch's work with his models. Colonel Agersted was of the same opinion as Major Borking, that Rasch as a brilliant scientist should be allowed full freedom to work with his theories. In 1955 the Danish Institute for Educational Research, DPI, was founded with Dr. E. Thomsen as director. Based on his reputation, not least at the MPA/MPT, Rasch was hired as a statistical consultant. At DPI Rasch was given the same freedom to work on his theory, while working on real data from educational tests. In addition he was provided with several students to make the necessary computations after the data had been collected.

In 1957 Rasch was asked to give a series of lectures at DPI. These lectures contained the main results of his work since 1952 and they represented a thoroughly prepared version of the way Rasch discovered his models and the associated statistical techniques. The lectures were recorded on tape by Rasch's daughter, Lotte, who later transcribed them as a birthday present to her father. Director Thomsen at DPI was convinced that Rasch's work was extremely important and ought to be published. DPI not only published the notes, but also paid Rasch for writing them. In addition the five heads of sections at DPI met every Friday morning, where they discussed various parts of the transcribed lecture notes.

In the duplicated notes from 1958 Rasch suggested three models: the multiplicative Poisson model (1.3), the gamma model (1.5), and the dichotomous Rasch model (1.6). Although each model was proposed as a way to analyze a specific data set, an important part of the notes was that within the framework of each model it should be possible to *measure* independently the attainment of the students and the degree of difficulty of the test items. Rasch was thus asking fundamental questions regarding measurement in psychology. It was also fundamental to Rasch that one should be able to estimate the item parameters independently of the given population.

Rasch made a living as a statistical consultant and under normal circumstances he would not be given time to do research. It is therefore beyond question that the work at DPI and MPA in the 1950s, the immense support from Director Thomsen, and the opportunity to publish his findings were crucial factors for the publication of the Rasch model.

8 The Conversation with Ragnar Frisch in 1959

According to the Andrich-interviews and other sources, a conversation Rasch had with Professor Ragnar Frisch in 1959 made him realize how important it was that a statistical model fulfilled a separation property. While Professor Frisch was in Copenhagen in 1959 to receive an honorary doctorate, Rasch visited him by appointment, primarily to renew their acquaintance from Rasch's stay with Frisch in the Spring of 1935. Frisch asked Rasch what he had been doing in the 25 years since he had been in Oslo. Rasch told him about his work with the intelligence tests. As an example, Rasch showed Frisch the multiplicative Poisson model (1.3). Rasch pointed out the important fact that the distribution of one person's responses to two different reading tests *conditional* on the sum of the two responses only depended on the two item parameters while *the person parameter was eliminated.* If $a_{\nu i}$ and $a_{\nu j}$ are the two responses, the formula Rasch showed Frisch was

$$P(a_{\nu i}, a_{\nu j} \mid a_{\nu i} + a_{\nu j}) = \binom{a_{\nu .}}{a_{\nu i}} \left(\frac{\delta_i}{\delta_i + \delta_j}\right)^{a_{\nu i}} \left(\frac{\delta_j}{\delta_i + \delta_j}\right)^{a_{\nu j}} . \qquad (1.7)$$

Rasch later wrote about the reaction of Frisch to this property (Rasch 1977, p. 65):

> *On seeing [(1.7)] Frisch opened his eyes widely and exclaimed: "it (the person parameter) was eliminated, that is most inter-*

esting!" And this he repeated several times during our further conversation. To which I of course agreed every time—while I continued reporting the main result of the investigation and some of my other work.

Only some days later I all of a sudden realized what in my exposition had caused this reaction from Ragnar Frisch. And immediately I saw the importance of finding an answer to the following question: "Which class of probability models has the property in common with the multiplicative Poisson model, that one set of parameters can be eliminated by means of conditional probabilities while attention is concentrated on the other set, and vice versa?"

Rasch told about the conversation with Frisch on many occasions, and basically in the same way, for example, several times to one of the authors (E.B.A.) and also prominently in the Andrich-interviews.

9 Two Important Publications (1960 and 1961)

Rasch's two most important publications, which contain the main results of his models, were published in 1960 and 1961. The publication from 1960 is Rasch's famous book *Probabilistic Models for Some Intelligence and Attainment Tests.* It consists of two parts: Part A is actually just a translation of the notes from 1958, and Part B treats the mathematical as well as the statistical properties of the three models. In Part B the main point is that the models fulfill what Rasch calls *a separability theorem.* This theorem can be formulated as follows:

It is possible to estimate the item parameters $\delta_1, \ldots, \delta_k$ in a distribution where the person parameters $\theta_1, \ldots, \theta_n$ have been eliminated. And symmetrically it is possible to estimate the person parameters in a distribution where the item parameters have been eliminated. Finally, the control of the model can be based on a distribution where both the item and the person parameters have been eliminated.

On many later occasions Rasch claimed that the development of the separability theorem was induced by his conversation with Ragnar Frisch.

Also in 1960 Rasch was invited by Jerzy Neyman to speak at the Fourth Berkeley Symposium of Probability Theory and Mathematical Statistics. At that time it was regarded as a great honor to be invited to speak at these symposia. Although Rasch knew Jerzy Neyman, who organized the event, from his time in London in 1935 to 1936, the invitation indicated that Rasch's work on statistical analysis of intelligence tests was recognized among mathematical statisticians as an important development. His paper published in the proceedings carries the title "On General Laws and the Meaning of Measurement in Psychology" (Rasch, 1961). In this paper Rasch gave his first amazingly complete answer to the question he asked himself after the conversation with Ragnar Frisch. He argued that the separability theorem strongly depended on the existence of sufficient statistics for the parameters and he presented a model which according to the Neyman criterion had sufficient statistics for the person parameters as well as for the item parameters. This model is in our notation as follows.

Let $X_{\nu i}$ be the response of individual number ν on item number i. The probability of response h, $h = 0, \ldots, m$, is then

$$P(X_{\nu i} = h) = \frac{\exp[\phi_h \theta_\nu + \psi_h \delta_i + \chi_h \theta_\nu \delta_i + \omega_h]}{\sum_{l=0}^{m} \exp[\phi_l \theta_\nu + \psi_l \delta_i + \chi_l \theta_\nu \delta_i + \omega_l]}, \tag{1.8}$$

where $(\phi_1, \ldots, \phi_m)$ and $(\psi_1, \ldots, \psi_m)$ are vectors of constants that can be interpreted as scoring functions for the categories. Both the person parameter θ_ν and the item parameter δ_i are regarded as unidimensional. However, in later articles Rasch relaxes this assumption and allows the parameters to be multidimensional. Finally, $(\chi_1, \ldots, \chi_m)$ and $(\omega_1, \ldots, \omega_m)$ are vectors of constants. When presenting the model (1.8) Rasch gave few explanations. However, the Andrich-interviews leave no doubt that his aim was not to interpret the model but to propose a large class of models that fulfills a separability theorem.

In the Berkeley paper Rasch quickly decides that the case $\chi_h = 0$ is the only interesting one, since only in this situation do there exist sufficient statistics for the person parameters that do not depend on the item parameters and vice versa. There is no interpretation of what the restriction $\chi_h = 0$ means. Thus it is beyond question that Rasch imposed the restriction $\chi_h = 0$ to ensure that a *separability theorem* holds. Rasch's separability theorem for the model (1.8) with $\chi_h = 0$ can be formulated as follows.

Consider independent random variables $X_{\nu i}$, $\nu = 1, \ldots, n$, $i = 1, \ldots, k$, and assume that the $X_{\nu i}$ follow the distribution (1.8) with $\chi_h = 0$; then:

(i) the distribution of the item marginals $(X_{.1}, \ldots, X_{.k})$ conditional on the person marginals $(X_{1.}, \ldots, X_{n.})$ only depends on the item parameters, while the person parameters have been eliminated;

(ii) the distribution of the person marginals $(X_{1.}, \ldots, X_{n.})$ conditional on the item marginals $(X_{.1}, \ldots, X_{.k})$ only depends on the person parameters, while the item parameters have been eliminated;

(iii) the distribution of all responses conditional on both the person and item marginals does not depend on any of the two sets of parameters.

Whether the model satisfied a separability theorem was not the only question Rasch tried to answer in the Berkeley paper. His main concern was *measurement*, which according to Rasch essentially was a question of comparisons. In the Berkeley paper he defined the very concept of measurements as follows (Rasch, 1961, pp. 331–332).

> *In formulating a law of the type (1.8), we have so far replaced the observations by quantitative parameters, but that does not imply that we have a proper measurement ...*
>
> *Now the requirements are as follows.*
> *The comparison between two stimuli should be independent of which particular individuals were instrumental for the comparison; and it should also be independent of which other stimuli within the considered class were or might also have been compared.*
> *Symmetrically, a comparison between two individuals should be independent of which particular stimuli within the class considered were instrumental for the comparison; and it should also be independent of which other individuals were also compared, on the same or on some other occasion.*

It is important, when trying to follow the development of the Rasch model, that the considerations about measurement originally led Rasch to choose a multiplicative parameterization when proposing the model. In the Berkeley paper it has now led to *requirements to comparisons.* Rasch went on to analyze which deterministic function $\mu(\theta, \delta)$ satisfied his now well-formulated requirements to comparisons. The analysis was strictly mathematical, and he concluded that for a parametric function of the type that appears in his general model (1.8), that is,

$$\mu(\theta_\nu, \delta_i) = \phi_h \theta_\nu + \psi_h \delta_i + \chi_h \theta_\nu \delta_i + \omega_h ,$$

the measurement requirements implied (in the case $\chi_h = 0$) that the two "scoring functions" $(\phi_1, \ldots, \phi_m)$ and $(\psi_1, \ldots, \psi_m)$ must be equal. The model accordingly simplified to

$$P(X_{\nu i} = h) = \frac{\exp[\phi_h(\theta_\nu + \delta_i) + \omega_h]}{\sum_{l=0}^{m} \exp[\phi_l(\theta_\nu + \delta_i) + \omega_l]} . \tag{1.9}$$

The mathematical derivation of the parametric structure of (1.9) was typical for Rasch who primarily thought of himself as a mathematician.

The model (1.9) is today known as the polytomous Rasch model, or the rating scale model (Andrich, 1978). In the years since 1965 much work has been done examining the statistical properties of the model. Rasch himself only examined a few of these. In the rating scale model the category scores $(\phi_1, \ldots, \phi_m)$ are assumed to be equidistant. Since this assumption is not derived from Rasch's measurement requirements, it is doubtful whether Rasch would have accepted equidistant scoring of the test categories.

10 Last Years (1962–1980)

About the time Rasch presented his results at the Berkeley Symposium a chair in statistics became vacant in Denmark. Rasch was persuaded by friends to apply, in spite of his age (60 years), and was in fact appointed professor in statistics at the Faculty of Social Science, much to the surprise of many statisticians in Denmark. Rasch's appointment did not stop him from devoting a considerable amount of time to statistical consulting and he continued to work both at MTP and at DPI. At these two institutions Rasch put a lot of energy into teaching students and staff to understand his models and to use them properly. In addition, Rasch gave lectures on some of the basic properties of his models at the university as part of his teaching duties. When teaching he began to call his models *Models for Measurement*, undoubtedly to stress that the models' main property was to measure, and that they therefore could be used to solve all sorts of measurement problems within the social sciences.

After 1962 Rasch published some 10 articles and conference papers and a few textbooks. Basically all the material in the articles is covered by an account of some free seminars Rasch gave in 1965 (Rasch, 1965). From a statistical point of view Rasch made few substantially new discoveries about the properties of his models. However, from a philosophical point of view much was achieved. As described, Rasch originally arrived at his first models in connection with the statistical analysis of data sets. But when Rasch gave his talk at the Berkeley symposium his requirements for the models were that they should fulfill a separability theorem and that the deterministic function of the parameters $\mu(\theta_\nu, \delta_i)$, determining the model, must allow for measurement. In the following years Rasch became more and more fascinated by this method of deriving his models. Based on the Berkeley paper Rasch thus formulated a new concept which he called *specific objectivity*. Obviously Rasch's aim was to formulate this concept as a generalization of the two model-theoretical requirements he presented at Berkeley.

Excluding details, Rasch's requirements for specific objectivity correspond to the second quotation in Section 9 from the Berkeley paper. In that paper these requirements are primarily analyzed in connection with a deterministic response function. But when Rasch formulated the concept of specific objectivity he also analyzed comparisons based on a probability distribution. Rasch interpreted the fact that the item parameters in the Rasch model can be estimated independently of the person parameters as showing that the item parameters can be estimated in a specifically objective way.

A main interest of Rasch after 1962 was to find necessary and sufficient conditions for specific objectivity. Rasch's early considerations about *when* it is possible to interpret the parameters of a statistical model as measurement was carried out as a strictly mathematical analysis of whether comparisons based on a deterministic function $\mu(\theta_\nu, \delta_i)$ are specifically objective. Rasch solved this problem for unidimensional parameters (Rasch, 1972). He showed that comparisons are specifically objective if and only if the function $\mu(\theta_\nu, \delta_i)$ can be transformed by continuous monotonic functions to a linear expression in θ_ν and δ_i. During the years Rasch focused on the deterministic situation and sometimes presented specific objectivity without references to a statistical model. But he also wanted to find necessary and sufficient conditions for comparisons based on a statistical model to be specifically objective, that is, to find necessary and sufficient conditions for a statistical model to fulfill a separability theorem. By 1965 Rasch had shown that when only two response categories are considered then the dichotomous Rasch model (1.6) is the only model with this property (Rasch, 1965, pp. 67–70). Rasch showed a similar result for models with k response categories, but in this case the definition of specific objectivity had to be strengthened. Rasch's theorem states that the model (1.9) with multidimensional parameters is the only model that allows for what Rasch called *complete specific objectivity* (which basically means that the model also must fulfill a separability theorem if one or more of the response categories is ignored). Having shown that the model (1.9) is the only model that allows for complete specific objectivity Rasch (1968, pp. 27–28) concluded:

> *Thus, if a set of empirical data cannot be described by that model then complete specifically objective statements cannot be derived from them.*
> *Firstly, the failing of specific objectivity means that the conclusions about, say, any set of person parameters will depend on which other persons are also compared. As a parody we might think of the comparison of the volumes of a glass and a bottle as being influenced by the heights of some of the books on a shelf. Secondly, the conclusion about the persons would depend on just which items were chosen for the comparison, a situation to which a parallel would be, that the relative height of two persons would depend on whether the measuring stick was calibrated in inches or in centimeters.*
> *Avoiding such irrelevant dependencies is just my reason for recommending the use of the models for measuring whenever they may utilized.*

During the years specific objectivity almost became a passion of Rasch's. He wanted people to understand the importance of the concept and he had little patience with those who failed to see that specific objectivity was the most important point in model building in the social sciences.

11 Epilogue

In our minds there are absolutely no doubts that Rasch developed the Rasch model as described. It all started with some empirical analyses and some basic techniques but over 10 to 20 years Rasch developed it into the Rasch model and the concept of specific objectivity. However, one may ask whether Rasch's development was original or whether he was influenced by the work of other statisticians of his time. We cannot give a complete answer here, but it is a characteristic of the writings of Rasch, that he quotes few papers, and if he does they are often very early works of a philosophic nature. This has to do with a peculiar fact, observed by the senior author (E.B.A.) over many years after he started working for Rasch in the summer of 1961, and documented by Rasch's own words in the Andrich-interviews: Rasch actually seldom read newly published statistical or psychometric papers or books. He was, therefore, unaware of many key papers in the psychometric literature. That some key papers, for example, Lord (1953), were known to him is evident from footnotes in Rasch (1960), but we do not believe he read them in detail and that they influenced his thinking. The same was true when the important paper by Luce and Tukey (1964) on the concept of conjoint measurement and the important book by Lord and Novick (1968), in which Allan Birnbaum wrote about models very similar to the Rasch model, were published. It must therefore be concluded that the development of the Rasch model is a truly original scientific contribution. We shall never know, of course, if readings of more key papers would have influenced his thinking.

Acknowledgments: The authors would like to thank David Andrich for giving us copies of the tapes and transcript of his interview with Rasch. We are also grateful to Peter Allerup, Michael Weiss Bentzon, Hans Brøns, Anders Hald, Eggert Petersen, and Børge Prien for letting L.W.O. interview them on matters related to Georg Rasch and his work.

References

Andrich, D. (1978). A rating formulation for ordered response categories. *Psychometrika*, *43*, 357–374.

Lord, F.M. (1953). An application of confidence intervals and of maximum likelihood to the estimation of an examinee's ability. *Psychometrika*, *18*, 57–75.

Lord, F.M., & Novick, M.R. (1968). *Statistical theories of mental test scores.* Reading, MA: Addison-Wesley.

Luce, R.D., & Tukey, J.M. (1964). Simultaneous conjoint measurement: A new type of fundamental measurement. *Journal of Mathematical Psychology. 1*, 1–27.

Rasch, G. (1947). *Om vurdering af intelligensprøer* [On evaluation of intelligence tests]. (Duplicated notes.) Copenhagen: Department of Psychology, University of Copenhagen.

Rasch, G. (1958). *Om anvendelse af et almindeligt måleprincip til brobygning mellem ensartede psykologiske prøver* [On the use of a general measurement principle to bridge-building between similar psychological tests]. Copenhagen: The Danish Institute of Educational Research.

Rasch, G. (1960). *Probabilistic models for some intelligence and attainment tests.* Copenhagen: The Danish Institute of Educational Research. (Expanded edition, 1980. Chicago: The University of Chicago Press.)

Rasch, G. (1961). On general laws and the meaning of measurement in psychology. *Proceedings of the Fourth Berkeley Symposium on Mathematical Statistics and Theory of Probability* (Vol. IV, pp. 321–333). Berkeley: University of California Press.

Rasch, G. (1965). *Målingsmodellerne og deres principielle baggrund* [Models for measurements and their fundamental background]. (Notes taken by J. Stene at the statistical seminar.) Copenhagen: Department of Statistics, University of Copenhagen.

Rasch, G. (1968). *A mathematical theory of objectivity and its consequences for model construction.* Paper presented at the European Meeting on Statistics, Econometrics and Management Sciences, Amsterdam.

Rasch, G. (1972). Objektivitet i samfundsvidenskaberne et metodeproblem [Objectivity within the social sciences: A methodological problem]. *Nationaløkonomisk Tidsskrift, 110,* 161–196.

Rasch, G. (1977). On specific objectivity: An attempt at formalizing the request for generality and validity of scientific statements. In M. Blegvad (Ed.), *The Danish Yearbook of Philosophy* (pp. 58–94). Copenhagen: Munksgaard.

2

The Growing Family of Rasch Models

Jürgen Rost[1]

ABSTRACT A family of Rasch models is defined in terms of the prominent properties of all Rasch models, that is, separability, sufficiency, specific objectivity, and latent additivity. It is argued that concepts such as item homogeneity, person homogeneity, and unidimensionality do not hold for all generalizations of the Rasch model (RM). Four directions of generalizing the model are discussed: the multidimensional, the ordinal polytomous, the linear logistic, and the mixture distribution generalization. A hierarchical system of generalized Rasch models is presented. Four out of these eight models are discussed in the literature and can be applied by means of reliable computer software.

1 Introduction

In many areas of statistical modeling, like structural equation, log-linear and item response models, a dramatic increase of new models can be observed that help to solve old problems better than before, or new problems that were not recognized before. Progress in the development of statistical models means to *blow up* the number of different models including their derivations, restrictions, justifications, and also their specific names reflecting the glory of their inventors. For students and applied researchers the amount of alternatives for analyzing data becomes hard to follow.

On the other side, progress means *reduction* of complexity by hierarchically ranking all these models, that is, by identifying which model is a sub- or supermodel of which other one. The game "Who has the most general model," often played at conferences and in scientific journals, has the essential function of finding links among bunches of puzzle pieces. Model development at its best, therefore, means reducing the manifold different models to only one hierarchical system that defines the logical structure among models, already invented or not.

---

[1]Institute for Science Education, University of Kiel, Olshausenstr. 62, D-24098 Kiel, Federal Republic of Germany; e-mail: *rost@ipn.uni-kiel.de*

However, the one and only most general model, if we ever find it, tells us nothing about the usefulness, beauty, and mathematical elegance of all its submodels and not much more about how to find the model that best fits a special set of data. Those "most general models" do not prevent us from studying their submodels and special cases. However, they make this job easier and may protect us from "getting lost in the models."

The general framework of *log-linear modeling* (Bishop, Fienberg, & Holland, 1975) is a good candidate for being the most general model structure for many or even all Rasch models. Since the articles by Kelderman (1984) and Tjur (1982) it has been known that the Rasch model (RM) (Rasch, 1960) can be formalized as a log-linear model and from papers such as Kelderman and Rijkes (1994) or Meiser (1996) it can be concluded that this is true for generalized Rasch models as multidimensional models for polytomous data. Even generalizations of the Rasch model to a discrete mixture distribution model (Rost, 1990) do not really leave the framework of log-linear modeling, but only require the inclusion of *latent* categorical variables in the log-linear structure (e.g., Haberman, 1988; Vermunt, 1996).

But of what use is the consideration of a general statistical framework such as the log-linear to be the supermodel for a family of item response models (IRT models)? It covers—besides Rasch models—a variety of other models that have nothing to do with Rasch models and it does not give any impression of the structure of the family of Rasch models. The present chapter tries to define the properties that all members of the Rasch family have in common, and it traces some historical lines generalizing the Rasch model. These considerations end up with a hierarchical system of generalized Rasch models, the most general of which never have been applied before and, which are probably far away from being applied in the future.

It is easy to imagine a model structure that combines all the nice features of existing models and, hence, covers the latter as submodels. It is difficult, however, to solve all problems related to applying such a complex model successfully and to take advantage of its statistical properties.

2 What Is a Rasch Model?

It is hard to give an abstract definition or even an axiomatic derivation of the family of Rasch models. Therefore, different properties that discriminate between Rasch models and non-Rasch models are discussed.

Looking into the North American and European literature on the Rasch model reveals some differences in terminology that may or may not correspond to differences in substance. In the American literature (e.g., Wright & Masters, 1982) *sufficiency* and *separability* are often mentioned as the essential properties of the Rasch model, while European authors stress the principle of *specific objectivity* (Fischer, 1988). Additivity or *latent addi-*

tivity comes into play when the Rasch model is treated from the tradition of measurement theory, where the relationships to the model of conjoint measurement become obvious (Luce & Tukey, 1964). The property of sample free item calibration (USA) or sample independence of measurement (German) seems to be as ubiquitous as it is misleading.

In contrast to these philosophical terms, it is easier to define the property of intersection-free and parallel item-characteristic curves (ICCs) or of local independence and accept them as Rasch-specific features. Others use the concepts of unidimensionality, item homogeneity, and person homogeneity as basic assumptions of Rasch measurement.

Not all of these properties also hold for generalized Rasch models and, hence, only some of them can be used to define the family of Rasch models. A preliminary response to the question, "What is a Rasch model?" would be: a Rasch model is an item response model aimed at measuring one or more quantitative latent variables on a metric level of measurement, and that has the properties of sufficiency, separability, specific objectivity, and latent additivity. It has to be noted that the specification "on a metric level of measurement" is crucial for not subsuming the Mokken (1971) model of double monotonicity (DM) in the family of Rasch models. Sufficiency, separability, and specific objectivity also hold—in a nonmetric sense—for the Mokken model (Hemker, Sijtsma, Molenaar, & Junker, 1996; Molenaar, 1997). It is only the different level of measurement (i.e., metric vs. ordinal) that discriminates between these two approaches to item analysis.

Hereafter, these four basic properties are discussed with respect to Rasch models and it is shown that their semantics overlap; that is, they only shed a different light on the same subject. It is further discussed which features are specific for the basic (dichotomous) model and are not shared by all of its generalizations.

Sufficiency means that the sum of responses in a certain category is a sufficient statistic for estimating the person's ability or trait value to respond in that category (this definition also holds for polytomous models). The importance of this statement becomes obvious only if it is paraphrased in different ways: if the sum of responses in a questionnaire is sufficient for estimating the trait parameter to be measured, then the model does exactly what is done in naive test analysis; that is, it takes the "number correct" or the "sum of positive answers" as an indicator of individual test performance. The only difference is that sufficiency is not a matter of definition but a consequence of the fact that the Rasch model holds for the data. Sufficiency also means that the sum score covers all the information about the person's trait provided by the test. The pattern of responses contributes no further information about the person. In particular, sufficiency means that *which* items a person has solved does not count but only how many.

Although it is possible to treat sufficiency as an axiom for deriving the Rasch model, it is far more interesting to see sufficiency as a necessary consequence of certain properties of the Rasch model, for example, constant item discriminations. The Rasch model (and its generalizations) is the only parametric item response model with this property and, hence, fits best with the usual practice of test analysis, where the "number correct" is treated as an interval scaled measure of the ability level.

Separability denotes the property that person and item effects on the response behavior can be isolated from each other. This can be seen as an analogy to analysis of variance where the separation of two factors in a main effects model implies that there are no interaction effects between these factors. It follows from this analogy that intersecting ICCs would contradict separability, since it is a kind of interaction: intersecting ICCs means that the "effect size" of the person factor with respect to the response probability varies from item to item.

There is, however, an even stronger notion of separability in Rasch models: it is possible to estimate the item parameters without estimating the person parameters and moreover without even knowing their distribution. Since the sum of responses in a certain category is a sufficient statistic for the person's tendency to respond in that category, it is possible to condition out the person parameters from the estimation equations. This is the basis for conditional maximum likelihood (CML) estimation of the item parameters, the only estimation procedure that is free of any parameter of the latent trait distribution and which can be used for any member of the Rasch family.

Rasch introduced the term *specific objectivity* to address the important property of measurement models to assure that the measure of one sort of objects (e.g., persons) is independent of the other objects (e.g., items) involved in the procedure of measurement. This requirement is self-evident for any kind of physical measurement; for example, everybody would expect that the weight of a package is the same when it is measured at home or at the post office. This is not always true for psychological measurement; for example, the level of intelligence measured by an intelligence test may differ per test. Rasch's demand for specific objective measurement means that the measure of a person's ability must be independent of which items were used. This kind of independence, however, can only be reached if the population of items is well defined and homogeneous with respect to the measured trait.

Fischer (1988) showed that the family of Rasch models is the only model type with a metric level of measurement that fulfills this requirement of specific objectivity. Moreover, he showed that the property of specific objective measurement is related to the property of *latent additivity*. Whenever specific objectivity holds for a set of items and a set of persons, then there exists a representation of the model where the person and the item parameter are connected by addition or subtraction.

It can be considered as a practical consequence of the latent additivity of Rasch models that they can be formalized as log-linear models (Kelderman, 1984; Tjur, 1982), which is not possible for models where item and person parameters are related in both ways, additively and multiplicatively, as in the Birnbaum model.

These features of Rasch models (i.e., sufficiency, separability, specific objectivity, and latent additivity), which contribute to the philosophy of science arguments for measurement, are valid for all generalized Rasch models. This is different in the case of more concrete features such as parallel ICCs, and person and item homogeneity. Parallel item response functions are an important feature of the dichotomous model but vanish when Rasch models for ordinal polytomous item responses are considered. In these models the form and, in particular, the slope of the ICCs depend on the threshold distances of the items (Rost, 1988, 1999).

Item homogeneity is an important feature of the dichotomous unidimensional model, but is given up in multidimensional generalizations, where the items are heterogeneous as they address different traits with different weights. Even *person homogeneity* is not a sine qua non, for example, when mixture distribution models are employed to unmix a heterogeneous population into homogeneous subpopulations. These issues are addressed in detail in Section 3.

Local stochastic independence, however, is not a typical property of Rasch models, but a common feature of all latent trait and latent class models. Local independence is the goal of many probabilistic models that use a latent variable which causes the manifest variables to be locally independent, that is, independent if the latent variable is held constant (cf. Andrich, 1991). Moreover, there are attempts to develop Rasch models where the response probabilities depend on which items have been solved before (Verhelst & Glas, 1993, 1995a), and hence, the item responses are—even locally—dependent on earlier responses.

In terms of a mathematical structure, the family of Rasch models can be formalized as follows. The probability of a person's (S_v) response on item I_i in category x ($x = 0, 1, \ldots, m$), p_{vix}, is an exponential function with an exponent that is additively decomposed into some person, item, and category parameters:

$$p_{vix} = \frac{\exp(\lambda_{vix})}{d_{vi}} , \tag{2.1}$$

where

$$d_{vi} = \sum_x \exp(\lambda_{vix}) . \tag{2.2}$$

Notice that the denominator d_{vi} does not depend on the data (i.e., on the response x), but is the sum of the numerator over all categories x. Hence, it plays the role of a normalizing factor ensuring that all response probabilities have values between 0 and 1 and sum to unity for a fixed person and item. In this notation, the original dichotomous Rasch model is given by

$$\lambda_{vix} = x(\theta_v - \beta_i) , \tag{2.3}$$

where θ_v is the trait parameter of person S_v and β_i is the difficulty parameter of item I_i. Category x does not have its own parameter, but enters the function as a fixed coefficient.

3 Some Historic Tracks of Generalizing the Rasch Model

Georg Rasch himself proposed some important generalizations of the "simple" logistic model for dichotomous data; in particular he analyzed models for multicategorical or polytomous data. Other generalizations go back to the work of P. Lazarsfeld (Lazarsfeld & Henry, 1968) and D.R. Cox (1970). The aim of the present chapter is not to present pieces of historic research on which author followed the ideas of which others. The aim is rather to mark some stages of model development that are not always straightforward but sometimes show features of a stochastic process. In particular, four different directions of generalizing the Rasch model are distinguished and discussed: multidimensional Rasch models, ordinal Rasch models, linear logistic models, and latent class models.

3.1 Multidimensional Rasch Models

Rasch's (1961) multidimensional model was based on the idea that the response in one out of $m+1$ response categories depends on as many latent traits as there are response categories (only m of which are independent). That is, the person's tendency to give a response in category x is a function of a trait parameter and an item difficulty parameter, both specific for that response category:

$$\lambda_{vix} = \theta_{vx} - \beta_{ix} \ . \qquad (2.4)$$

It should be noted here that, due to the normalizing factor d_{vi} (see Equation 2.2), the response probability depends on *all* trait parameters of the person; that is, the higher the trait for one of the other response categories $y \neq x$, the lower the response probability for category x. An important property of the data is that the sum scores of the response categories for each person add to the number of items k, and hence are ipsative measures. This property is preserved in the trait parameters θ_{vi}, which also show all the nice features of ipsative measures (Hicks, 1970).

Ipsative measures are not necessarily given in other multidimensional models where the traits do not correspond to the response categories. An early attempt of formalizing such a multidimensional Rasch model was presented by Stegelmann (1983). He started with an item component model like the linear logistic test model (LLTM) by Fischer (1972; see Equation 2.8 in Section 3.3) and introduced component-specific latent traits. Unfortunately, the component difficulties η_j (see the *linear logistic Rasch model* in Figure 2.1) turned out to be inseparable from the normalizing constant of the component-specific person parameters θ_{vj}. Hence, a multidimensional model was obtained that did not belong to the Rasch family, because no separation between item and person effects on the response probabilities was modeled.

Rost (1996; see also Rost & Carstensen, in press) followed Stegelmann's idea of a *multidimensional item component model* but left the item parameters undecomposed in order to keep them in the model:

$$\lambda_{vix} = \sum_{g=1}^{G} a_{ig}\theta_{vg} - \beta_{ix} \ . \tag{2.5}$$

Since the a priori fixed coefficients a_{ig} define the slope of the ICC of an item I_i with respect to dimension g, model (2.5) is also a multidimensional generalization of the OPLM by Verhelst and Glas (1995b).

However, this is by far not the most general multidimensional model of the Rasch family. Kelderman and Rijkes (1994) and Meiser (1996) discuss multidimensional Rasch models which, for example, can also model latent dimensions that correspond to the response categories. Figure 2.1 shows this extended model structure where the weights a_{igx} have an additional index x. As a consequence, Rasch's multidimensional model (2.4) is a submodel of this multidimensional structure. These authors refer to the work of Wilson, Adams, Wang, and Wu, who developed the most general multidimensional model, which was named after their computer program ConQuest (Wilson & Wang, 1995; Adams, Wilson, & Wang, 1997; Wu, Adams, & Wilson, 1997). In contrast to Kelderman and Rijkes (1994), and Meiser (1996), they use different weight matrices for the trait and item parameters (see Figure 2.1), so that their model is also a generalization of the linear logistic model of Fischer (1972).

The weight matrices in all of these models are a priori defined and, hence, part of the model. As compared to factor analytic models, they are analogues not to exploratory but to confirmatory factor models (Hoijtink & Rooks, 1999). Attempts at introducing multiplicative parameters as weights in a multidimensional structure (Muraki & Carlson, 1995; Bartholomew, 1987) are beyond the latent additive framework of Rasch models. However, the multidimensional Rasch model provides a very general framework for specifying subtypes of models with more than one latent dimension. Examples are models with an additional learning ability parameter (Klauer & Sydow, 1992, 2001) or models for facet-designed tests (Rost & Carstensen, in press).

3.2 Ordinal Rasch Models

Rasch himself (1961) investigated the conditions for reducing the multidimensional structure (2.4) to a unidimensional model for ordered cate-

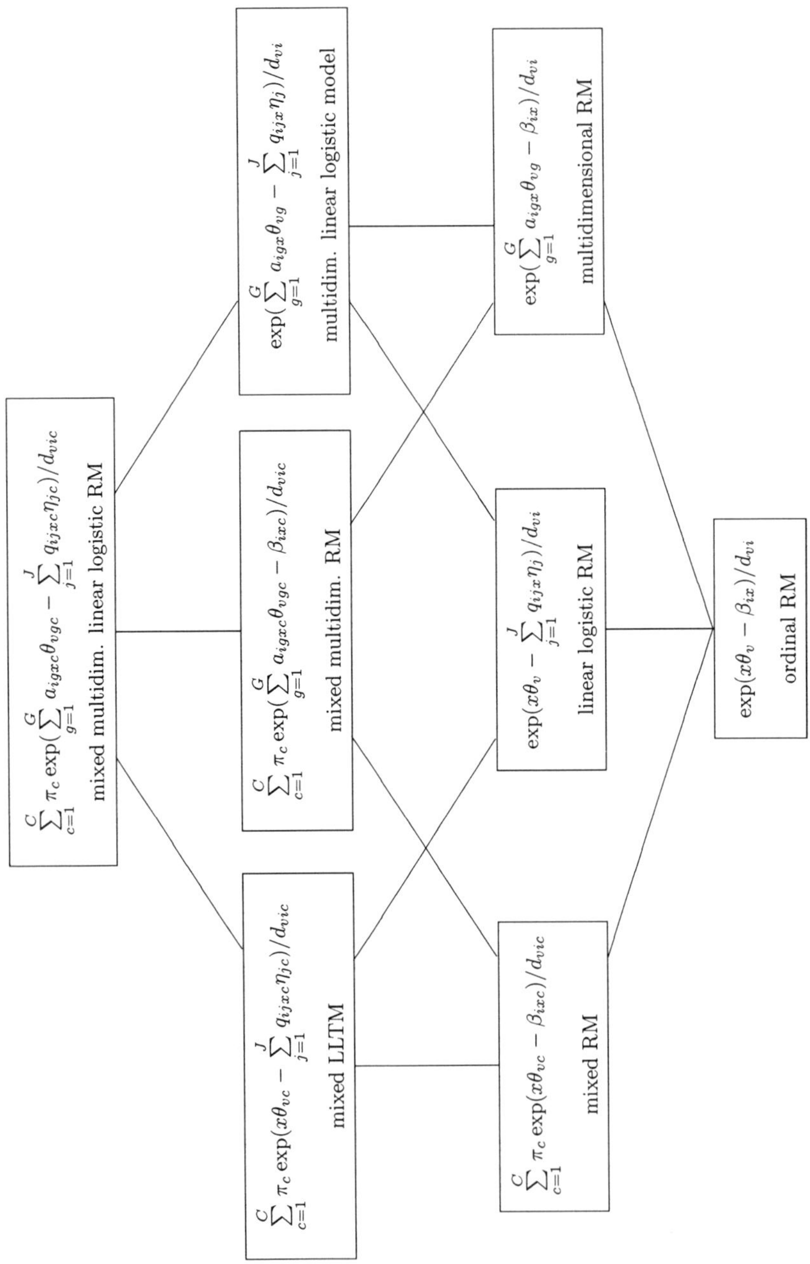

FIGURE 2.1. A hierarchy of generalized Rasch models.

gories. He proved that the vector parameters θ_{vx} and β_{ix} have to be a linear function of unidimensional (scalar) parameters in order to preserve the properties of specific objective measurement:

$$\lambda_{vix} = \phi_x(\theta_v - \beta_i) + \psi_x \ . \tag{2.6}$$

Andersen (1977) has shown that the multiplicative parameter ϕ_x has to have equidistant values in order to keep the property of sufficiency. Together with the necessary normalization restrictions (i.e., $\phi_0 = 0$ and $\sum \phi_x = c$), it followed that the multiplicative coefficient is not a parameter but a fixed term. Following these arguments, Andrich (1978) replaced the ϕ_x parameters by the category number x and proved that this number represents the number of thresholds passed by the respondent, when the dichotomous Rasch model describes the threshold probabilities. In this case, the additive category parameter ψ_x turns out to be defined in terms of the cumulative threshold locations. The resulting model has become known as the *rating scale model*

$$\lambda_{vix} = x(\theta_v - \beta_i) + \psi_x \quad \text{and} \quad \psi_x = \sum_{s=1}^{x} -\tau_s \ , \tag{2.7}$$

where τ_s is the location or difficulty of threshold s.

However, to derive a unidimensional Rasch model for ordered categories it is not necessary to restrict the item category parameters of the multidimensional model at all. The model that leaves the β_{ix} parameters as they are but replaces θ_{vx} by $x\theta_v$, was discussed by Masters (1982) as the *partial credit model* (PCM). Again, the β_{ix} parameters can be interpreted as the cumulative threshold locations for item I_i.

The partial credit model is the most general Rasch model for ordinal data and, hence, should be addressed as the ordinal Rasch model (see Figure 2.1). It has a series of submodels that result when certain restrictions are imposed on the threshold distances. Whereas the rating scale model assumes constant distances between any two thresholds τ_x and τ_{x+1} across all items I_i, the *dispersion model* by Andrich (1982) assumes equidistant thresholds within each item and introduces a distance parameter for the items. A combination of both models is discussed by Rost (1988). Müller (1987) generalized the dispersion model to a model for continuous item responses.

Although the class of Rasch models for ordinal data seems to be well represented by these models, there still is a debate about the property of these models of not being invariant against collapsing adjacent response categories when analyzing data (Roskam & Jansen, 1989). Maybe Andrich's (1995) argument will be accepted in the long run: the violation of the "joining assumption" is not a fault of ordinal Rasch models but, on the contrary, an implication of strong measurement models.

3.3 Linear Logistic Models

At an early stage of Rasch model development, Fischer (1972, 1973) presented a model that replaces the item parameters of the dichotomous model by a weighted sum of (a smaller number of) basic parameters η_j, resulting in

$$\lambda_{vix} = x\left(\theta_v - \sum_{j=1}^{J} q_{ij}\eta_j\right) . \qquad (2.8)$$

This linear logistic test model (LLTM) turned out to be a rather general model structure that covers many other models developed later on as submodels. The most prominent example probably is the three-mode extension of the Rasch model that was already discussed by Micko (1970) and Fischer (1974), but can also be traced back to unpublished work by Rasch. Later on it was published by Linacre (1989) as the multifaceted Rasch model.

For defining this generalization of the Rasch model, an extended data structure has to be introduced, that is, a data cube instead of a data matrix with item responses x_{vit} as entries. The three-mode Rasch model then assumes separable effects of three factors, that is, persons, items, and time points, situations, or judges. In the model a third parameter δ_t is specified defining the influence (main effect) of, for example, time point t:

$$p_{vitx} = \frac{\exp(\lambda_{vitx})}{d_{vit}} \quad \text{and} \quad \lambda_{vitx} = x(\theta_v - \beta_i + \delta_t) . \qquad (2.9)$$

where d_{vit} is the normalizing factor; cf. (2.2).

Due to the fact that the third type of parameters δ_t are structural parameters like the β_i and not incidental parameters like the θ_v, the model can be specified as a LLTM for a two-dimensional data matrix with new "virtual" items for each slice of the former data cube (cf. Rost & Langeheine, 1997). In the same way, the LLTM may be applied to data structures with four or even more modes. In cases where a test is constructed according to a facet design, but separate traits for the components of each facet are assumed, a multidimensional Rasch model has to be applied (Rost & Carstensen, in press).

There are many possible specifications of the LLTM that model assumptions about learning within a test and between two or more test applications, about cognitive operations during test performance, or about cognitive development (Fischer, 1995). The LLTM thus has been proven to be a flexible and very general model structure and it defines one of the three directions of generalizing the Rasch model in Figure 2.1. Although most applications of this model have been done with the dichotomous version of the model, the model has, in the mean time, been generalized for ordinal response data (Fischer & Parzer, 1991; Fischer & Ponocny, 1994).

3.4 Latent Class Analysis and Rasch Model

Latent class analysis (Lazarsfeld & Henry, 1968) and Rasch models have often been treated as alternative approaches of analyzing item response data, one using discrete and the other continuous latent variables (Langeheine & Rost, 1988). Nevertheless, in the 1980s some research was done on their relationship, yielding interesting models that combine ideas from both types of models. The driving force, again, was the question "What is the more general model?" This question was answered by Lindsay, Clogg, and Grego (1991) and Formann (1989) who have shown that the Rasch model can be represented as a restricted latent class model with a certain number of classes.

This relationship is important as it allows us to apply and to test certain restricted Rasch models within the framework of latent class modeling. In order to impose appropriate restrictions on the latent class parameters, it is necessary to use a version with logistic parameters (instead of probability parameters) that can be made subject to linear restrictions (Formann, 1985). In Formann's linear logistic latent class model, Lazarsfeld's idea of located classes can be specified for the Rasch model:

$$p_{vix} = \sum_{c=1}^{C} \pi_c \frac{\exp[x(\theta_c - \beta_i)]}{d_{ic}} , \tag{2.10}$$

where c is the subscript for latent classes and the normalization factor d_{ic} is again the sum of the numerator over all categories x: $d_{ic} = \sum_x \exp[x(\theta_c - \beta_i)]$. As mentioned before, with a sufficiently large number of classes ($C \geq (k-1)/2$), the model is algebraically equivalent to the Rasch model. With a smaller number of latent classes special assumptions about the latent distribution can be modeled, for example, the assumption of mastery testing in educational contexts, where only two classes (i.e., masters and non-masters) are assumed.

Quite a different way of combining the ideas of latent class and Rasch analysis is given by the assumption that the person population is a mixture of latent classes but the Rasch model holds within each class (with different sets of parameters, of course). This is the idea of the mixed Rasch model proposed by Rost (1990, 1991) and—in a modified form—by Mislevy and Verhelst (1990):

$$p_{vix} = \sum_{c=1}^{C} \pi_c \frac{\exp(x\theta_{vc} - \beta_{ixc})}{d_{vic}} . \tag{2.11}$$

The model becomes more flexible and general if the type of the Rasch model (if a Rasch model at all) may vary from class to class. A combination of a Rasch class and one or more ordinary latent classes (i.e., those with fixed values of θ) was presented by Yamamoto (1989) who named those models hybrid models. In the same way, other combinations of Rasch-type and non-Rasch models can be specified (Von Davier, 1995).

Examples would be a two-class model with the ordinal Rasch model in one, and some equidistance model in the other class, or a two-class model for separating a masters-class (with high solution probabilities for all items) from the rest of the students showing variation of their abilities according to the Rasch model.

Nevertheless, (2.11) represents the most general form of this kind of extending Rasch's measurement theory, since all other hybrid or located classes' models can be specified by appropriate restrictions. An ordinary latent class is obtained by fixing $\theta_{vc} = \theta_c$ and the model with located classes requires the further restriction of constant item parameters in all classes; that is, $\beta_{ixc} = \beta_{ix}$. The ordinary Rasch model follows from $\theta_{vc} = \theta_v$.

4 A Hierarchical Structure of Generalized Rasch Models

The four directions of Rasch family growth form the basis of the system of Rasch models shown in Figure 2.1. The straightforward way of generalizing the dichotomous Rasch model to a model for ordinal data, however, can be neglected here, since all other generalizations have already been formalized for such data. Hence, Figure 2.1 starts off with the ordinal Rasch model at the bottom and differentiates to its mixture, linear logistic, and multidimensional generalizations. The models at these two lower levels were introduced in the preceding sections. Test analyses according to these models can be performed by the programs WINMIRA (Von Davier, 1998), LPCM-WIN (Fischer & Ponocny-Seliger, 1998), and MULTIRA (Carstensen & Rost, 1998) for each type of generalization, respectively.

On the third level, each two of these generalizations are combined resulting, again, in three models. The multidimensional linear logistic model was investigated by Adams, Wilson, and Wang (1997). These authors provide the computer program ConQuest, which is based on the marginal likelihood approach to the estimation of model parameters. Although there are quite a few experiences with specifying and applying submodels of this general model structure, the range of possible specifications is far from being explored completely.

A mixed multidimensional Rasch model (cf. the middle of the third level of Figure 2.1) has not yet been developed and applied. It would provide the frame for testing hypotheses about differences of the multidimensional structure of a test among latent subpopulations. It would be possible, for example, to analyze whether a personality questionnaire measures the latent trait under consideration in one subpopulation, and social desirability *and* the target trait in a second subpopulation. Another example from the assessment of school achievement would be the hypothesis that fewer ability dimensions are needed for students who gained a deeper understanding of the basic concepts in a knowledge domain than for students who still are on the level of reproducing pieces of knowledge they have just learned. At the moment, however, there is little knowledge about which kinds of model specifications are identifiable, and which problems could arise and be solved with parameter estimation and model tests.

The third model at this level, that is, the mixture generalization of the linear logistic model, also belongs to the yet unexplored model structures. In fact, the appeal of that combination relates to the demand for testing a particular item component structure to be invariant across people. In some sense, this model can be motivated by the same argument that let Stegelmann (1983) and others develop multidimensional item component models (see Section 3): to extend the LLTM in a way that allows us to model individual differences with respect to the assumed item components. In one case, this is done by introducing component-specific dimensions, and in the other case, by introducing item component parameters that have different values in different subpopulations or even component structures that are specific for the expected subpopulations.

The model combining all three generalizations of the Rasch model (at the top of Figure 2.1) represents the supermodel in this family of Rasch models. The future will show whether this model is only a virtual completion of this systematic approach or useful for the practice of test analysis. In any case, the required sample size for a stable estimation of all parameters would be enormous.

The family of Rasch models seems to have reached a stage where the basic problems of generalizing the "simple logistic" to an analytic framework for applied researchers and practitioners of test construction have been solved. Moreover, even the lines to a joint supermodel of this family can be recognized. The challenge now is to derive results on uniqueness and identifiability, to develop methods of parameter estimation and efficient computer algorithms, model tests and fit indices for items and persons, and to provide user-friendly programs for the practitioners who have to specify the input and interpret the output of those programs.

Acknowledgments: The author would like to thank the two anonymous reviewers for their helpful comments and suggestions.

References

Adams, R.J., Wilson, M., & Wang, W. (1997). The multidimensional random coefficients multinomial logit model. *Applied Psychological Measurement, 21*, 1–23.

Andersen, E.B. (1977). Sufficient statistics and latent trait models. *Psychometrika, 42*, 69–81.

Andrich, D. (1978). A rating formulation for ordered response categories. *Psychometrika, 43*, 561–573.

Andrich, D. (1982). An extension of the Rasch model for ratings providing both location and dispersion parameters. *Psychometrika, 47*, 105–113.

Andrich, D. (1991). [Review of the book *Latent trait and latent class models*]. *Psychometrika, 56*, 155–168.

Andrich, D. (1995). Models for measurement, precision, and the nondichotomization of graded responses. *Psychometrika, 60*, 7–26.

Bartholomew, D.J. (1987). *Latent variable models and factor analysis.* London: Griffin.

Bishop, Y.M.M., Fienberg, S.E., & Holland, P.W. (1975). *Discrete multivariate analysis: Theory and practice.* Cambridge, MA: MIT Press.

Carstensen, C., & Rost, J. (1998). *MULTIRA: Ein Programmsystem für mehrdimensionale Rasch-Modelle* [MULTIRA: A program system for multidimensional Rasch models]. Kiel: IPN.

Cox, D.R. (1970). *The analysis of binary data.* Methuen: London.

Fischer, G.H. (1972). A measurement model for the effects of mass-media. *Acta Psychologica, 36*, 207–220.

Fischer, G.H. (1973). The linear logistic test model as an instrument in educational research. *Acta Psychologica, 37*, 359–374.

Fischer, G. (1974). *Einführung in die Theorie psychologischer Tests: Grundlagen und Anwendungen* [Introduction to mental test theory: Foundations and applications]. Bern: Huber.

Fischer, G.H. (1988). Spezifische Objektivität [Specific objectivity]. In K.D. Kubinger (Hrsg.), *Moderne Testtheorie* (pp. 87–111). Weinheim: Beltz.

Fischer, G.H. (1995). The linear logistic test model. In G.H. Fischer & I.W. Molenaar (Eds.), *Rasch models: Foundations, recent developments, and applications* (pp. 131–155). New York: Springer-Verlag.

Fischer, G.H., & Parzer, P (1991). An extension of the rating scale model with an application to the measurement of treatment effects. *Psychometrika, 56*, 637–651.

Fischer, G.H., & Ponocny, I. (1994). An extension of the partial credit model with an application to the measurement of change. *Psychometrika, 59*, 177–192.

Fischer, G.H., & Ponocny-Seliger, E. (1998). LPCM-WIN: For structural Rasch modeling [Software manual]. Groningen: ProGAMMA.

Formann, A.K. (1985). Constrained latent class models: Theory and applications. *British Journal of Mathematical and Statistical Psychology, 38*, 87–111.

Formann, A.K. (1989). Constrained latent class models: Some further applications. *British Journal of Mathematical and Statistical Psychology, 42*, 37–54.

Haberman, S.J. (1988). A stabilized Newton–Raphson algorithm for loglinear models for frequency tables derived by indirect observation. In C.C. Clogg (Ed.), *Sociological methodology 1988* (pp. 193–211). Washington, DC: American Sociological Association.

Hemker, B.T., Sijtsma, K., Molenaar, I.W., & Junker, B.W. (1996). Polytomous IRT models and monotone likelihood ratio of the total score. *Psychometrika, 61*, 679–693.

Hicks, L.E. (1970). Some properties of ipsative, normative, and forced-choice normative measures. *Psychological Bulletin, 74*, 167–184.

Hoijtink, H., & Rooks, G. (1999). Confirmatory factor analysis of items with a dichotomous response format using the multidimensional Rasch model. *Psychological Methods, 4*, 300–314.

Kelderman, H. (1984). Loglinear Rasch model tests. *Psychometrika, 49*, 223–245.

Kelderman, H., & Rijkes, C.P.M. (1994). Loglinear multidimensional IRT models for polytomously scored items. *Psychometrika, 59*, 149–176.

Klauer, K.C., & Sydow, H. (1992). Interindividuelle Unterschiede in der Lernfähigkeit: Zur Analyse von Lernprozessen bei Kurzzeitlerntests [Interindividual differences in learning ability: Analyzing learning by means of short-term learning tests]. *Zeitschrift für Differentielle und Diagnostische Psychologie, 13*, 175–190.

Klauer, K.C., & Sydow, H. (2001). Modeling learning in short-term learning tests. In A. Boomsma, M.A.J. van Duijn, & T.A.B. Snijders (Eds.), *Essays on item response theory* (pp. 69–87). New York: Springer-Verlag.

Langeheine, R., & Rost, J. (1988). *Latent trait and latent class models.* New York: Plenum.

Lazarsfeld, P.F., & Henry, N.W. (1968). *Latent structure analysis.* Boston: Houghton Mifflin.

Linacre, J.M. (1989). *Many-faceted Rasch measurement.* Chicago: MESA Press.

Lindsay, B., Clogg, C.C., & Grego, J. (1991). Semiparametric estimation in the Rasch model and related exponential response models, including a simple latent class model for item analysis. *Journal of the American Statistical Association, 86*, 96–107.

Luce, R.D., & Tukey, J.W. (1964). Simultaneous conjoint measurement: A new type of fundamental measurement. *Journal of Mathematical Psychology, 1*, 1–27.

Masters, G.N. (1982). A Rasch model for partial credit scoring. *Psychometrika, 47*, 149–174.

Meiser, T. (1996). Loglinear Rasch models for the analysis of stability and change. *Psychometrika, 61*, 629–645.

Micko, H.-C. (1970). Eine Verallgemeinerung des Meßmodells von Rasch mit einer Anwendung auf die Psychophysik der Reaktionen [A generalization of Rasch's measurement model with an application to the psychophysics of reactions]. *Psychologische Beiträge, 12*, 4–22.

Mislevy, R.J., & Verhelst, N. (1990). Modeling item responses when different subjects employ different solution strategies. *Psychometrika, 55*, 195–215.

Mokken, R.J. (1971). *A theory and procedure of scale analysis: With applications in political research.* The Hague: Mouton.

Molenaar, I.W. (1997). Nonparametric models for polytomous responses. In W.J. van der Linden & R.K. Hambleton (Eds.), *Handbook of modern item response theory* (pp. 3–14). New York: Springer-Verlag.

Müller, H. (1987). A Rasch model for continuous ratings. *Psychometrika, 52*, 165–181.

Muraki, E., & Carlson, J.E. (1995). Full-information factor analysis for polytomous item responses. *Applied Psychological Measurement, 19*, 73–90.

Rasch, G. (1960). *Probabilistic models for some intelligence and attainment tests.* Copenhagen: The Danish Institute of Educational Research. (Expanded edition, 1980. Chicago: The University of Chicago Press.)

Rasch, G. (1961). On general laws and the meaning of measurement in psychology. *Proceedings of the Fourth Berkeley Symposium on Mathematical Statistics and Theory of Probability* (Vol. IV, pp. 321–333). Berkeley: University of California Press.

Roskam, E.E., & Jansen, P.G.W. (1989). Conditions for Rasch-dichotomizability of the unidimensional polytomous Rasch model. *Psychometrika, 54*, 317–333.

Rost, J. (1988). Measuring attitudes with a threshold model drawing on a traditional scaling concept. *Applied Psychological Measurement, 12*, 397–409.

Rost, J. (1990). Rasch models in latent classes: An integration of two approaches to item analysis. *Applied Psychological Measurement, 14*, 271–282.

Rost, J. (1991). A logistic mixture distribution model for polychotomous item responses. *British Journal of Mathematical and Statistical Psychology*, 44, 75–92.

Rost, J. (1996). *Lehrbuch Testtheorie, Testkonstruktion* [Textbook test theory, test construction]. Bern: Huber.

Rost, J. (1999). Was ist aus dem Rasch-Modell geworden? [Where has the Rasch model gone?] *Psychologische Rundschau, 50*, 140–156.

Rost, J., & Carstensen, C. (in press). Multidimensional Rasch measurement via item component models and faceted designs. *Applied Psychological Measurement.*

Rost, J., & Langeheine, R. (1997). A guide through latent structure models for categorical data. In J. Rost & R. Langeheine (Eds.), *Applications of latent trait and latent class models in the social sciences* (pp. 13–37). Münster: Waxmann.

Stegelmann, W. (1983). Expanding the Rasch model to a general model having more than one dimension. *Psychometrika, 48*, 259–267.

Tjur, T. (1982). A connection between Rasch's item analysis model and a multiplicative Poisson model. *Scandinavian Journal of Statistics, 9*, 23–30.

Verhelst, N.D., & Glas, C.A.W. (1993). A dynamic generalization of the Rasch model. *Psychometrika, 58*, 395–415.

Verhelst, N.D., & Glas, C.A.W. (1995a). Dynamic generalization of the Rasch model. In G.H. Fischer & I.W. Molenaar (Eds.), *Rasch Models: Foundations, recent developments, and applications* (pp. 181–201). New York: Springer-Verlag.

Verhelst, N.D., & Glas, C.A.W. (1995b).The one parameter logistic model. In G.H. Fischer & I.W. Molenaar (Eds.), *Rasch models: Foundations, recent developments, and applications* (pp. 215–237). New York: Springer-Verlag.

Vermunt, J.K. (1996). LEM: A general program for the analysis of categorical data [Computer software]. Tilburg University, Department of Methodology, The Netherlands.

Von Davier, M. (1995). Mixtures of polytomous Rasch models and latent class models for ordinal variables. In F. Faulbaum & W. Bandilla (Eds.), *Softstat 95—Advances in Statistical Software 5.* Stuttgart: Lucius & Lucius.

Von Davier, M. (1998). WINMIRA: A WINDOWS program for mixed Rasch models [Computer software]. Kiel: IPN.

Wilson, M., & Wang, W. (1995). Complex composites: Issues that arise in combining different modes of assessment. *Applied Psychological Measurement, 19*, 51–71.

Wright, B.D., & Masters, G.N. (1982). *Rating scale analysis: Rasch measurement.* Chicago: Mesa Press.

Wu, M., Adams, R.J., & Wilson, M. (1997). ConQuest—Generalised item response modelling software (Draft Release 2) [Software manual]. Camberwell: Australian Council for Educational Research.

Yamamoto, K. (1989). *A hybrid model of IRT and latent class models* (ETS Research Report RR–89–41). Princeton, NJ: Educational Testing Service.

3

Gain Scores Revisited Under an IRT Perspective

Gerhard H. Fischer[1]

ABSTRACT For the measurement and statistical assessment of individual gain scores based on item sets that satisfy the assumptions of the Rasch, Rating Scale, or Partial Credit Models, a conditional maximum likelihood estimator, Clopper–Pearson confidence intervals, uniformly most accurate confidence intervals, and uniformly most powerful unbiased tests of the hypothesis of no change are presented. All methods are grounded on the exact conditional distribution of the gain score, given the total score for both time points, so that no asymptotic approximations are required. Typical applications of the methods are mentioned.

1 Introduction

The measurement of change in testees between testing occasions or conditions of observation and the statistical evaluation of such measures is a central topic in psychological research and assessment. Developmental and educational psychologists, for instance, ask how much children gain when certain educational conditions are improved, in other words, what the effects of educational conditions or treatments are with respect to abilities or traits of children. Clinical psychologists investigate the effects of treatments on patients; that is, they ask to what degree patients' psychic disorders or their subjective feelings of illness are assuaged by certain treatment conditions. Applied psychologists study the effects of working conditions or of training on the performance or job satisfaction of employees; in other words, they ask what the amount of change is in certain dimensions of attitude or achievement induced by changes of the working conditions or by training. All these questions can be asked meaningfully both in a group-oriented way (i.e., with respect to an average amount of change in a group of testees) or focusing on the individual. This chapter deals with the measurement of

[1]Department of Psychology, University of Vienna, Liebiggasse 5, A-1010 Wien, Austria; e-mail: *gh.fischer@univie.ac.at*

individual change between two time points and its statistical significance, exploiting the possibilities inherent in a family of Rasch models.

The problems of the measurement of change have been notorious within the framework of classical psychometrics (cf., e.g., Bereiter, 1963; Cronbach & Furby, 1970; Holtzman, 1963; Huber, 1977; Willett, 1989; Williams & Zimmerman, 1996). Modern IRT, and particularly a family of Rasch models, allows interesting new approaches to the problem. Many authors have realized these possibilities: mainly, however, in regard to group-oriented research; work on IRT models for the measurement and statistical testing of individual change, on the other hand, still seems to be scarce (Embretson, 1991; Fischer, 1995; Fischer & Ponocny-Seliger, 1998; Hoijtink & Boomsma, 1996; Mellenbergh & Van den Brink, 1998; Ponocny & Ponocny-Seliger, 1999). It must be said, though, that some work on person fit is also highly relevant to the issue (e.g., Klauer, 1991a, 1995; Liou, 1993; Liou & Chang, 1992; Molenaar & Hoijtink, 1990; Ponocny, 2000; for a recent review, see Meijer & Sijtsma, in press), because intraindividual change can be viewed as a violation of the assumption of a constant person parameter and/or of the generalizability of item parameters over persons.

The following questions are considered for a pretest–posttest design (two testing occasions where raw scores r_{v1} and r_{v2}, respectively, are observed on testee S_v), within the framework of a family of Rasch models.

1. How can change be measured?
2. What are the metric (or scale) properties of measurements of change?
3. How can the change measurements be assessed statistically?

The advantage of the chosen family of IRT models—defined by the Partial Credit Model (PCM; Masters, 1982), which comprises the Rating Scale Model (RSM; Rasch, 1965; Andrich, 1978) and the Rasch Model (RM; Rasch, 1960) as special cases—is that inference can be based on the exact conditional distribution of the gain score, given the item parameters of the test and the observed total raw score of the testee. As shown, this implies that no questionable asymptotics are needed.

Before going into technical details, a typical area of application of the methods proposed below is mentioned. One subfield of psychology where the quest for measuring individual change invariably arises is psychological assessment. Let us suppose that testees are tested under neutral laboratory conditions and again under some sort of strain or load and that the difference in performance between the two testing conditions is used as a predictor, for instance, of stress resistance. This approach to psychological assessment has been studied extensively and termed "ergopsychometry" by G. Guttmann (cf. Guttmann & Etlinger, 1991). Applications have been done particularly in sports psychology. Recently, a doctoral student of the author (Prieler, 2000) carried out a joint study with the Department of

Military Psychology of the Österreichisches Bundesheer (Austrian army), in which candidates for the career of noncommissioned officer were subjected to a battery of tests both before and after a long and strenuous night march. The research question was to what extent the difference in test performance between normal conditions and after the night march was a measure of stress resistance and thus a useful predictor of aptitude for the intended career.

In that study, a psychometric approach proposed by Fischer (1995) for the measurement and the statistical assessment of individual change in test performance was used, based on the assumption that the RM was previously applied to the test and found to fit sufficiently well. The method is implemented in the software LPCM-WIN 1.0 (Fischer & Ponocny-Seliger, 1998). Within the framework of dichotomous Rasch tests, this approach answers all three questions raised above: measurement can be realized by means of conditional maximum likelihood (CML) estimation of the individual change parameter; under suitable assumptions about the item universe, the scale of the change measure is a ratio scale (cf. Fischer, 1987, 1995); the statistical evaluation of change can be carried out by means of a Clopper–Pearson confidence interval for the individual change parameter, based on the exact conditional null distribution of the gain score, given the item parameters and the total score of the testee.

The present chapter deals with a generalization of this method to tests consisting of ordered-response items where either the RSM or the PCM is assumed to apply. First, the CML method for the estimation of a change parameter is extended to the present class of models. Then, in addition to the Clopper–Pearson intervals which are known to be somewhat conservative, uniformly most accurate (UMA) confidence intervals and uniformly most powerful (UMP) unbiased tests are presented. Thereby, special attention is paid to questions concerning the existence and uniqueness of the boundaries of such intervals. Methods described in detail by Witting (1985, Chap. 2) in connection with UMA confidence intervals and UMP unbiased hypothesis tests for the binomial distribution are employed, following Klauer (1991a, 1991b) who was the first to apply these techniques to the (dichotomous) RM. Finally, an artificial example with four-categorical ordered-response items is given to demonstrate what results can be expected from an application of the proposed methods.

The substantive motivation for the present generalization is that, in clinical psychology, rating scales are popular for the assessment of changes in the severity of psychopathological disorders of patients or, in applied psychology, for the measurement of changes in dimensions like self-assertiveness or job satisfaction, under certain training or working conditions, respectively; similarly, in social psychology attitude scales with ordered-response items are often used to monitor changes of attitude. Researchers then try to find answers to the questions listed above, namely, how much patient or testee S_v has changed on the trait measured by the scale, what the status of the "measurement" of change is, and whether the observed change is statistically significant. Therefore it is important to make sound psychometric methods available that can be employed for these purposes.

2 Measuring Change on the Basis of a PCM

Throughout this chapter, the following notation and definitions are used. Let the item sample consist of a set of items $\mathcal{I} = I_1, \ldots, I_k$. It is assumed that the PCM fits these items,

$$P(X_{vij} = 1 \mid \theta_v, \beta_{i0}, \ldots, \beta_{im_i}) = \frac{\exp(j\theta_v + \beta_{ij})}{\sum_{l=0}^{m_i} \exp(l\theta_v + \beta_{il})}, \tag{3.1}$$

where

X_{vij} denotes the random response variable with realizations $x_{vij} = 1$ if person S_v's response to item I_i belongs to response category C_{ij}, and $x_{vij} = 0$ otherwise;

m_i is the number of item I_i's response categories minus 1 (the response categories being numbered $0, \ldots, m_i$);

θ_v is testee S_v's person parameter;

β_{ij}, for $j = 0, \ldots, m_i$, are item I_i's item $\times$ category parameters.

To make the model identifiable, $k + 1$ normalization conditions have to be imposed on the parameters; for instance,

$$\beta_{i0} = 0, \quad \text{for all } i, \text{ and} \quad \sum_{i=1}^{k}\sum_{j=0}^{m_i} \beta_{ij} = 0. \tag{3.2}$$

In psychometric literature, the PCM is often presented in a somewhat different but equivalent form: the β_{ij} are written as sums of "threshold" parameters τ_{ij}, where $\beta_{ij} = \sum_{l=0}^{j} \tau_{ij}$, with $\tau_{i0} = 0$. For the following, however, the introduction of threshold parameters is neither necessary nor convenient. Rather, it is advantageous to reparameterize (3.1) as

$$P(X_{vij} = 1 \mid \xi_v, \epsilon_{i0}, \ldots, \epsilon_{im_i}) = \frac{\xi_v^j \epsilon_{ij}}{\sum_{l=0}^{m_i} \xi_v^l \epsilon_{il}}, \tag{3.3}$$

where $\xi_v = \exp(\theta_v)$, and $\epsilon_{ij} = \exp(\beta_{ij})$. The normalization conditions (3.2) now become

$$\epsilon_{i0} = 1, \quad \text{for all } i, \text{ and} \quad \prod_{i=1}^{k}\prod_{j=0}^{m_i} \epsilon_{ij} = 1. \tag{3.4}$$

Note that the RSM and the RM are special cases of the PCM: let all items have the same $m+1$ response categories $C_0,\ldots,C_m$, and let the item $\times$ category parameters β_{ij} satisfy the restrictions

$$\beta_{ij} = j\beta_i + \omega_j \ , \tag{3.5}$$

where β_i is an easiness parameter of item I_i, and ω_j is an attractiveness parameter of response category C_j; then (3.1) becomes

$$P(X_{vij} = 1 \mid \theta_v, \beta_i, \omega_0, \ldots, \omega_m) = \frac{\exp[j(\theta_v + \beta_i) + \omega_j]}{\sum_{l=0}^{m} \exp[l(\theta_v + \beta_i) + \omega_l]} \ , \tag{3.6}$$

which is the RSM. Three restrictions are required. Two conditions are implied by (3.2), $\omega_0 = 0$ and $\sum_j j \sum_i \beta_i + k \sum_j \omega_j = 0$. Thus we may put $\sum_i \beta_i = 0$ and $\sum_j \omega_j = 0$, or alternatively,

$$\omega_0 = \omega_1 = 0 \quad \text{and} \quad \sum_i \beta_i = 0 \ . \tag{3.7}$$

Finally, the RM is easily obtained from (3.6) and (3.7) by setting $m = 1$,

$$P(X_{vi1} = 1 \mid \theta_v, \beta_i) = \frac{\exp(\theta_v + \beta_i)}{1 + \exp(\theta_v + \beta_i)} \ , \tag{3.8}$$

where, however, the β_i are interpreted as easiness (instead of the more familiar difficulty) parameters.

All three models, PCM, RSM, and RM, have in common that the raw score is a sufficient statistic for the person parameter, given the item and category parameters. The raw score is of the same form in all three models, namely,

$$r_v = \sum_{i=1}^{k} \sum_{j=0}^{m_i} j x_{vij} \ , \tag{3.9}$$

with $m_i = m$ for the RSM, and $m_i = 1$ for the RM.

As regards the topic of this chapter, it suffices to treat the case of the PCM: all results immediately apply also to the RSM and the RM.

For the following considerations it is assumed that some (or all) of the items $\mathcal{I} = I_1, \ldots, I_k$ are presented at time point T_1, henceforth called the "pretest," denoted $\mathcal{I}_1$, and some (or all) items are presented at time point T_2, henceforth called the "posttest," denoted $\mathcal{I}_2$. The nonempty item subsets $\mathcal{I}_1$ and $\mathcal{I}_2$ may be completely different, may overlap, or may be identical. For convenience, however, a notation is adopted where $\mathcal{I}_1$ and $\mathcal{I}_2$ are considered as disjoint subsets of $\mathcal{I}$, namely, $\mathcal{I}_1 = \{I_1, \ldots, I_h\}$ and $\mathcal{I}_2 = \{I_{h+1}, \ldots, I_k\}$. But cases where $\mathcal{I}_1$ and $\mathcal{I}_2$ overlap are implicitly covered; it suffices to let some $I_a \in \mathcal{I}_1$ have the same parameters as some $I_b \in \mathcal{I}_2$.

Note that the perfect scores attainable in $\mathcal{I}_1$ and $\mathcal{I}_2$ are $s_1 = \sum_{i=1}^{h} m_i$ and $s_2 = \sum_{i=h+1}^{k} m_i$, respectively.

TABLE 3.1. Pretest–posttest design and notation.

	T_1	T_2
Occasions	T_1	T_2
Subtests	$\mathcal{I}_1$	$\mathcal{I}_2$
Items	$I_1, \ldots, I_h$	$I_{h+1}, \ldots, I_k$
Responses	$\mathbf{x}_1, \ldots, \mathbf{x}_h$	$\mathbf{x}_{h+1}, \ldots, \mathbf{x}_k$
Response Patterns	$\mathbf{X}_1$	$\mathbf{X}_2$
Numbers of Items	h	$k-h$
Raw Scores	r_1	$r_2 = r - r_1$
Perfect Scores	s_1	s_2
Item × Category Parameters of the Real Items	$\epsilon_{10}, \ldots, \epsilon_{hm_h}$	$\epsilon_{h+1,0}, \ldots, \epsilon_{km_k}$
Parameter Vector	$\boldsymbol{\epsilon}$	
Item × Category Parameters of the Virtual Items	$\epsilon_{10}, \ldots, \epsilon_{hm_h}$	$\delta^0\epsilon_{h+1,0}, \ldots, \delta^{m_k}\epsilon_{km_k}$

The model in (3.1) and (3.3) is assumed to hold for all items if they are presented on a single occasion. Therefore, the model holds for the items of pretest $\mathcal{I}_1$. For posttest $\mathcal{I}_2$, however, the model has to be rewritten since it is assumed that the person parameter θ_v may change between the two testing occasions; so θ_v has to be replaced by $\theta_v + \eta_v$, and ξ_v by $\xi_v\delta_v$, where η_v is a parameter representing the amount of change in person S_v, and $\delta_v = \exp(\eta_v) > 0$. Therefore, the model for the posttest becomes

$$P(X_{vij} = 1 \mid \xi_v, \epsilon_{i0}, \ldots, \epsilon_{im_i}, \delta_v) = \frac{(\xi_v\delta_v)^j \epsilon_{ij}}{\sum_{l=0}^{m_i} (\xi_v\delta_v)^l \epsilon_{il}} \,. \tag{3.10}$$

As is immediately clear, the model in (3.3), for $\mathcal{I}_1$, and the model in (3.10), for $\mathcal{I}_2$, can be jointly interpreted as *one* PCM for all items of $\mathcal{I} = I_1, \ldots, I_k$ given at a single time point to a person with a constant person parameter ξ_v, where, however, the item × category parameters for all $I_i \in \mathcal{I}_2$ have changed from ϵ_{ij} to $\delta_v^j\epsilon_{ij}$. Since items characterized by the latter parameters do not exist in reality, they can be denoted "virtual items."

The symbol $\boldsymbol{\epsilon}$ is used for the entire vector of item × category parameters $(\epsilon_{10}, \ldots, \epsilon_{1m_1}; \ldots; \epsilon_{k0}, \ldots, \epsilon_{km_k})$. Dropping index v henceforth for convenience, the design and notation can be summarized as in Table 3.1.

For the measurement of the individual amount of change δ, a straightforward approach is to maximize the conditional likelihood of S's response pattern, $\mathbf{X} = (\mathbf{x}_1, \ldots, \mathbf{x}_k)$, given S's total raw score $r = r_1 + r_2$, where $\mathbf{x}_i$ denotes the response vector $(x_{vi1}, \ldots, x_{vim_i})'$. As is well known, this conditional likelihood is independent of the person parameter ξ and thus depends only on δ (the item $\times$ category parameters $\boldsymbol{\epsilon}$ being treated as known constants). This conditional probability (likelihood) is

$$P(\mathbf{X} \mid r) = \frac{\prod_{i=1}^{k} \prod_{j=0}^{m_i} \epsilon_{ij}^{x_{ij}} \delta^{r_2}}{\gamma_r(\mathcal{I}_1, \mathcal{I}_2, \delta)} , \tag{3.11}$$

where $\gamma_r(\mathcal{I}_1, \mathcal{I}_2, \delta) = \gamma_r(\epsilon_{10}, \ldots, \epsilon_{hm_h}; \delta^0 \epsilon_{h+1,0}, \ldots, \delta^{m_k} \epsilon_{km_k})$ is a combinatorial expression resulting from the summation over all patterns $\mathbf{Y}$ of responses compatible with the fixed total score r, denoted by $\mathbf{Y} \mid r$ under the summation sign in (3.12), namely,

$$\gamma_r(\mathcal{I}_1; \mathcal{I}_2, \delta) = \sum_{\mathbf{Y} \mid r} \left(\prod_{i=1}^{h} \prod_{j=0}^{m_i} \epsilon_{ij}^{y_{ij}} \right) \left(\prod_{i=h+1}^{k} \prod_{j=0}^{m_i} (\delta^j \epsilon_{ij})^{y_{ij}} \right) \tag{3.12}$$

(cf. Andersen, 1995, p. 277). Denoting that part of response pattern $\mathbf{Y}$ which pertains to $\mathcal{I}_1$, as $\mathbf{Y}_1$, and the rest, which pertains to $\mathcal{I}_2$, as $\mathbf{Y}_2$, with respective raw scores r_1 and r_2 satisfying $r_1 + r_2 = r$, we can rewrite (3.12) as

$$\gamma_r(\mathcal{I}_1, \mathcal{I}_2, \delta) = \sum_{\substack{r_1, r_2 \\ r_1 + r_2 = r}} \left(\sum_{\mathbf{Y}_1 \mid r_1} \prod_{i=1}^{h} \prod_{j=0}^{m_i} \epsilon_{ij}^{y_{ij}} \right) \left(\sum_{\mathbf{Y}_2 \mid r_2} \prod_{i=h+1}^{k} \prod_{j=0}^{m_i} \epsilon_{ij}^{y_{ij}} \right) \delta^{r_2}, \tag{3.13}$$

where the summation is taken over all r_1 and r_2 sufficing $r_1 + r_2 = r$. This again can be written in a more compact form as

$$\gamma_r(\mathcal{I}_1, \mathcal{I}_2, \delta) = \sum_{l=a}^{b} \gamma_{r-l}(\mathcal{I}_1) \gamma_l(\mathcal{I}_2) \delta^l . \tag{3.14}$$

The limits of this summation, a and b, are determined by the following obvious inequality constraints for the summation index l: $l \geq 0$, $l \geq r - s_1$, $l \leq s_2$, and $l \leq r$. These inequalities imply

$$a = \max(0, r - s_1) , \tag{3.15}$$
$$b = \min(r, s_2) . \tag{3.16}$$

Similarly, the conditional probability of posttest score r_2, given r, is

$$\begin{aligned} p_\delta(r_2) = P(r_2 \mid r) &= P(r_2 - r_1 \mid r) = \sum_{\mathbf{Y} \mid (r_1, r_2)} \frac{\prod_{i=1}^{k} \prod_{j=0}^{m_i} \epsilon_{ij}^{y_{ij}} \delta^{r_2}}{\gamma_r(\mathcal{I}_1, \mathcal{I}_2, \delta)} \\ &= \frac{\gamma_{r-r_2}(\mathcal{I}_1) \gamma_{r_2}(\mathcal{I}_2) \delta^{r_2}}{\gamma_r(\mathcal{I}_1, \mathcal{I}_2, \delta)} , \end{aligned} \tag{3.17}$$

the summation being taken over all response patterns $\mathbf{Y}$ with fixed scores r_1 and r_2 in $\mathcal{I}_1$ and $\mathcal{I}_2$, respectively. Note that the conditional distribution of the gain score variable R_2 is one-parametric, with parameter δ.

It is now an easy matter to find the conditional maximum likelihood (CML) estimator for the change parameter δ (and thus also for $\eta = \ln \delta$). One way to this goal would be to equate the derivative of the logarithm of (3.11) with respect to δ with zero. Another way is to apply results from the theory of exponential families: knowing that the PCM is an exponential family (Andersen, 1990, p. 416), it follows from standard results that the conditional model defined by (3.11) is also an exponential family and that the CML estimation equation for δ (and $\eta = \ln \delta$) is therefore obtained by equating the observed posttest score r_2 with its conditional expectation,

$$r_2 = \sum_{l=a}^{b} p_\delta(l)l = \sum_{l=a}^{b} \frac{\gamma_{r-l}(\mathcal{I}_1)\gamma_l(\mathcal{I}_2)\delta^l l}{\gamma_r(\mathcal{I}_1, \mathcal{I}_2, \delta)} . \tag{3.18}$$

If (3.18) is solved for $\hat{\delta}$, an estimate $\hat{\eta} = \ln \hat{\delta}$ of S's amount of change on the latent scale θ obtains. Presuming that the PCM, under suitable assumptions about the item universe, can be shown to yield measures of the latent trait θ on an interval scale (here, however, without proof), the estimator $\hat{\eta}$ lies on a ratio scale (η being a difference of values of θ). This basically answers Question 1, how to measure change, and Question 2, what the metric properties of the measurements are (cf. Fischer, 1987, 1995).

3 Some Technical Considerations

In this section, the following questions are answered: when does a finite CML estimator—a positive root of (3.18)—exist, when is it unique, what is its behavior when the gain score is extreme (i.e., falls on the boundary of the support of the exponential family), and how can it be computed numerically? First, we rewrite the estimation equation (3.18) as

$$F(\delta) = \sum_{l=a}^{b} (r_2 - l)\gamma_{r-l}(\mathcal{I}_1)\gamma_l(\mathcal{I}_2)\delta^l = 0 . \tag{3.19}$$

Function F is a polynomial in δ, with coefficients $(r_2 - l)\gamma_{r-l}(\mathcal{I}_1)\gamma_l(\mathcal{I}_2)$. Notice that, since the γ's are positive numbers, the signs of the coefficients depend only on $r_2 - l$, for $l = a, \ldots, b$.

Obviously, cases where $r = 0$ or $r = s_1 + s_2 = \sum_{i=1}^{k} m_i$ have to be entirely excluded, because then the conditional likelihood (3.11) would be 1 and thus uninformative. (Even intuitively it is clear that a measurement of change is not feasible when the total score is extreme, because then the scores on both the pretest and posttest must also be extreme.) For all other cases, the above questions are answered by the following Theorem 3.1.

Theorem 3.1 Let $0 < r < s_1 + s_2$. For $0 < r_1 < s_1$ and $0 < r_2 < s_2$, the CML equation (3.19) has a unique positive root $\hat{\delta}$; that is, a finite CML estimator $\hat{\delta}$ exists and is unique. For $r_1 = 0$ or $r_2 = s_2$, the CML estimator $\hat{\delta}$ diverges to $+\infty$ (implying $\hat{\eta} \to +\infty$), and for $r_1 = s_1$ or $r_2 = 0$, $\hat{\delta}$ converges to 0 (implying $\hat{\eta} \to -\infty$).

Proof First observe that $a = b$ is not feasible, because then, owing to (3.15) and (3.16), one of the following relations would have to hold: (a) $r = 0$; (b) $s_1 = 0$; (c) $s_2 = 0$; (d) $r - s_1 = s_2$, implying $r = s_1 + s_2$. All four contradict the assumptions made. Therefore it can be concluded that index l in (3.19) must take on at least two different values.

Next consider those cases where either r_1 or r_2 is extreme, which implies that the gain score $r_2 - r_1$ lies on the boundary of the support of the exponential family. The consequences for the behavior of the estimator $\hat{\delta}$ are seen from the following considerations.

(a) Case $r_1 = 0$, which implies $r_2 = r > 0$ and $r_2 = b$. If $\hat{\delta}$ were to satisfy (3.18), the term $\gamma_0(\mathcal{I}_1)\gamma_{r_2}(\mathcal{I}_2)\hat{\delta}^{r_2}$ would have to be infinitely large in comparison with all other terms, which entails that $\hat{\delta}$ (and therefore also $\hat{\eta}$) would diverge to $+\infty$.

(b) Case $r_2 = s_2$, implying $b = \min(r, r_2) = r_2$. The consequences are the same as in (a).

(c) Case $r_1 = s_1$. It follows that $a = \max(0, r - r_1) = r_2$. If $\hat{\delta}$ were a root of (3.18), the term $\gamma_{r_1}(\mathcal{I}_1)\gamma_{r_2}(\mathcal{I}_2)\hat{\delta}^{r_2}$ would have to be infinitely large in comparison with all other terms with $l > r_2$; that is, $\hat{\delta}$ would go to 0 (and $\hat{\eta}$ to $-\infty$).

(d) Case $r_2 = 0$, implying $r = r_1$, $a = \max(0, r_1 - s_1) = 0$, and $b > 0$. Then the term $\gamma_0(\mathcal{I}_1)\gamma_r(\mathcal{I}_2)\delta^0$ would have to become infinitely large in comparison with all other terms, with consequences as in (c).

So only cases with $0 < r_1 < s_1$ and $0 < r_2 < s_2$ remain; in these, a unique CML estimator exists by virtue of the well-known convexity of the loglikelihood function in exponential families. (The same can be concluded from (3.19) by applying Descartes' "sign rule": only one sign change occurs in the sequence of coefficients of $F(\delta)$.) □

The next technical question is how to compute the γ-functions needed in (3.19). Recursion formulae for these combinatorial functions have been described in the literature (e.g., Andersen, 1995, p. 279; Fischer & Ponocny, 1995, pp. 361–363). For the sake of completeness, a simple and very effective recursion—in fact, a special case of (3.14), first used by Andersen (1972)—is repeated here: to compute all $\gamma_r(I_1, \ldots, I_t)$ for $r = 0, \ldots, \sum_{i=1}^{t} m_i$, apply the following equation repeatedly, for $j = 2, \ldots, t$:

$$\gamma_r(I_1, \ldots, I_j) = \sum_{l=u}^{w} \gamma_{r-l}(I_1, \ldots, I_{j-1})\gamma_l(I_j) , \tag{3.20}$$

where $u = \max(0, r - \sum_{i=1}^{j-1} m_i)$, $w = \min(r, m_j)$, $\gamma_0(I_1, \ldots, I_j) = 1$ for all j, and $\gamma_l(I_j) = \epsilon_{jl}$. The method is fast and numerically stable, since only multiplications and additions of positive numbers are involved.

The last question is how to solve the polynomial (3.19) for $\hat{\delta}$. This is not entirely trivial because F is a polynomial of high order: if, for instance, the same 20 items with four categories each are given at T_1 and T_2, then $h = 20$, $k - h = 20$, $m_i = m = 3$, and the order of the polynomial may go up to 120; moreover, some coefficients of the polynomial become very large. Therefore, the computations can be affected by numerical error.

As it turned out, the Newton method is not suitable for solving the present polynomial equation, owing to the occurrence of so-called "spider-web phenomena." The Regula Falsi is also ineffective for the following reason. It is based on a weighted averaging of approximate values δ_1 and δ_2 sufficing $F(\delta_1) < 0$ and $F(\delta_2) > 0$, whereby $|F(\delta_1)|$ and $|F(\delta_2)|$ are employed as weights of δ_2 and δ_1, respectively. If, for instance, $|F(\delta_2)|$ is much larger than $|F(\delta_1)|$, the weighted average, δ_3, will be very near to δ_1; in other words, the convergence will become extremely slow, or—owing to numerical error—the iterations may even get permanently caught at δ_1.

Fortunately, there is a simple and effective solution to the problem. Instead of the weighted average, take $\delta_3 = \sqrt{\delta_1 \delta_2}$, which is equivalent to using the unweighted average $\eta_3 = (\eta_1 + \eta_2)/2$, and proceed with the pair (δ_1, δ_3) if $F(\delta_3) > 0$, or with (δ_3, δ_2) if $F(\delta_3) < 0$, and so on. This bisection algorithm converges only linearly, but it has been seen to be quite efficient for the present task because only the sign of F is needed.

To summarize this section, the computation of the combinatorial functions γ and of the root of the polynomial equation (3.19) poses no serious problems. This answers the practical side of Question 1, namely, how to measure change within the specified framework.

4 Statistical Assessment of Change Under a One-Sided H_1

Question 3 still remains: how to assess individual change statistically. One-sided alternative hypotheses, which are both the simpler and the more typical ones in applications, are treated first.

Suppose the researcher expects an improvement of test performance after some treatment or training, so that the null hypothesis H_0: $\delta = 1$ (or H_0: $\eta = 0$, the "no change" hypothesis) is to be tested against the alternative hypothesis H_1: $\delta > 1$ (or H_1: $\eta > 0$). Such a test is again conveniently based on the conditional distribution of R_2, given r, characterized by the probability function (3.18). Under H_0: $\delta = \delta_0 = 1$, the probability of observing raw scores $R_2 \geq r_2$ is

$$P(R_2 \geq r_2 \mid \epsilon, \delta_0, r) = \sum_{l=r_2}^{b} p_{\delta_0}(l) \,. \tag{3.21}$$

Hence, the H_0 is rejected on significance level α if the probability in (3.21) is $\leq \alpha$. This test of significance has some attractive properties. First, it does not require any asymptotic approximations because it is based on the exact conditional distribution of the raw score R_2. Second, it is an intuitively plausible approach analogous to familiar tests in other discrete distributions, for example, binomial tests.

Another, related approach to the statistical assessment of individual change consists in the computation of a so-called Clopper–Pearson confidence interval $[\delta_*, \infty)$ for the change parameter δ, again using the same conditional distribution. Its lower boundary δ_* is that value $\delta > 0$ for which

$$P(R_2 \geq r_2 \mid \epsilon, \delta, r) = \alpha \tag{3.22}$$

(cf. Santner & Duffy, 1989). Equation 3.22 is again conveniently rewritten as a polynomial: define dummy constants $g_l = 0$, for $l = a, \ldots, r_2 - 1$, and $g_l = 1$, for $l = r_2, \ldots, b$. Then the left-hand side of (3.22) can be written as $\sum_{l=a}^{b} g_l p_\delta(l)$, the right-hand side as $\alpha \sum_{l=a}^{b} p_\delta(l)$, and rearranging terms yields a polynomial equation in δ,

$$G(\delta) = \sum_{l=a}^{b} (g_l - \alpha)\gamma_{r-l}(\mathcal{I}_1)\gamma_l(\mathcal{I}_2)\delta^l = 0 \,. \tag{3.23}$$

The boundary δ_* is then a positive root of (3.23).

By symmetry, under alternative hypothesis H_1: $\delta < 1$, the upper boundary of a one-sided Clopper–Pearson interval $(-\infty, \delta^*]$ is also obtained as a root $\delta^* > 0$ of the polynomial equation (3.23), however, with dummy constants $g_l = 1$ for $l = a, \ldots, r_2$ and $g_l = 0$ for $l = r_2 + 1, \ldots, b$.

Equation 3.23 can be solved in the same manner as the polynomial (3.19) above. The uniqueness questions connected with (3.23) are answered by the following theorem.

Theorem 3.2 The lower boundary δ_* of a one-sided Clopper–Pearson confidence interval $[\delta_*, \infty)$ is unique if $a < r_2 \leq b$; for $r_2 = a$, no root $\delta_* > 0$ of (3.23) exists. The upper boundary δ^* of the one-sided Clopper–Pearson interval $(0, \delta^*]$ is unique if $a \leq r_2 < b$; for $r_2 = b$, no finite root $\delta^* > 0$ of (3.23) exists.

Note For the interval $[\delta_*, \infty)$ with $r_2 = a$, technically put $\delta_* = 0$ and $\eta_* = -\infty$; for the interval $(-\infty, \delta^*]$ with $r_2 = b$, put $\delta^* = \infty$ and $\eta^* = \infty$.

Proof Both in the polynomial equation for the lower boundary δ_* and in that of the upper boundary δ^*, one and only one sign change occurs in the sequence of coefficients (except for the obvious boundary cases: $r_2 = a$ for δ_*, $r_2 = b$ for δ^*). Hence, the uniqueness of the respective positive roots follows from Descartes' sign rule. □

The null hypothesis H_0: $\delta = 1$ against H_1: $\delta > 1$ is rejected if the point $\delta = 1$ lies outside the confidence interval. The result is in accordance with that obtained by means of the probability (3.21). Testing H_0: $\delta = 1$ against H_1: $\delta < 1$ is completely analogous.

The disadvantage of the Clopper–Pearson confidence intervals and of the probabilities (3.21) as means of hypothesis testing is that they lead to somewhat conservative decisions. The construction of a UMP unbiased test of the H_0 of no change and of a UMA confidence interval for δ (or η), on the other hand, requires transforming the discrete random variable R_2 into a continuous variable $W = R_2 + U$, where R_2 and U are stochastically independent and U has a rectangular distribution on support $[0, 1)$ (cf., e.g., Witting, 1985, Chap. 2). The result of this randomization is a continuous distribution function of W, strictly monotone decreasing in δ,

$$A_\delta(w) = \sum_{l=a}^{[w]-1} p_\delta(l) + (w - [w]) p_\delta([w]) \, , \tag{3.24}$$

where $[w]$ denotes the largest integer $\leq w$ and the sum is defined to be zero if $a > [w] - 1$. Both the construction of a hypothesis test and of a confidence interval for δ can be based on this continuous distribution.

The lower boundary of the one-sided UMA confidence interval $[\delta_*, \infty)$ under H_1: $\delta > 1$ is obtained by solving

$$A_{\delta_*}(w) = 1 - \alpha \tag{3.25}$$

for δ_*. Analogously, the upper boundary δ^* of the one-sided confidence interval $(0, \delta^*]$ under H_1: $\delta < 1$ is obtained by solving

$$A_{\delta^*}(w) = \alpha \tag{3.26}$$

for δ^*. The question of the uniqueness of these boundaries and of their behavior when r_2 is extreme is answered in Theorem 3.3.

Theorem 3.3 A unique root $\delta_* > 0$ of Equation 3.25 exists for $a + 1 - \alpha < w < b + 1 - \alpha$, and a unique root $\delta^* > 0$ of Equation 3.26 exists for $a + \alpha < w < b + \alpha$.

Note In the case of the interval $[\delta_*, \infty)$, technically put $\delta_* = 0$ ($\eta_* = -\infty$) for $a \leq w \leq a + 1 - \alpha$, and $\delta_* = \infty$ ($\eta_* = \infty$) for $b + 1 - \alpha \leq w < b + 1$. In the case of the interval $(-\infty, \delta^*]$, put $\delta^* = 0$ ($\eta^* = -\infty$) for $a \leq w \leq a + \alpha$, and $\delta^* = \infty$ ($\eta^* = \infty$) for $b + \alpha \leq w < b + 1$.

Proof The uniqueness of the roots of (3.25) and (3.26) follows from the monotonicity in δ of the distribution function $A_\delta(w)$. The boundary cases can be found by inspection of (3.24): it is seen that

$$\lim_{\delta \to 0} A_\delta(w) = \begin{cases} 1 \, , & \text{for } w \geq a + 1 \, , \\ w - [w] \, , & \text{for } a \leq w < a + 1 \, , \end{cases} \tag{3.27}$$

and

$$\lim_{\delta\to\infty} A_\delta(w) = \begin{cases} 0\,, & \text{for } w < b\,, \\ w - [w]\,, & \text{for } b \le w < b+1\,. \end{cases} \tag{3.28}$$

The limits (3.27) imply that, for $w - a = 1 - \alpha$, that is, for $w = a + 1 - \alpha$, $\delta_* \to 0$, and that for $w < a+1-\alpha$ there is no solution $\delta_* > 0$ of (3.25). From (3.28) it can be concluded that, for $w - b = 1 - \alpha$, that is, for $w = b+1-\alpha$, $\delta_* \to \infty$, and that for $w > b + 1 - \alpha$ there is no finite solution $\delta_* > 0$. Analogously, (3.27) implies that, for $w = a + \alpha$, $\delta^* \to 0$, and that for $w < a + \alpha$ there is no solution $\delta^* > 0$ of (3.26). Finally, (3.28) implies that for $w = b + \alpha$, $\delta^* \to \infty$, and that for $w > b + \alpha$ there is no finite solution $\delta^* > 0$. □

The essential disadvantage of the UMA confidence intervals is that the boundary in each individual case depends on the actual realization u of the random component U. Therefore, the confidence intervals for the change parameters will vary slightly between persons who attained exactly the same raw scores r_1 and r_2. This may be considered undesirable by applied researchers.

A UMP unbiased test of H_0: $\delta = 1$ against H_1: $\delta > 1$ can be carried out in two (equivalent) ways. One method is to reject the H_0 if the parameter $\delta = 1$ lies outside the confidence interval $[\delta_*, \infty)$. The other is to put $\delta = 1$ in $A_\delta(w)$ and to compute the numerical value of the latter. If this probability exceeds $1 - \alpha$, then H_0 is rejected on significance level α. The case of the UMP unbiased test against H_1: $\delta < 1$ is analogous.

It should be mentioned that for the special case of the dichotomous RM, one-sided randomized tests have already been suggested by Liou (1993), as person-fit indices under the alternative hypothesis of learning during testtaking.

The UMP hypothesis tests have the attractive property that they are always at least as powerful as the test procedures based on Clopper–Pearson confidence intervals; however, they share with the UMA confidence intervals the problem that the outcome depends on the random component U and may therefore differ between individuals with the same raw scores.

5 Statistical Assessment of Change Under a Two-Sided H_1

The construction of a two-sided confidence interval and of a test of H_0: $\delta = \delta_0$ against H_1: $\delta \neq \delta_0$ is a somewhat more complicated matter than the one-sided case. A "naive" method is to apply two-sided Clopper–Pearson confidence intervals and related hypothesis tests, with $\alpha/2$ on either side. This makes sense, however, only if the conditional distribution is symmetric (e.g., if the pretest and posttest are composed of the same items); but even then these confidence intervals are not the smallest possible and the tests not most powerful. Moreover, if pretest and posttest comprise different samples of items, the conditional distribution will generally not be symmetric, so that this simple method lacks a rigorous justification.

A UMA two-sided confidence interval and a UMP unbiased two-sided test can be constructed as follows. According to the theory of UMP unbiased tests in one-parametric exponential families, such a test can be written as a function ϕ of the observation w,

$$\phi(w) = \begin{cases} 1 \, , & \text{for } w \leq c_1 \text{ or } w \geq c_2 \, , \\ 0 \, , & \text{otherwise} \, , \end{cases} \tag{3.29}$$

where W is the randomized score variable, c_1 and c_2 are cutoff scores, $\phi = 1$ means rejecting, and $\phi = 0$, retaining the H_0. The cutoff scores are determined by the system

$$E_{\delta_0}[\phi(W)] = \alpha \, , \tag{3.30}$$

$$E_{\delta_0}[R_2\phi(W)] = \alpha E_{\delta_0}(R_2) \, , \tag{3.31}$$

with $\delta_0 = 1$ (and $\eta_0 = 0$) under the present H_0 of no change (cf., e.g., Witting, 1985, Chap. 2).

Evaluation of the expectations in (3.30) and (3.31) yields

$$A_{\delta_0}(c_1) + B_{\delta_0}(c_2) = \alpha \, , \tag{3.32}$$

$$\tilde{A}_{\delta_0}(c_1) + \tilde{B}_{\delta_0}(c_2) = \alpha \, , \tag{3.33}$$

where $A_\delta(c_1)$ is the distribution function (3.24), for $w = c_1$, and

$$B_\delta(c_2) = ([c_2] + 1 - c_2)p_\delta([c_2]) + \sum_{l=[c_2]+1}^{b} p_\delta(l) \, , \tag{3.34}$$

$$\tilde{A}_\delta(c_1) = \frac{\sum_{l=0}^{[c_1]-1} p_\delta(l)l + (c_1 - [c_1])p_\delta([c_1])[c_1]}{\sum_{l=a}^{b} p_\delta(l)l} \, , \tag{3.35}$$

$$\tilde{B}_\delta(c_2) = \frac{([c_2] + 1 - c_2)p_\delta([c_2])[c_2] + \sum_{l=[c_2]+1}^{b} p_\delta(l)l}{\sum_{l=a}^{b} p_\delta(l)l} \, . \tag{3.36}$$

The Equations 3.32 and 3.33 can be used for the construction of both a UMA confidence interval $[\delta_*, \delta^*]$ and a two-sided UMP unbiased test of H_0: $\delta = \delta_0$ (cf. Mood, Graybill, & Boes, 1974, p. 464). For the first of these two aims, the following procedure has been adopted.

To obtain the lower and upper limits of the confidence interval, w is in turn equated with c_2 and with c_1, respectively. Since both cases are completely analogous, only one of them is described. Consider, for example, the second one. Choose an initial approximate value for δ, say, δ_1, insert $c_1 = w$ and $[c_1] = [w] = r_2$ into (3.32) and solve it for $c_2 = c_2(c_1)$.

This can be done by successively inserting trial values $[c_2] = b, b-1, \ldots$, computing the corresponding numerical values of $B_{\delta_1}(c_2)$ and, from these, the corresponding terms $[c_2]+1-c_2$, using Equation 3.34, until the condition $0 < [c_2]+1-c_2 \leq 1$ is satisfied. This yields that unique value for c_2 which fits to δ_1 and $c_1 = w$ in (3.32).

Now insert both δ_1 and c_2 into the left-hand side of (3.33) and compare the outcome with its right-hand side: if both are equal, δ_1 is the desired solution δ^*. If the left-hand side of (3.33) is too large, then δ_1 was too small, so that δ_1 has to be replaced by a larger trial value δ_2. If, however, the left-hand side of (3.33) is too small, then δ_1 was too large and has therefore to be replaced by a smaller δ_2. The bisection algorithm can again be used to obtain convergence of the sequence $\delta_1, \delta_2, \ldots$ to the solution δ^*.

Similar methods for analogous cases of the hypothesis of person invariance and for a UMA confidence interval for the person parameter in the dichotomous RM are described in Klauer (1991a, 1991b).

Typically this procedure will result in unique boundaries $[\delta_*, \delta^*]$. That uniqueness is not always attainable, however, can be seen from the following theorem.

Theorem 3.4 Both boundaries of the two-sided UMA confidence interval $[\delta_*, \delta^*]$ are unique if the following inequalities are satisfied: $[c_1] < [c_2]$ and $a+2-\alpha < w < b-1+\alpha$. For $a+2-\alpha < w < b+1-\alpha$, a finite boundary $\delta_* > 0$ exists, but for $w \leq a+2-\alpha$ or $w \geq b+1-\alpha$, there is no finite solution $\delta_* > 0$. Similarly, for $a+\alpha < w < b-1+\alpha$, a finite boundary $\delta^* > 0$ exists, but for $w \leq a+\alpha$ or $w \geq b-1+\alpha$, there is no finite solution $\delta^* > 0$.

Note For $w \leq a+2-\alpha$, technically put $\delta_* = 0$ $(\eta_* = -\infty)$, and for $w \geq b+1-\alpha$, put $\delta_* = \infty$ $(\eta_* = \infty)$. Similarly, for $w \leq a+\alpha$, put $\delta^* = 0$ $(\eta^* = -\infty)$, and for $w \geq b-1+\alpha$, put $\delta^* = \infty$ $(\eta^* = \infty)$.

Proof First observe that the continuous functions $A_\delta(c_1)$ and $\tilde{A}_\delta(c_1)$ are strictly monotone increasing in c_1 and decreasing in δ; similarly, $B_\delta(c_2)$ and $\tilde{B}_\delta(c_2)$ are strictly monotone decreasing in c_2 and increasing in δ.

To see why the condition $[c_1] < [c_2]$ is necessary, assume that a solution exists with $[c_1] = [c_2]$, so that $[c_1] \leq c_1 < c_2 < [c_1]+1$ holds. Consider a pair of numbers $(\bar{c}_1, \bar{c}_2)$ satisfying $\bar{c}_1 = c_1 + d$, $\bar{c}_2 = c_2 + d$, and $[c_1] \leq c_1 < c_1 + d < c_2 < c_2 + d < [c_1]+1$. Inserting $(\bar{c}_1, \bar{c}_2)$ in (3.32) and (3.33) shows that this pair is also a solution. Hence, no solution with $[c_1] = [c_2]$ can be unique. (Notice, however, that $[c_1] = [c_2]$ is a rather unlikely case.)

To verify the necessity of the other inequality conditions, insert δ in (3.32) and (3.33) and let in turn $\delta \to 0$ and $\delta \to \infty$. For $\delta \to 0$, (3.32) becomes $(c_1 - [c_1])p_\delta([c_1]) = \alpha$ together with $p_\delta([c_1]) = 1$, which implies $c_1 - a = \alpha$. This shows that for $c_1 = w$ and $w = a + \alpha$ the boundary δ^* becomes zero. Hence it follows from the strict monotonicity of c_1 in δ (cf. Witting, 1985,

Chap. 2) that there cannot exist a solution $\delta^* > 0$ for $w \leq a+\alpha$. Similarly, for $\delta \to 0$ and $c_2 = w$, with $[c_2] = a+1$, it is seen that (3.33) converges to $a+2-w = \alpha$. From this it can be concluded that at $w = a+2-\alpha$ the boundary δ_* becomes zero, so that for $w \leq a+2-\alpha$ no $\delta_* > 0$ can exist. The other two inequalities follow from analogous arguments.

Now we are ready to show that, if the above inequalities are satisfied, there exists a solution. Owing to the condition $a+\alpha < w$ for the boundary δ^*, there exists a number $c_1 = c_x$ such that $A(c_x) = \alpha$ and $B[c_2(c_x)] = 0$, where $c_2(c_x)$ is that value of c_2 which corresponds to c_x in (3.32). This implies, however, $\tilde{A}(c_x) \leq A(c_x) = \alpha$ and $\tilde{B}[c_2(c_x)] = 0$; that is, $\tilde{A}(c_x) + \tilde{B}[c_2(c_x)] \leq \alpha$. Symmetrically, because of the condition $w < b+1-\alpha$ for δ_*, it is possible to find a value $c_1 = c_y$ such that $A(c_x) = 0$ and $B[c_2(c_y)] = \alpha$, which implies $\tilde{A}(c_x) = 0$ and $\tilde{B}[c_2(c_y)] \geq B[c_2(c_y)] = \alpha$; that is, $\tilde{A}(c_y) + \tilde{B}[c_2(c_y)] \geq \alpha$. Hence, continuity implies that there exists at least one c with $a+\alpha < c < b+1-\alpha$ such that $A(c) + B[c_2(c)] = \alpha$ and $\tilde{A}(c) + \tilde{B}[c_2(c)] = \alpha$. This proves the existence of a solution $[c, c_2(c)] = (c_1, c_2)$.

Next, we prove the uniqueness of a solution (c_1, c_2), given that the inequalities in Theorem 3.4 are satisfied. Suppose there also existed another solution $(\bar{c}_1, \bar{c}_2) \neq (c_1, c_2)$; then it would follow from (3.32) and (3.33) that

$$A_{\delta_0}(\bar{c}_1) - A_{\delta_0}(c_1) - [B_{\delta_0}(c_2) - B_{\delta_0}(\bar{c}_2)] = 0 \ , \tag{3.37}$$

$$\tilde{A}_{\delta_0}(\bar{c}_1) - \tilde{A}_{\delta_0}(c_1) - [\tilde{B}_{\delta_0}(c_2) - \tilde{B}_{\delta_0}(\bar{c}_2)] = 0 \ . \tag{3.38}$$

Equation 3.38, however, is a contradiction of (3.37), because the terms in $\tilde{A}_{\delta_0}(c_1)$ and $\tilde{B}_{\delta_0}(c_2)$ are weighted with l, so that—if (3.37) holds true—the left-hand side of (3.38) has to be > 0. Hence, for fixed δ_0 and α, there can be only one pair (c_1,c_2) with $[c_1] < [c_2]$ satisfying both (3.32) and (3.33).

Finally, from the fact that to determine cutoff scores c_1 and c_2 for a UMP unbiased test of H_0: $\delta = \delta_0$ against H_1: $\delta \neq \delta_0$ is a task equivalent to finding a two-sided confidence interval $[\delta_*, \delta^*]$ for δ and checking whether δ_0 is inside or outside the interval (cf. Mood, Graybill, & Boes, 1974, Chap. 9), it can be concluded that uniqueness of the cutoff scores implies uniqueness of the confidence interval. □

Next, we construct a UMP unbiased test for H_0: $\delta = \delta_0$. (Typically, δ_0 will be 1 and, therefore, $\eta_0 = 0$). Instead of fixing beforehand a significance level α in (3.32) and (3.33), we compute α as that unknown significance level at which the test (3.29) becomes just significant. If this is the case for a certain value of α, the observed realization w must lie exactly at one of the cutoff scores, c_1 or c_2. Therefore, three cases can be distinguished: if $r_2 = E_{\delta_0}(R_2)$, then obviously no significance is possible, and the probability α is equal 1; if $r_2 < E_{\delta_0}(R_2)$, then put $c_1 = r_2 + u = w$, so that $[c_1] = r_2$; else if $r_2 > E_{\delta_0}(R_2)$, then put $c_2 = r_2 + u = w$, so that $[c_2] = r_2$. The second and the third cases are obviously completely analogous, so that it suffices to treat one of them in detail. Suppose, for example, that $r_2 < E_{\delta_0}(R_2)$; then elimination of α from (3.32) and (3.33) yields $[c_2]+1-c_2<1$ explicitly,

$$[c_2]+1-c_2 = -\frac{A_\delta(c_1)-\tilde{A}_\delta(c_1)+\sum_{l=[c_2]+1}^{b} p_\delta(l)[1-l/E_\delta(R_2)]}{p_\delta([c_2])\{1-[c_2]/E_{\delta_0}(R_2)\}}. \tag{3.39}$$

The problem, however, is that $[c_2]$ is not known. But since, for given c_1, there is a unique solution (c_2, α) of Equations 3.32 and 3.33, it suffices to compute the right-hand side of formula (3.39) in turn for $[c_2] = r_2, r_2+1$, $\ldots$, until its left-hand side satisfies the condition $0 \le [c_2]+1-c_2 < 1$. This procedure yields both $[c_2]$ and $[c_2]+1-c_2$, so that c_2 is fully determined. Inserting finally both $c_1 = w$ and c_2 into (3.32) gives the probability α, and H_0 is rejected if α is less than the chosen significance level.

These procedures for the computation of a UMA confidence interval and for a UMP unbiased test look more formidable than they actually are. All necessary computations can be carried out on a PC in a fraction of a second even for long tests. We may therefore summarize this section by stating that the problem of the construction of UMA confidence intervals and of making UMP unbiased tests of change in test performance of individuals is solved satisfactorily within the specified framework.

Alternative methods for hypothesis testing in the RM, using Monte Carlo simulations for test lengths $k > 30$, are implemented in the software "T-Rasch" by Ponocny and Ponocny-Seliger (1999).

6 An Example

An example (with artificial item parameters) is now given as an illustration of the use of the methods presented in this chapter. Suppose a scale of "Social Competence" has been developed to measure the effects of a training seminar for personality development. Let the scale consist of a total of $k = 20$ questionnaire items with $m + 1 = 3$ response categories each. A previous calibration is assumed to have shown that the RSM applies reasonably well; the item easiness parameters β_i, for $i = 1, \ldots, 20$, and the category parameters ω_j, for $j = 0, \ldots, 2$, are given in Table 3.2.

Since it is assumed that the seminar increases the social competence of the trainees, a somewhat easier selection of items ("easier" in a purely technical sense) is used for the pretest in comparison with the posttest (see Table 3.2).

To measure the individual amounts of change and to test the H_0: $\eta = 0$ against the one-sided H_1: $\eta > 0$ for an individual with observed raw scores r_1 and r_2, the methods outlined above are now used. The complete

TABLE 3.2. Item and category parameters for the example of Section 6.

Item Easiness Parameters β_i									
2.90	2.30	1.90	1.60	1.50	0.80	0.70	0.50	0.20	0.10
−0.20	−0.20	−0.50	−0.60	−0.90	−1.30	−1.40	−2.00	−2.40	−3.00
Category Parameters ω_j									
0.00	0.00	−3.50							
Pretest Items									
1	1	1	1	0	1	1	0	1	1
1	0	1	0	0	0	1	0	1	0
Posttest Items									
0	1	0	0	1	0	1	1	0	0
0	1	0	1	1	1	1	1	1	1

Note: $k = 20$, $m = 2$; by normalization, $\sum_i \beta_i = 0$ and $\omega_0 = \omega_1 = 0$. The entries "1" vs. "0" denote that an item belongs vs. does not belong to the pretest or posttest, respectively.

results comprise, for each score combination r_1 and r_2, (a) the change parameter estimate $\hat{\eta}$, (b) the lower boundaries of a semi-open Clopper–Pearson and of a UMA randomized confidence interval $[\eta_*, +\infty)$, and (c) the conditional probabilities $P(R_2 \geq r_2 \mid r)$ and $P(W \geq w)$, respectively, under H_0. Since the total tables of these results, for all combinations of r_1 and r_2, are much too large to be reproduced here, only a selection of r_1- and r_2-values is shown in Tables 3.3 and 3.4. The significance level assumed for the confidence intervals and tests is always $\alpha = .05$.

Several observations can be made. First, the two test forms (pre- and posttest) are of a different difficulty level: the difference between the average item easiness parameters of the two test forms is 1.06. Therefore, the same raw scores on both tests represent different levels of test performance. Accordingly, the estimates of the amounts of change (i.e., improvements) for equal raw scores $r_1 = r_2$, as far as they are given in Tables 3.3 and 3.2, are 1.06 for $r_1 = r_2 = 4$, 1.09 for $r_1 = r_2 = 10$, 1.10 for $r_1 = r_2 = 16$, and 1.08 for $r_1 = r_2 = 22$. Intuitively, this seems to make sense since the estimates $\hat{\eta}$ are almost of the same size as the difference between average difficulty parameters of the two test forms. If both forms had been equal, their parameter difference would have been 0, and the estimators $\hat{\eta}$ for $r_1 = r_2$ would always have been exactly zero, as can be deduced from the estimation equation (3.19).

Second, the comparison of Tables 3.3 and 3.4 shows that in some cases the lower limits of one-sided UMA confidence intervals were not estimable where those of the Clopper–Pearson intervals could be computed, and vice versa. This illustrates the different conditions for existence of a finite boundary δ_* under the two approaches. Notice that rerunning the computation of the UMA confidence intervals would probably yield estimates of some of the missing boundaries; however, some other estimates of boundaries that now are finite, might then become infinite. This once more demonstrates the effect of the randomization.

Third, the minimum score differences leading to significance of the change parameter estimate $\hat{\eta}$ are remarkably constant over the entire range of raw scores. This can be seen from Tables 3.5 and 3.6 which give the two-sided nonrandomized and randomized levels of significance, respectively. (Limitation of space precludes the reproduction of more tables.) The pairs of raw scores (r_1, r_2), where r_1 is a given pretest score, and r_2 the smallest posttest score for which an increase is significant at $\alpha = .05$, read from Table 3.5, are (0,3), (1,4), (2,5), (3,6), ..., (19,22), (20,23), (21,24); in other words, the just significant gain scores in this example are equal to 3 over the entire range. This is not to say, however, that significant gain scores are so constant in *all* tests, of course, but a surprising regularity has been observed in other examples as well. A similar picture is shown by Table 3.6; however, there is slightly more irregularity in this table, owing to the randomization. Nevertheless, the minimal significant gain scores are smaller now: the respective pairs of scores are (0,3), (1,4), (2,4), (3,6), ..., (18,20), (19,21), (20,22), (21,23); that is, in many cases gain scores as small as 2 are already significant. Moreover, some of them even are significant on the $\alpha = .01$ level. (Similar results would be found in tables for the one-sided alternative hypothesis H_1: $\eta > 0$, with a tendency to still smaller significant gain scores. In interpreting these results, however, it should be kept in mind that the significance of such small score differences in this example is mainly caused by the difficulty difference of the two test forms and does not generalize to other cases.)

Fourth, in a recomputation of Table 3.6, it could happen that the level of significance changes in some cells of the table. This uncertainty is the price to be paid for the higher power of the UMP significance tests.

Finally, there appear to occur nonmonotonicities in the boundaries of some randomized confidence intervals and in the respective significance levels: in Table 3.4, compare the probabilities for raw score pairs (24,18) and (24,20); in Table 3.6, see the significances for raw score pairs (2,6) and (2,7), or (23,16), (23,17), and (23,18), or (24,19) and (24,20). At first sight, this looks like a formal contradiction to monotonicity properties of the UMA confidence intervals and UMP tests, but in fact it is not: monotonicity of the significance levels in the gain scores holds only *within* each conditional distribution of R_2 with fixed total score r. This means that monotonicity must hold for entries parallel to the codiagonal of Tables 3.4 and 3.6, which is actually the case. From a substantive point of view, however, one would wish to have "double monotonicity" within rows and within columns, as in Tables 3.3 and 3.5. For the empirical researcher, this may be an argument

TABLE 3.3. Effect parameters, one-sided Clopper–Pearson confidence intervals, and levels of significance for the example of Section 6.

	r_2											
r_1	0	2	4	6	8	10	14	16	18	20	22	24
	···	∞	∞	∞	∞	∞	∞	∞	∞	∞	∞	∞
0	···	-0.28	1.29	2.28	3.07	3.75	5.01	5.66	6.35	7.11	8.01	9.35
	···	0.07	0.00	0.00	0.00	0.00	0.00	0.00	0.00	0.00	0.00	0.00
	-∞	1.83	3.03	3.90	4.62	5.27	6.53	7.20	7.92	8.73	9.76	∞
1	-∞	-1.12	0.49	1.50	2.29	2.98	4.25	4.89	5.58	6.34	7.24	8.57
	1.00	0.22	0.02	0.00	0.00	0.00	0.00	0.00	0.00	0.00	0.00	0.00
	-∞	-0.13	1.06	1.92	2.63	3.29	4.56	5.23	5.96	6.79	7.87	∞
4	-∞	-2.65	-0.99	0.04	0.84	1.53	2.80	3.45	4.13	4.89	5.78	7.11
	1.00	0.75	0.27	0.05	0.00	0.00	0.00	0.00	0.00	0.00	0.00	0.00
	-∞	-1.33	-0.14	0.72	1.44	2.09	3.36	4.03	4.76	5.60	6.69	∞
7	-∞	-3.75	-2.07	-1.04	-0.24	0.45	1.72	2.37	3.06	3.81	4.70	6.02
	1.00	0.97	0.73	0.34	0.09	0.01	0.00	0.00	0.00	0.00	0.00	0.00
	-∞	-2.33	-1.14	-0.28	0.44	1.09	2.36	3.03	3.76	4.60	5.70	∞
10	-∞	-4.72	-3.04	-2.01	-1.21	-0.52	0.76	1.40	2.09	2.85	3.73	5.06
	1.00	1.00	0.96	0.78	0.45	0.16	0.00	0.00	0.00	0.00	0.00	0.00
	-∞	-3.28	-2.10	-1.24	-0.52	0.13	1.41	2.08	2.81	3.65	4.75	∞
13	-∞	-5.67	-4.00	-2.97	-2.16	-1.47	-0.20	0.45	1.13	1.89	2.78	4.10
	1.00	1.00	1.00	0.97	0.86	0.59	0.08	0.01	0.00	0.00	0.00	0.00
	-∞	-4.26	-3.07	-2.21	-1.50	-0.85	0.43	1.10	1.83	2.67	3.77	∞
16	-∞	-6.67	-5.00	-3.98	-3.18	-2.49	-1.22	-0.57	0.12	0.87	1.77	3.10
	1.00	1.00	1.00	1.00	0.99	0.93	0.45	0.17	0.04	0.00	0.00	0.00
	-∞	-5.34	-4.16	-3.31	-2.59	-1.94	-0.67	0.00	0.73	1.57	2.67	∞
19	-∞	-7.82	-6.20	-5.19	-4.40	-3.72	-2.45	-1.80	-1.11	-0.35	0.55	1.90
	1.00	1.00	1.00	1.00	1.00	1.00	0.89	0.67	0.35	0.10	0.01	0.00
	-∞	-6.88	-5.73	-4.89	-4.18	-3.53	-2.27	-1.59	-0.86	-0.02	1.08	∞
22	-∞	-9.71	-8.29	-7.35	-6.59	-5.93	-4.66	-4.01	-3.31	-2.54	-1.61	-0.18
	1.00	1.00	1.00	1.00	1.00	1.00	1.00	0.98	0.92	0.72	0.37	0.06
	-∞	-∞	-∞	-∞	-∞	-∞	-∞	-∞	-∞	-∞	-∞	···
24	-∞	-∞	-∞	-∞	-∞	-∞	-∞	-∞	-∞	-∞	-∞	···
	1.00	1.00	1.00	1.00	1.00	1.00	1.00	1.00	1.00	1.00	1.00	···

Note: The table gives effect parameter estimates $\hat{\eta}$ (first row), lower boundaries of one-sided Clopper–Pearson confidence intervals η_*, under H_0: $\eta = 0$ against H_1: $\eta > 0$ (second row), and levels of significance $P(R_2 \geq r_2 \mid r)$ under H_0 (third row).

TABLE 3.4. Effect parameters, one-sided UMA confidence intervals, and levels of significance of the UMP test for the example of Section 6.

	r_2											
r_1	0	2	4	6	8	10	14	16	18	20	22	24
0	···	∞	∞	∞	∞	∞	∞	∞	∞	∞	∞	∞
	···	-0.24	1.32	2.48	3.19	3.79	5.48	6.41	∞	7.67	12.15	11.09
	···	0.07	0.00	0.00	0.00	0.00	0.00	0.00	0.00	0.00	0.00	0.00
1	$-\infty$	1.83	3.03	3.90	4.62	5.27	6.53	7.20	7.92	8.73	9.76	∞
	$-\infty$	-0.37	0.64	1.94	2.63	3.85	4.50	5.11	5.93	6.48	7.42	9.93
	0.96	0.08	0.01	0.00	0.00	0.00	0.00	0.00	0.00	0.00	0.00	0.00
4	$-\infty$	-0.13	1.06	1.92	2.63	3.29	4.56	5.23	5.96	6.79	7.87	∞
	$-\infty$	-1.76	-0.43	0.47	1.38	1.55	3.41	3.60	4.65	4.93	6.02	8.48
	0.97	0.41	0.11	0.02	0.00	0.00	0.00	0.00	0.00	0.00	0.00	0.00
7	$-\infty$	-1.33	-0.14	0.72	1.44	2.09	3.36	4.03	4.76	5.60	6.69	∞
	$-\infty$	-3.17	-2.04	-0.34	-0.19	0.97	1.98	2.46	3.07	3.91	4.70	6.05
	1.00	0.86	0.71	0.11	0.08	0.00	0.00	0.00	0.00	0.00	0.00	0.00
10	$-\infty$	-2.33	-1.14	-0.28	0.44	1.09	2.36	3.03	3.76	4.60	5.70	∞
	$-\infty$	-3.92	-3.02	-2.01	-1.03	-0.11	1.10	1.51	2.10	2.91	4.02	5.15
	1.00	0.98	0.96	0.78	0.33	0.07	0.00	0.00	0.00	0.00	0.00	0.00
13	$-\infty$	-3.28	-2.10	-1.24	-0.52	0.13	1.41	2.08	2.81	3.65	4.75	∞
	$-\infty$	-4.75	-3.97	-2.50	-2.14	-1.38	0.37	0.79	1.70	2.40	3.39	4.13
	1.00	1.00	1.00	0.89	0.84	0.52	0.02	0.01	0.00	0.00	0.00	0.00
16	$-\infty$	-4.26	-3.07	-2.21	-1.50	-0.85	0.43	1.10	1.83	2.67	3.77	∞
	-8.84	-6.66	-4.99	-3.95	-2.69	-2.29	-1.18	-0.33	0.53	1.24	2.10	∞
	1.00	1.00	1.00	1.00	0.94	0.85	0.43	0.10	0.01	0.00	0.00	0.00
19	$-\infty$	-5.34	-4.16	-3.31	-2.59	-1.94	-0.67	0.00	0.73	1.57	2.67	∞
	$-\infty$	-7.42	-5.86	-5.08	-3.74	-3.54	-2.45	-1.45	-0.80	0.37	0.62	2.89
	1.00	1.00	1.00	1.00	1.00	0.99	0.89	0.46	0.21	0.02	0.01	0.00
22	$-\infty$	-6.88	-5.73	-4.89	-4.18	-3.53	-2.27	-1.59	-0.86	-0.02	1.08	∞
	$-\infty$	-8.75	-8.27	-6.95	-6.39	-5.32	-3.76	-3.30	-3.16	-2.53	-1.47	∞
	1.00	1.00	1.00	1.00	1.00	1.00	0.98	0.90	0.85	0.71	0.31	0.00
24	$-\infty$	$-\infty$	$-\infty$	$-\infty$	$-\infty$	$-\infty$	$-\infty$	$-\infty$	$-\infty$	$-\infty$	$-\infty$	···
	$-\infty$	$-\infty$	$-\infty$	-10.70	-8.91	$-\infty$	$-\infty$	$-\infty$	$-\infty$	$-\infty$	$-\infty$	···
	1.00	1.00	1.00	1.00	1.00	1.00	1.00	1.00	0.97	0.99	0.75	···

Note: The table gives effect parameter estimates $\hat{\eta}$ (first row), lower boundaries of semi-open UMA confidence intervals η_*, under H_0: $\eta = 0$ against H_1: $\eta > 0$ (second row), and UMP levels of significance $P(W \geq w \mid r)$ under H_0, based on a randomization of the scores (third row).

TABLE 3.5. Two-sided levels of significance (nonrandomized) for the example of Section 6.

											1										2					
	0	1	2	3	4	5	6	7	8	9	0	1	2	3	4	5	6	7	8	9	0	1	2	3	4	
0				s	S	T	T	T	T	T	T	T	T	T	T	T	T	T	T	T	T	T	T	T	T	0
1					s	S	S	T	T	T	T	T	T	T	T	T	T	T	T	T	T	T	T	T	T	1
2						s	S	S	T	T	T	T	T	T	T	T	T	T	T	T	T	T	T	T	T	2
3							s	S	S	T	T	T	T	T	T	T	T	T	T	T	T	T	T	T	T	3
4							.	s	S	S	T	T	T	T	T	T	T	T	T	T	T	T	T	T	T	4
5								.	s	S	S	T	T	T	T	T	T	T	T	T	T	T	T	T	T	5
6	.								.	s	S	S	T	T	T	T	T	T	T	T	T	T	T	T	T	6
7	s									.	s	S	S	T	T	T	T	T	T	T	T	T	T	T	T	7
8	S	.									.	s	S	S	T	T	T	T	T	T	T	T	T	T	T	8
9	S	s										.	s	S	S	T	T	T	T	T	T	T	T	T	T	9
10	T	S	s										.	s	S	S	T	T	T	T	T	T	T	T	T	10
11	T	S	s	.										.	s	S	S	T	T	T	T	T	T	T	T	11
12	T	T	S	s											.	s	S	S	T	T	T	T	T	T	T	12
13	T	T	S	s	s											.	s	S	S	T	T	T	T	T	T	13
14	T	T	T	S	s	.											.	s	S	S	T	T	T	T	T	14
15	T	T	T	T	S	s	.											.	s	S	S	T	T	T	T	15
16	T	T	T	T	T	S	s	.											.	s	S	S	T	T	T	16
17	T	T	T	T	T	S	S	s	.											.	s	S	S	T	T	17
18	T	T	T	T	T	T	S	S	s	.											.	s	S	T	T	18
19	T	T	T	T	T	T	T	S	S	s	.											.	s	S	T	19
20	T	T	T	T	T	T	T	T	T	S	s	s												s	S	20
21	T	T	T	T	T	T	T	T	T	T	S	s	s	.											s	21
22	T	T	T	T	T	T	T	T	T	T	T	S	S	s	.											22
23	T	T	T	T	T	T	T	T	T	T	T	T	T	S	s	s	.									23
24	T	T	T	T	T	T	T	T	T	T	T	T	T	T	S	s	s	.								24
											1										2					
	0	1	2	3	4	5	6	7	8	9	0	1	2	3	4	5	6	7	8	9	0	1	2	3	4	

Note: Rows correspond to scores r_1, columns to scores r_2. The symbols ".", "s", "S", and "T" mean significance at $\alpha = 0.10$, 0.05, 0.01, and 0.001, respectively; blank spaces denote nonsignificant gain scores.

in favor of the less powerful nonrandomized Clopper–Pearson confidence intervals and the significance levels based on them.

7 Conclusion

The change parameter estimate $\hat{\eta}_v$ of testee S_v can be used as a measure of S_v's amount of change between two testing occasions, and this measure

TABLE 3.6. Levels of significance (randomized) for the example of Section 6.

	0	1	2	3	4	5	6	7	8	9	10	11	12	13	14	15	16	17	18	19	20	21	22	23	24	
0				S	T	T	T	T	T	T	T	T	T	T	T	T	T	T	T	T	T	T	T	T	T	0
1					s	S	T	T	T	T	T	T	T	T	T	T	T	T	T	T	T	T	T	T	T	1
2					s	S	T	S	T	T	T	T	T	T	T	T	T	T	T	T	T	T	T	T	T	2
3						.	s	S	T	T	T	T	T	T	T	T	T	T	T	T	T	T	T	T	T	3
4						.	.	s	S	S	T	T	T	T	T	T	T	T	T	T	T	T	T	T	T	4
5	.							s	s	S	T	T	T	T	T	T	T	T	T	T	T	T	T	T	T	5
6	s								.	s	S	T	T	T	T	T	T	T	T	T	T	T	T	T	T	6
7	s	.								.	S	S	T	T	T	T	T	T	T	T	T	T	T	T	T	7
8	S	s									.	s	S	T	T	T	T	T	T	T	T	T	T	T	T	8
9	S	s										s	S	S	T	T	T	T	T	T	T	T	T	T	T	9
10	T	T	s	s								.	s	s	S	S	T	T	T	T	T	T	T	T	T	10
11	T	T	s	.								.	.	s	s	S	S	T	T	T	T	T	T	T	T	11
12	T	T	S	s	.										.	s	S	S	T	T	T	T	T	T	T	12
13	T	T	T	S	s	.										s	s	T	T	T	T	T	T	T	T	13
14	T	T	T	T	S	s	s										s	s	S	S	T	T	T	T	T	14
15	T	T	T	T	S	s	s	s	.									.	s	S	S	T	T	T	T	15
16	T	T	T	T	T	S	S	s	.									.	.	S	S	S	T	T	T	16
17	T	T	T	T	T	T	S	S	s											s	s	T	T	T	T	17
18	T	T	T	T	T	T	T	S	S	s											s	S	S	T	T	18
19	T	T	T	T	T	T	T	T	S	S	s										.	s	s	S	T	19
20	T	T	T	T	T	T	T	T	T	S	S	S	s	s									s	s	S	20
21	T	T	T	T	T	T	T	T	T	T	T	S	s	.										s	S	21
22	T	T	T	T	T	T	T	T	T	T	T	T	S	s	s	.									.	22
23	T	T	T	T	T	T	T	T	T	T	T	T	T	S	S	S	.		.							23
24	T	T	T	T	T	T	T	T	T	T	T	T	T	T	T	S	S	s	s		.					24
	0	1	2	3	4	5	6	7	8	9	10	11	12	13	14	15	16	17	18	19	20	21	22	23	24	

Note: Rows correspond to scores r_1, columns to scores r_2. The symbols ".", "s", "S", and "T" mean significance at $\alpha = 0.10$, 0.05, 0.01, and 0.001, respectively; blank spaces denote nonsignificant gain scores.

can be assessed statistically by means of confidence intervals and hypothesis tests based on the exact conditional distribution of the raw score variable R_{v2}, given S_v's observed total raw score r_v. Slightly conservative but intuitively appealing methods are available, as well as UMA randomized confidence intervals and UMP tests of no change. The latter, however, have some peculiar properties and may therefore not seem equally acceptable to empirical researchers. None of these methods requires any asymptotic results about person parameter or change parameter estimates.

The practical application of these methods is quite simple. Once a table of all possible raw score combinations (r_1, r_2) with $r_1 + r_2 = r$, for $r = 1, \ldots, \sum_{i=1}^{k} m_i - 1$, has been computed, containing the change parameter estimates, their one- or two-sided confidence intervals, and the respective significances, the user may simply look up the raw scores of a given testee and will find both the measure of change and an indication of its significance or nonsignificance.

It is the opinion of the author that such tables should be part of the manual of any commercial test whenever one of the models PCM, RSM, RM has been applied and found to fit sufficiently well. They would prove useful even when no repeated testing is done: the tables can immediately be interpreted also in terms of the measurements of two testees and as a statistical test for the significance of their score difference, affording a general guideline for the interpretation of test results.

In situations where many different subsets of items are drawn from an item pool and are used as test forms, as in computerized adaptive testing, however, such tables will soon become impractical. In such cases, a computer program would have to be implemented in the testing software to estimate change and to compute significance levels for each individual.

Acknowledgments: The author thanks Dr. G. Schuhfried Ges.m.b.H. for financial support of this work, and the two anonymous referees for valuable suggestions.

References

Andersen, E.B. (1972). The numerical solution of a set of conditional estimation equations. *Journal of the Royal Statistical Society, Series B, 34*, 42–54.

Andersen, E.B. (1990). *The statistical analysis of categorical data.* Heidelberg: Springer-Verlag.

Andersen, E.B. (1995). Polytomous Rasch models and their estimation. In G.H. Fischer & I.W. Molenaar (Eds.), *Rasch models: Foundations, recent developments, and applications* (pp. 271–291). New York: Springer-Verlag.

Andrich, D. (1978). A rating formulation for ordered response categories. *Psychometrika, 43*, 561–573.

Bereiter, C. (1963). Some persisting dilemmas in the measurement of change. In C.W. Harris (Ed.), *Problems in measuring change* (pp. 3–20). Madison, WI: University of Wisconsin Press.

Cronbach, L.J., & Furby, L. (1970). How should we measure change, or should we? *Psychological Bulletin, 74*, 68–80.

Embretson, S.E. (1991). A multidimensional latent trait model for measuring learning and change. *Psychometrika, 56,* 495-515.

Fischer, G.H. (1987). Applying the principles of specific objectivity and generalizability to the measurement of change. *Psychometrika, 52,* 565–578.

Fischer, G.H. (1995). Some neglected problems in IRT. *Psychometrika, 60,* 459–487.

Fischer, G.H., & Ponocny, I. (1995). Extended rating scale and partial credit models for assessing change. In G.H. Fischer & I.W. Molenaar (Eds.), *Rasch models: Foundations, recent developments, and applications* (pp. 351–370). New York: Springer-Verlag.

Fischer, G.H., & Ponocny-Seliger, E. (1998). Structural Rasch modeling: Handbook of the usage of LPCM-WIN 1.0 [Software manual]. Groningen: ProGAMMA.

Guttmann, G., & Etlinger, S.C. (1991). Susceptibility to stress and anxiety in relation to performance, emotion, and personality: The ergopsychometric approach. In Ch. Spielberger, I.G. Sarason, J. Strelau, & J.M.T. Brebner (Eds.), *Stress and anxiety* (pp. 23–52). New York: Hemisphere.

Hoijtink, H., & Boomsma, A. (1996). Statistical inference based on latent ability estimates. *Psychometrika, 61,* 313–330.

Holtzman, W.H. (1963). Statistical models for the study of change in the single case. In C.W. Harris (Ed.), *Problems in measuring change* (pp. 199–211). Madison, WI: The University of Wisconsin Press.

Huber H. (1977). Zur Planung und Auswertung von Einzelfalluntersuchungen [On the planning and analysis of single case studies]. In L.J. Pongratz (Ed.), *Handbuch der Psychologie: Vol. 8. Klinische Psychologie* (pp. 1153–1199). Göttingen: Hogrefe.

Klauer, K.C. (1991a). An exact and optimal standardized person test for assessing consistency with the Rasch model. *Psychometrika, 56,* 213–228.

Klauer, K.C. (1991b). Exact and best confidence intervals for the ability parameter of the Rasch Model. *Psychometrika, 56,* 535–547.

Klauer, K.C. (1995). The assessment of person fit. In G.H. Fischer & I.W. Molenaar (Eds.), *Rasch Models: Foundations, recent developments, and applications* (pp. 97–110). New York: Springer-Verlag.

Liou, M. (1993). Exact person tests for assessing model-data fit in the Rasch model. *Applied Psychological Measurement, 17,* 187–195.

Liou, M., & Chang, C.-H. (1992). Constructing the exact significance level for a person fit statistic. *Psychometrika, 47,* 169-181.

Masters, G.N. (1982). A Rasch model for partial credit scoring. *Psychometrika, 47,* 149–174.

Meijer, R.R., & Sijtsma, K. (in press). A review of methods for evaluating the fit of item score patterns on a test. *Applied Psychological Measurement.*

Mellenbergh, G.J., & Van den Brink, W.P. (1998). The measurement of individual change. *Psychological Methods, 3,* 470–485.

Molenaar, I.W., & Hoijtink, H. (1990). The many null distributions of person fit indices. *Psychometrika, 55,* 75–106.

Mood, A.M., Graybill, F.A., & Boes, D.C. (1974). *Introduction to the theory of statistics.* Singapore: McGraw-Hill.

Ponocny, I. (2000). Exact person fit indexes for the Rasch model for arbitrary alternatives. *Psychometrika, 65,* 29-42.

Ponocny, I., & Ponocny-Seliger, E. (1999). T-Rasch 1.0 [Software program]. Groningen: ProGAMMA.

Prieler, J. (2000). *Evaluation eines Ausleseverfahrens für Unteroffiziere beim Österreichischen Bundesheer* [Evaluation of a selection procedure for noncommissioned officers in the Austrian army]. Unpublished dissertation, University of Vienna, Department of Psychology.

Rasch, G. (1960). *Probabilistic models for some intelligence and attainment tests.* Copenhagen: The Danish Institute of Educational Research. (Expanded edition, 1980. Chicago: University of Chicago Press.)

Rasch, G. (1965). *Målingsmodellerne og deres principielle baggrund* [Models for measurements and their fundamental background]. (Notes taken by J. Stene at the statistical seminar.) Copenhagen: Department of Statistics, University of Copenhagen.

Santner, T.J., & Duffy, D.E. (1989). *The statistical analysis of discrete data.* New York: Springer-Verlag.

Willett, J.B. (1989). Some results on reliability for the longitudinal measurement of change: Implications for the design of studies of individual growth. *Educational and Psychological Measurement, 49,* 587–602.

Williams, R.H., & Zimmerman, D.W. (1996). Are simple gain scores obsolete? *Applied Psychological Measurement, 20,* 59–69.

Witting, H. (1985). *Mathematische Statistik I* [Mathematical statistics I]. Stuttgart: Teubner.

4 Modeling Learning in Short-Term Learning Tests

Karl Christoph Klauer[1] and Hubert Sydow[2]

ABSTRACT A new model-based method for analyzing learning processes in so-called short-term learning tests is proposed and applied. Learning is measured by means of a two-dimensional item response theory model that incorporates two latent factors: ability level and learning ability. The model allows one to assess the impact of both variables and their correlation for given data sets in a manner that implicitly corrects for statistical artifacts that arise in conventional analyses. The model is applied to four short-term learning tests administered to $N = 434$ children aged five to six years. Model tests and tests for empirical validity of the model parameters succeed in establishing construct validity for one of these tests.

1 Introduction

Learning tests aim at assessing both a person's ability level as well as his or her learning ability. There are short- and long-term learning tests (Guthke, Böttcher, & Sprung, 1991). Long-term learning tests realize a pretest-training-posttest design. Short-term learning tests are within-test procedures in which feedback and learning hints are provided for each item so that training and test coincide. A recent review of the concept, theoretical foundations, and applications of learning tests have been given by Guthke and Wiedl (1996; cf. also Guthke, 1982, 1990; Embretson, 1991; Flammer & Schmid, 1982; Hamers, Ruijssenaars, & Sijtsma, 1993; Klauer & Sydow, 1992; Klauer, Kauf, & Sydow, 1994, among others).

Important assumptions underlying the idea to assess learning potential by means of learning tests are the following.

1. The response patterns in learning tests are determined by two factors, ability level and learning ability.

[1]Psychologisches Institut, Rheinische Friedrich-Wilhelms-Universität Bonn, Römerstr. 164, 53117 Bonn, FR Germany; e-mail: *christoph.klauer@uni-bonn.de*

[2]Institut für Psychologie, Humboldt-Universität Berlin, Oranienburger Str. 18, 10178 Berlin, FR Germany; e-mail: *hubert.sydow@psychologie.hu-berlin.de*

2. There is significant variability between persons in learning ability.

3. Learning ability provides diagnostic information that is not contained in ability level; that is, ability level and learning ability are relatively independent and as a minimum requirement not correlated perfectly.

4. Learning ability may be more important for predicting future performance than current ability level, or that taking learning ability into account over and above ability status at least provides an increment in predictive validity.

The first three assumptions concern the factorial structure of learning tests. Testing these assumptions is fraught with considerable methodological problems that are subsumed under the label of "measurement of change" (e.g., Zimmermann & Williams, 1982a, 1982b). For short-term learning tests, Klauer and Sydow (1992) and Klauer et al. (1994) have proposed a model for analyzing learning and for testing the above structural assumptions. The model comprises a measurement model and a structural model. The measurement model is a probabilistic model and defines a latent scale relating manifest test performance to the latent variables learning ability and ability level. The structural model specifies the joint distribution of ability level and learning ability. In the general case, the structural model assumes a bivariate normal distribution with unknown variances and covariance. Figure 4.1 shows a graphical illustration of the resulting learning model. The relationship of the present model to previous dynamic generalizations of item response theory (IRT) is considered in the general discussion.

This setup suggests a three-step structural analysis to test the above assumptions. The first step tests whether the observed set of response patterns can be described by the learning model with two latent factors. If goodness of fit is satisfactory, the first assumption can be maintained. The second step tests whether the structural model can be simplified by omitting interindividual differences in learning ability; that is, it tests whether the variance of the latent variable learning ability can be set to zero. If this simplification must be rejected, the second assumption can be maintained according to which there are significant differences between persons in learning ability. Finally, it tests whether the correlation of both latent variables is smaller than one in absolute value, so that both variables contribute nonredundant diagnostic information (cf. Assumption 3). While this three-step analysis is focused on the internal structure of the test, evaluating Assumption 4 requires additional external data such as, for example, data on the tested persons' future performance.

Klauer and Sydow (1992) applied this three-step analysis to three short-term learning tests labeled analogy test (AT) 1, AT 2, and AT 3, respectively, on the basis of a sample of $N = 361$ five- to six-year-old children. The items of AT 3 are figural analogy tasks, while those of AT 1 and AT 2

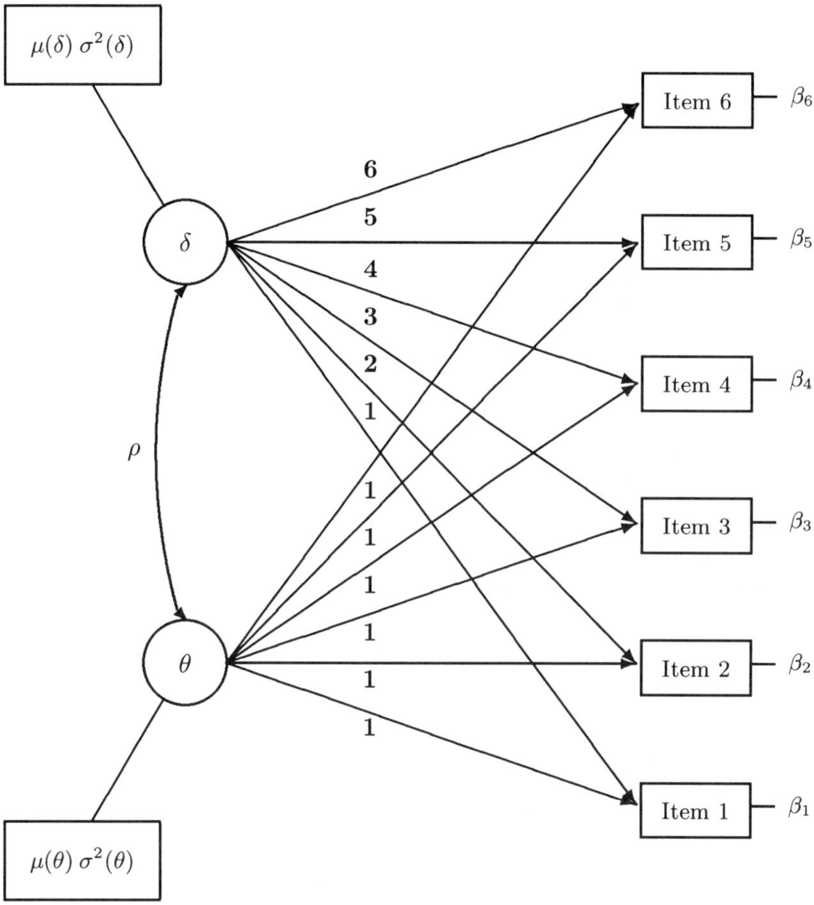

FIGURE 4.1. The learning model: θ is the ability level, δ the learning ability, μ and σ^2 the respective means and variances of the latent variables, while β_i are item difficulty parameters. The path weights connecting latent variables and item response variables are proportional to the weights v_i and w_i of the latent variables discussed below. A test with six items is assumed.

test component tasks of figural analogies. Klauer and Sydow (1992) showed that in the easiest test, AT 1, learning could not be identified, while in the second test, AT 2, there were no significant differences between children in learning ability. In AT 3, there were such differences that correlated almost perfectly with ability level, however.

Klauer et al. (1994) focused on the most promising AT 3 and implemented a number of modifications that aimed at increasing the impact of differential learning ability on test performance. In addition, there was a control group in which the short-term learning test was administered without feedback and learning hints. Based on a new sample of $N = 226$ five- to six-year-old children, the modified short-term learning test AT 3 was found to reveal significant interindividual differences in learning ability that were not correlated perfectly with ability level (cf. Assumptions 2 and 3).

The present work replicates and extends these results. Based on yet a new sample of $N = 434$ five- to six-year-old children, the AT 3 items were again used along with three short-term learning tests, called Vorschullerntests (VLT-tests), developed by Roether (1983). All tests were administered either with feedback and learning aids, that is, as learning tests, or in a control group in the traditional fashion without feedback and learning hints. The objective of this investigation was to provide another validation of the short-term learning test AT 3, to analyze the well-established VLT-tests by means of the new method, and to test for relationships, at the level of the latent variables, between AT 3 and the VLT-tests.

Attempts to identify stable and consistent interindividual differences in learning performance are of both practical and theoretical importance, specifically for the sampled age range of preschool children. Possible deficits in general learning ability should be diagnosed as early as possible so that special interventions can be attempted to improve learning ability before school entry. A number of diagnostic procedures (e.g., Guthke, 1983; Katzbeck, Kastner-Koller, & Deimann, 1998; Roether, 1983) as well as intervention programs (K.J. Klauer, 1989; Sydow & Meincke, 1994) have been developed for this purpose.

2 The Learning Model

The learning model is composed of a measurement model that sets up a scale for the latent variables, and a structural model mapping the relationships between the latent variables.

2.1 The Measurement Model

Klauer and Sydow's (1992) learning model assumes that the items $I_1, \ldots, I_k$ of the test under scrutiny can be described by a simple item response theory model when there is no learning. That is, it is assumed that conditional independence holds and that the probability $P(X_i = 1|\theta)$, that an examinee with ability θ solves item I_i, can be described by a simple item characteristic curve (ICC) $f_i(\theta)$.

For the AT 3 items, in particular, it turned out that a logistic ICC with a guessing parameter r that could be set equal for all items was adequate.

Thus, in previous analyses

$$f_i(\theta) = r + (1-r)\frac{\exp(\theta - \beta_i)}{1+\exp(\theta - \beta_i)} ,$$

where β_i is an item-specific difficulty parameter.

Furthermore, when the set of items is administered as a learning test, that is, with learning hints, it is assumed that each item with accompanying feedback and learning aids presents an opportunity to learn that can be modeled by a learning gain δ in ability value θ. The learning gain may be a function of the tested person, the particular item, and of whether the item was solved without learning hints or only after learning hints had been given.

As a first simplifying assumption it was postulated that the learning gain may be person-specific, but that it is not a function of the particular item or the examinee's solution behavior vis à vis the item. Under this simplifying assumption, the probability of a spontaneous solution is given by

$$P(X_i = 1 \mid \theta, \delta) = f_i(v_i\theta + w_i\delta) ,$$

where $v_i = 1$ for all items I_i and where w_i models the $i-1$ learning opportunities that have accumulated up to item I_i: $w_i = (i-1)$.

Note that there are many alternative conceptions of the short-term learning process. For example, instead of a linear function of accumulating learning gains, a step-function could be assumed in which after some subject-dependent item i_0, a sudden subject-specific increase δ in ability occurs; this would mean that $w_i = 0$ for $i < i_0$ and $w_i = 1$ for $i \geq i_0$, where the number i_0 would depend on when the subject sees the light. Or, learning might depend on whether the answer to the last item was correct (Verhelst & Glas, 1995). In the absence of more definite information, a linear model is often assumed, but it is important to keep in mind that other models of the learning process exist and might be more adequate (Verhelst & Glas, 1995).

This measurement model defines both ability level θ as well as learning ability δ. Because $w_1 = 0$, the ability level is defined as the initial ability status. In order to make the variances of both latent variables comparable, a further refinement is that the vector of weights $\mathbf{w} = (w_1, \ldots, w_k)'$ and $\mathbf{v} = (v_1, \ldots, v_k)'$ of both variables should be of equal length: $|\mathbf{v}| = |\mathbf{w}|$. This is achieved by multiplying the weights w_i by a suitable factor

$$w_i^* = \sqrt{\frac{6}{(k-1)(2k-1)}}w_i .$$

2.2 The Structural Model

The structural model for the two latent variables is given by a bivariate normal distribution with unknown means $\mu(\theta)$ and $\mu(\delta)$, variances $\sigma^2(\theta)$ and $\sigma^2(\delta)$, and correlation $\rho(\theta, \delta)$ of the two variables.

3 Estimating the Model

The manifest probabilities for observing a given response vector $\mathbf{x} = (x_1, \ldots, x_k)'$ are given by

$$P(\mathbf{X} = \mathbf{x}) = \int \prod_{i=1}^{k} f_i^{x_i}(\theta + w_i\delta)[1 - f_i(\theta + w_i\delta)]^{1-x_i}\, d\mathcal{N}(\theta, \delta) ,$$

where integration is with respect to the bivariate normal distribution with means $\mu(\theta)$, $\mu(\delta)$ and variances $\sigma^2(\theta)$, $\sigma^2(\delta)$ as well as correlation $\rho(\theta, \delta)$. If Ω is the set of response vectors $\mathbf{x}$ observed in the sample, and $n_{\mathbf{x}}$ the frequency of occurrence, then the *likelihood* of the data is given by

$$L = \prod_{\mathbf{x} \in \Omega} P(\mathbf{X} = \mathbf{x})^{n_{\mathbf{x}}} .$$

The likelihood is a function of the model parameters, that is, of the item difficulties, the guessing parameter, if one is used, as well as of the structural parameters of the bivariate normal distribution. The maximum likelihood estimates of these parameters are those parameter values that maximize the likelihood function. Some of the parameters are not identified and can in fact be omitted without loss of generality. In particular, the parameter $\mu(\theta)$ for the mean of the ability-level distribution can be absorbed into the item difficulty parameters. Similarly, when there is only one group of examinees, the mean of the learning-ability distribution can also be omitted (cf. Klauer & Sydow, 1992).

The numerical integration required to compute the likelihood function is performed by means of the Gauss–Hermite algorithm (cf. Bock & Aitkin, 1981). For this purpose, the variables θ and δ are represented as linear combinations of two independent variables t and u, each following a standard normal distribution with mean zero and variance one:

$$\theta = \mu(\theta) + at \quad \text{and} \quad \delta = \mu(\delta) + bt + cu . \tag{4.1}$$

This means replacing the structural variance and covariance parameters by functions of the new parameters a, b, and c:

$$\sigma^2(\theta) = a^2 , \tag{4.2}$$
$$\sigma^2(\delta) = b^2 + c^2 , \tag{4.3}$$
$$\sigma(\theta, \delta) = ab . \tag{4.4}$$

4 Validating the Model

Validating mathematical models has two aspects: statistical fit and empirical validity. Statistical goodness of fit means testing the fit of the model to the data (D'Agostino & Stephens, 1986). Empirical validity goes above statistical goodness of fit and aims at testing whether the model parameters measure those variables and processes that they are intended to operationalize (Klauer & Wegener, 1998).

4.1 Statistical Goodness of Fit

In the present case, the model makes direct predictions for the manifest probabilities with which the different response vectors are observed. Unfortunately, there are very many different response vectors; for a test with 12 items, for example, there are $2^{12} = 4096$ response vectors. This means that the frequencies with which each response vector is observed will most often be zero, and traditional χ^2-based goodness-of-fit tests cannot be directly applied. Rather, it is necessary to group response vectors into categories with larger expected probabilities of occurrence. In the present case, the bivariate distribution of the test score variable X and the weighted score $Y = \sum_{i=1}^{k} w_i X_i$ is an important aspect of the model, because when there is no guessing, X and Y provide sufficient statistics for θ and δ, respectively. Note that Y assumes the largest values for response vectors for which correct answers cluster at the end of the test, that is, for which learning has likely occurred. For this reason, we recommend grouping response vectors into a contingency table that cross-tabulates (intervals of) X and Y, and to apply χ^2-tests to assess the fit of the model to this bivariate distribution as exemplified below. Maximum-likelihood estimation should be based on the grouped data in this test.

In addition, a traditional method of assessing consistency of the model with the data in the context of item response theory is to consider splits of the sample of examinees based on a variety of variables and to test whether the same model parameters can be used to describe both resulting subgroups of the sample in each case (Molenaar, 1983). This procedure can reveal model violations tied to specific examinee characteristics that may go unnoticed in general-purpose goodness-of-fit tests. Below, we illustrate this approach by means of splits based on gender and order of test presentation. When the test has many items, the procedure based on splits can be the only feasible one, because the general-purpose test sketched above in general requires repeated computations of the expected probability for each response vector, which quickly surpasses the limits of what can be computed within reasonable amounts of time as the number of items increases.

4.2 Experimental Validation

Statistical goodness of fit is a necessary condition for interpreting model-based analyses. Since several models with substantially different implications can often be fit to the same data set, goodness of fit alone does not guarantee that the substantive model implications are valid. In particular, parameters are often labeled by names that suggest that certain psychological processes are measured by these parameters: for example, there are difficulty parameters, guessing parameters, and ability level parameters. The confidence in such labels can be raised through empirical validations. The idea is to manipulate each psychological process through appropriate empirical manipulations and to check whether the differences between the different conditions can be mapped onto those model parameters assumed to measure the manipulated process and whether the other parameters remain unaffected, thereby establishing both convergent as well as discriminant validity (Campbell & Fiske, 1959) of the model parameters. In other words, the test is whether the parameters measure those processes they are intended to operationalize while they are unconfounded with other processes.

In the present study, we attempted an empirical validation of the learning parameters. For this purpose, the learning tests were administered without feedback and learning aids, that is, as traditional status tests in a control group. The learning model was then fit to both groups, the control group and the experimental group with learning aids. If the learning test successfully elicits and indeed accentuates differences in learning ability, the variance in the latent learning ability as well as the mean learning gain should be larger when it is administered as a short-term learning test rather than as a status test. The other model parameters should remain unaffected by the experimental manipulation, and thus it should be possible to fit the model to the data of both groups using the same parameters for the distribution of ability status as well as the same guessing and item difficulty parameters.

5 Application

5.1 The Analyzed Tests

The administered tests comprise the AT 3 previously analyzed by Klauer and Sydow (1992) and Klauer et al. (1994) as well as three short-term learning tests developed by Roether (1983). The tests aim at children of an age range of four to six years, and they test abilities in the domain of inductive reasoning (K.J. Klauer, 1989).

The AT 3 items are visual analogy items, in which a set of three figures must be completed so that a valid analogy arises. The items follow a multiple choice format with four response options. The items and the feedback and learning hints are described in more detail by Klauer et al. (1994). The AT 3 comprises 12 items.

The VLT by Roether (1983) consists of three short-term learning tests labeled forms, sequences, and colors. The items follow a multiple choice format with four response options in each case, and there are 21 items

TABLE 4.1. Composition of the sample.

		VLT, then AT 3		AT 3, then VLT	
Group	Age	Female	Male	Female	Male
CG	Five Years	16	13	19	17
	Six Years	17	16	21	19
CG Total		33	29	40	36
EG	Five Years	29	35	33	31
	Six Years	50	41	36	41
EG Total		79	76	69	72

grouped in three series of seven items in each test. In VLT-forms, the child must recognize the shape of a simple geometrical figure and discriminate it from a number of competing shapes presented as response options. The child must also learn that color is an irrelevant, distracting attribute from which it must abstract. In the case of false responses, the child is supported by means of a graded system of learning hints. In VLT-sequences, simple progressions of geometric shapes must be completed by picking the appropriate shape from four response options. In VLT-colors, the child must abstract from shapes and complete simple series of colored geometric shapes by picking the correctly colored figure from the set of response options, irrespective of shape.

The tests were administered in two different modes. In the experimental group (EG), the tests were presented as short-term learning tests with learning aids. In the control group (CG), there was neither feedback nor learning aids and thus, the tests were administered as traditional status tests. For each item, it was coded whether the item was solved spontaneously without learning hints or not. Since learning hints were given in the EG only if the item could not be solved, this performance measure can be defined equivalently for both groups.

Table 4.1 shows the joint distribution of the $N = 434$ children over the experimental conditions (CG versus EG), age groups (five years old versus six years old), gender (male versus female), and order of administration (VLT first, then AT 3 versus AT 3 first, then VLT). As can be seen, there are about twice as many participants in the condition with learning aids, reflecting our greater interest in estimating the model parameters in this condition than in the control condition. The other factors partition the sample into subgroups of about equal sizes in each case. The tests were administered by students as part of their course requirements. The children were recruited from, and the tests taken in, kindergartens that were usually located in the students' hometowns.

5.2 Goodness-of-Fit Tests and Experimental Validation

According to the baseline model outlined above, it should be possible (a) to describe the data from the control and experimental groups using *the same item parameters and parameters for the ability-level distribution,* and (b) to relegate all differences between the two groups to the structural parameters for the distribution of learning ability. This model has 19 functionally independent parameters for the AT 3 test.

- 12 item difficulties, β_i, $i = 1, \ldots, 12$.
- One guessing parameter, γ.
- One parameter for the variance in ability level, $\sigma^2(\theta)$.
- One parameter for the mean gain in ability level in the experimental group relative to the control group, $\mu_{\mathrm{EG}}(\delta)$.
- For each group GR = CG, EG, parameters for the variance in learning ability and its correlation with ability level, $\sigma^2_{\mathrm{GR}}(\delta)$, $\rho_{\mathrm{GR}}(\theta, \delta)$, totaling four parameters.

Note that the mean ability level as well as the mean gain in ability level through (incidental) learning in the control group can be absorbed into the item difficulty parameters and do not constitute independent parameters. The VLT tests comprise 21 items each and accordingly, use 28 parameters in this baseline model.

The goodness of fit of the model was evaluated by general-purpose goodness-of-fit tests (AT 3) as well as by the method based on splits of the sample of examinees (AT 3 and VLT). The sample of participants can be split by a number of variables, in particular into groups differing in the order of administration or in gender. For both splits, we do not expect differences in the model parameters. The model was separately fit to each resulting subgroup, yielding two maximum likelihoods L_1 and L_2 that can be compared with the maximum likelihood L for the total sample by $\Delta G^2 = -2[\ln(L) - \ln(L_1) - \ln(L_2)]$. The loglikelihood ratio statistic ΔG^2 is asymptotically chi-square distributed with 19 and 28 degrees of freedom for AT 3 and each VLT-test, respectively.

Furthermore, two additional splits were investigated: the split according to age group (children aged five years versus six years) and the split into control group and experimental group. Both followed the logic of experimental validation outlined above. For the split in terms of age, we do expect significant ΔG^2-values if the model is indeed sensitive to the many differences that can be expected in ability level, learning ability, propensity to guess, and so forth, as a function of age. Note that the model should be sensitive to such differences between subpopulations, because it comprises not only a probabilistic IRT measurement model, but also models the distribution of ability levels and learning abilities in the population under study.

TABLE 4.2. Split-based model tests for the baseline model.

	AT 3		VLT-Forms		VTL-Sequ.		VLT-Colors	
Split	*df*	ΔG^2	*df*	ΔG^2	*df*	ΔG^2	*df*	ΔG^2
Order	19	26.3	28	61.3**	28	30.5	28	41.2
Gender	19	11.3	28	26.0	28	24.0	28	26.2
CG/EG	13	10.6	22	26.2	22	68.7**	22	53.4**
Age	19	62.9**	28	99.7**	28	53.3**	28	80.4**

$^*p < .05$; $^{**}p < .01$

The split into control group and experimental group, on the other hand, allows us to test explicitly whether these groups do not differ in terms of item difficulties, guessing, and parameters of the ability-level distribution. Since different learning-ability parameters are allowed in the joint baseline model for the total sample, the degrees of freedom are somewhat smaller for this test, and any significant differences in the test performance of the two groups should already be absorbed by these baseline model parameters. This latter split should therefore yield a nonsignificant ΔG^2-value.

Table 4.2 shows the resulting split-based test statistics. As can be seen, AT 3 achieves satisfactory results in all of these analyses; that is, the log-likelihood ratio statistic is not significant for the splits based on order of administration, gender, and experimental condition, whereas significant differences are found, as expected, for the split based on age. On the other hand, each VLT-test has difficulties in at least one respect. The split according to the order of administration of AT 3 and VLT-tests revealed significant effects on the test VLT-forms. The split into control group and experimental group revealed significant model deviations of VLT-sequences as well as VLT-colors. Possible reasons for these latter deviations are considered in the following section. The effects of order of administration on VLT-forms might itself reflect learning effects so that participants in the condition "first AT 3, then VLT" might perform better on VLT-forms than participants in the condition "first VLT, then AT 3." This interpretation does not explain, however, why similar effects of the order of administration were not found on the other VLT-tests, nor on AT 3 itself. As discussed below, it is often difficult to pinpoint the cause of significant model deviations when they occur.

For AT 3, the relatively small number of items made it possible to conduct the general-purpose goodness of fit. In Table 4.3, the bivariate distribution of X and Y is shown separately for control group and experimental group. As discussed above, the bivariate distribution of the test score variable X and the weighted score Y, $Y = \sum_{i=1}^{k} w_i X_i$, is an important aspect

TABLE 4.3. Contingency table for AT 3. Expected frequencies in parentheses.

		Y-Interval					
Group	X	1		2		3	
CG	0	22	(17.8)				
	1	4	(9.7)	6	(7.0)	7	(6.5)
	2–3	11	(10.3)	15	(14.4)	15	(15.7)
	4–5	3	(3.0)	16	(11.3)	12	(14.3)
	6–7	2	(3.5)	7	(6.5)	8	(6.0)
	8–12	4	(4.8)	0	(2.0)	6	(4.9)
EG	0–1	5	(7.8)	6	(6.0)	36	(36.0)
	2	8	(12.5)	15	(9.4)	21	(20.9)
	3	26	(22.4)	7	(9.0)	15	(13.8)
	4	14	(11.0)	8	(9.1)	19	(22.4)
	5	13	(9.5)	12	(15.6)	9	(11.6)
	6	7	(8.5)	8	(11.4)	10	(9.4)
	7–8	18	(14.3)	14	(10.0)	12	(11.3)
	9–12	2	(3.5)	3	(4.0)	8	(6.6)

of the model, because when there is no guessing, X and Y provide sufficient statistics for θ and δ, respectively. To assess the population variances and the correlation of these latent variables, it therefore seemed to be of prime importance that the model validly describe the joint distribution of X and Y. For this reason, we grouped response vectors into a contingency table that cross-tabulates (intervals of) X and Y. To guarantee sufficient large cell frequencies, a number of adjacent score groups were aggregated, and for each resulting score group, the range of Y was divided into three intervals which had been determined on the basis of another sample of examinees by Klauer et al. (1994) so that approximately equal numbers of observations fell into each interval in that previous study. As a matter of fact, we used exactly the same groupings for X- and Y-values as Klauer et al. (1994) did, so that the interval boundaries are not functions of the data in the present analysis. Note, however, that even if they were, the chi-square test would not have been compromised (Moore & Spruill, 1975). The resulting contingency table has 40 cells yielding 38 degrees of freedom, because the two group totals are fixed. The model has 19 parameters so that the likelihood-ratio statistic for assessing goodness of fit has 19 degrees of freedom. Its value was $G^2 = 29.67$, indicating an acceptable, though not excellent, fit, $p = .06$.

5.3 A Modified Model for the VLT-Tests

While AT 3 achieved satisfactory goodness-of-fit results in the above analyses, the VLT-tests could not be described by the learning model. The reasons for model failure can be manifold. The model may be misspecified in the measurement part or in the structural part. In the first case, the analyzed test might still measure the two latent factors ability level and learning ability, but the assumed item characteristic curve may be misspecified, the linear growth function relating learning ability to the latent ability scale may be overly simplistic, and so forth. In the second, more serious case, the test might not even be governed by two latent factors. It is often difficult to pinpoint the cause of observed violations.

For the VLT-tests, we attempted a modification of the measurement model in which we allowed differential mean learning for each item. That is, we reasoned that transfer from previous items to subsequent items may not proceed in homogeneous equal-sized and accumulating steps for all items as assumed by the above baseline model, but that different items may be on average differentially susceptible to transfer. We maintained the assumption, however, that for each item I_i, the positions of individuals' learning gains, relative to the item-specific average transfer, would still follow a simple accumulation of itemwise increments as captured by the function $(i-1)\delta$. This led to a modified model in which different item difficulty parameters are used for all items (except the first one) in control versus experimental group to allow for potentially item-specific average amounts of transfer. The same parameters were used in both groups for the guessing tendency and for the ability distribution.

The modified model could be maintained in the split-based tests for the VLT-sequences and the VLT-colors, whereas it had to be rejected for the VLT-forms. VLT-forms still suffered from a significant effect under the split based on order of administration. Further investigations would be required to determine what the problem is with this subtest.

5.4 Structural Analyses

Departing from these results, further structural analyses were performed for AT 3 and VLT-sequences as well as VLT-colors. Table 4.4 shows the maximum likelihood estimates of the central structural parameters and structural tests for AT 3. As can be seen, the variance in learning ability, $\sigma^2(\delta)$, is about twice as large in the experimental group than in the control group supporting Assumption 2 of learning tests that interindividual differences in learning ability exist and are accentuated by learning tests. In addition, while learning ability and ability level seem perfectly correlated in the control group, they appear more independent in the experimental group supporting Assumption 3.

The structural tests for AT 3 corroborate this impression. The reported tests for variances and correlations are performed by setting appropriate

TABLE 4.4. Structural analysis of AT 3.

	$\hat{\sigma}^2(\theta)$	$\hat{\sigma}^2(\delta)$	$\hat{\rho}(\theta,\delta)$
CG	.63	.15	1.0
EG	.63	.32	−.07
	Structural Tests		
Test	*df*	ΔG^2	*p*
$\sigma^2_{\text{CG}}(\delta) = 0$	2	6.9	.03
$\lvert\rho_{\text{CG}}(\theta,\delta)\rvert = 1$	1	.0	.99
$\sigma^2_{\text{EG}}(\delta) = 0$	2	9.4	.01
$\lvert\rho_{\text{EG}}(\theta,\delta)\rvert = 1$	1	3.7	.06
$\mu_{\text{EG}}(\delta) = 0$	1	14.4	$< .01$

parameters a, b, or c equal to zero in the control group or experimental group (cf. Equations 4.2 through 4.4) and by conducting likelihood-ratio tests of such null hypotheses. As can be seen, there are significant amounts of variability in learning ability in both the control and experimental groups and a significant average learning gain in the experimental group relative to the control group. Learning ability and ability level are, however, redundant factors in the control group; that is, they can be assumed to correlate perfectly, whereas the test for nonredundancy approaches significance in the experimental group ($p = .06$).

Table 4.5 shows the parameter estimates for the VLT-sequences and VLT-colors. Since each item is allowed an item-specific amount of overall transfer in this model, the item parameters for experimental and control group are also shown. As can be seen, the items are generally easier in the experimental group, suggesting that the feedback and learning aids did in fact elicit overall learning gains. Wilcoxon tests comparing the parameter estimates for each test reveal in fact that the drop in item difficulty from control group to experimental group is significant in both tests (VLT-sequences: $Z = -2.39$, $p = .017$, and VLT-colors: $Z = -3.14$, $p = .002$). Although increased overall learning has taken place in the experimental group, the estimates of the parameters of the latent distributions reveal an interesting counterintuitive pattern: in both tests, the variability in learning ability is much larger in the control group than in the experimental group. In addition, ability level and learning ability are highly positively correlated in the control group and correlate lower or negatively in the experimental group. Clearly, the empirical validation of the learning model of the VLT learning tests has failed, because the variability in supposed

TABLE 4.5. Parameter estimates for two VLT-tests; item difficulties are multiplied by 100.

	VTL-Sequ.		VTL-Colors	
	CG	EG	CG	EG
$\sigma^2(\theta)$	2.79	2.79	1.25	1.25
$\sigma^2(\delta)$	1.85	.25	1.46	.25
$\rho(\theta, \delta)$	.89	-.87	.79	.14
β_1	−264	−264	−121	−121
β_2	17	6	−176	−139
β_3	−58	−89	−32	8
β_4	−109	−64	−103	−102
β_5	93	98	11	15
β_6	−53	−107	164	54
β_7	14	−51	210	79
β_8	−329	−258	−118	−111
β_9	−9	−87	−121	−117
β_{10}	−47	−207	−59	−158
β_{11}	−213	−247	−112	−176
β_{12}	113	25	−3	−60
β_{13}	−31	−18	−10	−114
β_{14}	151	8	223	−42
β_{15}	−269	−223	115	27
β_{16}	−46	−155	−62	−251
β_{17}	−62	−228	−211	−272
β_{18}	−279	−256	−56	−179
β_{19}	47	−75	34	−133
β_{20}	−145	−144	300	−15
β_{21}	163	−72	201	−92

learning ability is much larger when the test is administered as a traditional status test than when it is administered as a learning test. It seems that paradoxically, feedback and learning aids led to much more uniform amounts of individual learning gains than the traditional manner of administration instead of eliciting and enhancing existing differences in learning ability. A substantive interpretation of this pattern of structural results is given in the discussion.

In the instance of a more successful validation of the VLT-tests, we had planned to explore the joint multivariate distribution of the AT 3 latent variables and the latent factors tapped by the VLT-tests by fitting an appropriately extended supermodel to data from several tests simultaneously. This would have allowed us to assess the correlations between learning abilities in different domains. However, the pattern of results did not encourage this extension of the model for the present data.

6 Discussion

To summarize, the desired statistical and experimental validation of the AT 3 learning test was quite successful. The data from the control and experimental groups could be described by the same measurement model while differences between the groups could be mapped onto differences in the bivariate distribution of ability level and learning ability. In particular, test performance in the group without learning aids revealed relatively little interindividual variability in learning gains, and these gains correlated highly with ability level. In contrast, in modeling the experimental group's data, interindividual differences in learning gains became more important and were found to be less strongly correlated with ability level. These results were corroborated by the general-purpose goodness-of-fit test, by split-based model tests as well as in structural analyses of the latent factors' bivariate distribution. Taken together with earlier results by Klauer et al. (1994), these findings raise the confidence in the usefulness of AT 3 as a means of diagnosing learning ability in the domain of inductive reasoning for preschool children.

The results for the VLT-tests were less encouraging. The split-based model tests revealed violations of the baseline model. A modified model with relaxed assumptions could be maintained for two VLT-tests, but did not lead to a successful experimental validation of these tests. Specifically, more interindividual variability emerged in the learning gains of the control group, in which the tests were administered as traditional status tests, than in the experimental group with learning aids. In addition, learning gains were correlated highly with ability level in the control group and less strongly or even negatively in the experimental group.

This pattern can be interpreted in the context of so-called *aptitude treatment interaction* research (Snow, 1989). In this line of research, learning gains are found to correlate with general intelligence to the extent to which the learning environment is unstructured and does not provide learning aids. In other words, the more helpful and structured the learning opportunity, the more it helps low-ability learners in a compensatory fashion, while such opportunities to learn are quickly assimilated and exhausted by high-ability learners. Considering the VLT-tests, and the learning situation realized in the control group with an unstructured and unsupportive learning environment, these findings lead one to expect high correlations of learning gain and ability level in the control group. The more supportive learning opportunities realized in the experimental group, on the other hand, may have supported and helped low-ability learners more strongly than high-ability learners, thereby leveling the deficits exhibited in the control group by low-ability learners, explaining (a) the reduced variability in learning gains and (b) the lower correlations of learning gains and initial learning ability.

In contrast, for the AT 3 test, a larger variability of learning gains was revealed in the experimental group than in the control group. This suggests that the difficulty level of this test and the nature of the learning aids created learning opportunities that were helpful for learners over a wider range of initial ability levels.

From a test-theoretic point of view, the model-based approach to the analysis of proposed learning tests for the first time offers the possibility to identify and evaluate the contributions of two latent factors defined by the model. This makes it possible to circumvent the many statistical problems associated with conventional ad hoc indices of learning gains and ability level that are subsumed under the label "measurement of change" (e.g., Zimmerman & Williams, 1982a, 1982b). In conventional approaches, ability level is most often defined directly in terms of the test score and learning gains by means of difference scores between late and early sections of the test. In particular, note that the correlation estimated through the use of the model for the two latent factors is not only automatically corrected for attenuation, but also for bias that arises from the fact that both variables are estimated from the same data with shared measurement error (leading, for example, to the well-known initial values problem in conventional analyses; Wilder, 1965). This follows from the fact that the correlation is estimated consistently if the model correctly describes the data.

Previous dynamic generalizations of the Rasch model have been reviewed by Verhelst and Glas (1995). Many of them consider what Guthke et al. (1991) have called long-term learning tests in which a pretest-training-posttest design is used. An interesting family of models for learning during testtaking is considered by Verhelst and Glas (1995), in which learning decreases the item difficulty of each item by an amount that is defined by the participants' preceding responses and possible independent reinforcements. Since learning gain is not operationalized as a person parameter, this family of models does not appear to lend itself naturally to addressing the questions about interindividual differences in learning ability and about correlations between ability status and learning ability that are of central interest in the study of learning tests. It should be possible, however, to generalize this family of models so that learning gain is operationalized as a person parameter if marginal maximum likelihood estimation is used as in the present case.

Model-based analyses require, however, a statistically satisfactory fit of the model to the data and ideally additional empirical validations of the most important model parameters. Given the many simplifying assumptions that enter into IRT models in general and into the present learning model in particular, failure to validate the model may be a quite common outcome of appropriate studies. In these instances, it can be difficult to diagnose the exact nature and cause of the observed model deviations.

References

Bock, R.D., & Aitkin, M. (1981). Marginal maximum likelihood estimation of item parameters: Application of an EM algorithm. *Psychometrika, 46*, 443–459.

Campbell, D.T., & Fiske, D.W. (1959). Convergent and discriminant validation by the multitrait-multimethod matrix. *Psychological Bulletin, 56*, 81–105.

D'Agostino, R.B., & Stephens, M.A. (1986). *Goodness-of-fit techniques.* New York: Marcel Dekker.

Embretson, S.E. (1991). A multidimensional latent trait model for measuring learning and change. *Psychometrika, 56*, 495–515.

Flammer, A., & Schmid, H. (1982). Lerntests: Konzept, Realisierung, Bewährung. Eine Übersicht [Learning tests: Concept, realization, validation. A review]. *Schweizerische Zeitschrift für Psychologie, 41*, 114–138.

Guthke, J. (1982). The learning test concept—An alternative to the traditional static intelligence test. *German Journal of Psychology, 6*, 306–324.

Guthke, J. (1983). *Mengenfolgen-Test (MFT)* [Set-sequences test]. Berlin: Psychodiagnostisches Zentrum der Humboldt-Universität.

Guthke, J. (1990). Learning tests as an alternative or completion of intelligence tests: A critical review. *European Journal of Psychology of Education, 5*, 117–133.

Guthke, J., Böttcher, H.R., & Sprung, L. (1991). *Psychodiagnostik* (Band 2) [Psychodiagnostics (Vol. 2)]. Berlin: Deutscher Verlag der Wissenschaften.

Guthke, J., & Wiedl, K.H. (1996). *Dynamisches Testen. Zur Psychodiagnostik der intraindividuellen Variabilität* [Dynamic testing: The psychodiagnostics of intraindividual variability]. Göttingen: Hogrefe.

Hamers, J.H.M., Sijtsma, K., & Ruijssenaars, A.J.J.M. (1993). *Learning potential assessment: Theoretical, methodological and practical issues.* Amsterdam: Swets & Zeitlinger.

Katzbeck, C., Kastner-Koller, U., & Deimann, P. (1998). Konstruktion und erste Schritte zur Konstruktvalidierung eines treatment-orientierten Kurzzeitlerntests für das Vorschulalter [Construction and first steps in construct-validation of a treatment-oriented short-term learning test for preschool children]. *Diagnostica, 44*, 100–111.

Klauer, K.C., Kauf, H., & Sydow, H. (1994). Experimentelle Validierung eines Lernmodells für Kurzzeitlerntests [Experimental validation of a learning model for short-term learning tests]. *Diagnostica, 40*, 124–142.

Klauer, K.C., & Sydow, H. (1992). Interindividuelle Unterschiede in der Lernfähigkeit: Zur Analyse von Lernprozessen bei Kurzzeitlerntests [Interindividual differences in learning ability: Analyzing learning by means of short-term learning tests]. *Zeitschrift für Differentielle und Diagnostische Psychologie, 13*, 175–190.

Klauer, K.C., & Wegener, I. (1998). Unraveling social categorization in the "Who said what?"-paradigm. *Journal of Personality and Social Psychology, 75*, 1155–1178.

Klauer, K.J. (1989). *Denktraining für Kinder I* [Cognitive training for children I]. Göttingen: Hogrefe.

Molenaar, I.W. (1983). Some improved diagnostics for failure in the Rasch model. *Psychometrika, 48*, 49–72.

Moore, D.S., & Spruill, M.C. (1975). Unified large-sample theory of general chi-squared statistics for tests of fit. *The Annals of Statistics, 3*, 599–616.

Roether, D. (1983). *Vorschul-Lerntest (VLT)* [Preschool learning test]. Berlin: Psychodiagnostisches Zentrum der Humboldt-Universität.

Snow, R.E. (1989). Aptitude-treatment interaction as a framework for research on individual differences in learning. In P.L. Ackermann, R.J. Sternberg, & R. Glaser (Eds.), *Learning and individual differences: Advances in theory and research* (pp. 13–59). New York: Freeman.

Sydow, H., & Meincke, J. (1994). *DenkMit—Das Berliner Programm zur Förderung des Denkens und der Wahrnehmung von drei- bis sechsjährigen Kindern* [DenkMit—The Berlin program for training reasoning and perception in children aged three to six]. Simbach/Inn: ZAK.

Verhelst, N.D., & Glas, C.A.W. (1995). Dynamic generalizations of the Rasch model. In G.H. Fischer & I.W. Molenaar (Eds.), *Rasch models: Foundations, recent developments, and applications* (pp. 181–201). New York: Springer-Verlag.

Wilder, J. (1965). Pitfalls in the methodology of the law of initial values. *American Journal of Psychotherapy, 19*, 577–584.

Zimmerman, D.W., & Williams, R.H. (1982a). The relative error magnitude in three measures of change. *Psychometrika, 47*, 141–147.

Zimmerman, D.W., & Williams, R.H. (1982b). On the high predictive potential of change and growth measures. *Educational and Psychological Measurement, 42*, 961–968.

5

An IRT Model for Multiple Raters

Norman D. Verhelst[1] and Huub H.F.M. Verstralen[2]

ABSTRACT An IRT model for multiple ratings is presented. If it is assumed that the quality of a student performance has a stochastic relationship with the latent variable of interest, it is shown that the ratings of several raters are not conditionally independent given the latent variable. The model gives a full account of this dependence. Several relationships with other models appear to exist. The proposed model is a special case of a nonlinear multilevel model with three levels, but it can also be seen as a linear logistic model with relaxed assumptions (LLRA). Moreover, a linearized version of the model turns out to be a special case of a generalizability model with two crossed measurement facets (items and raters) with a single first-order interaction term (persons and items). Using this linearized model, it is shown how the estimated standard errors of the parameters are affected if the dependence between the ratings is ignored.

1 Introduction

The purpose of this chapter is to introduce and to discuss an item response model for discrete data collected in a design where the answers of subjects to questions or items are graded by raters. Feldt and Brennan (1989, pp. 128–129) use this measurement procedure to explain central concepts in generalizability theory. Items and raters are called measurement facets. If all raters grade all responses of all subjects, the design is called a crossed design. The model discussed here is restricted to such a two-facet crossed design.

While developing the model, it appeared that the main problem was related closely to a central concept in item response theory (IRT), namely, the assumption of local independence. Before presenting the model, this assumption is discussed briefly for the case of a single facet design (items only, say) and a simple IRT model, the Rasch model.

[1]National Institute for Educational Measurement (CITO), P.O. Box 1034, 6801 MG Arnhem, The Netherlands; Faculty of Educational Science and Technology, University of Twente, The Netherlands; e-mail: *norman.verhelst@cito.nl*

[2]National Institute for Educational Measurement (CITO), P.O. Box 1034, 6801 MG Arnhem, The Netherlands; e-mail: *huub.verstralen@cito.nl*

When a student responds to a number of items, it is assumed that his latent ability θ is constant during testtaking and that each response X_i is a independent draw from some distribution. For example, if the Rasch model is used, the probability of a correct response is given by

$$P(X_i = 1 \mid \theta) = \frac{\exp(\theta - \delta_i)}{1 + \exp(\theta - \delta_i)} ,$$

where δ_i is the difficulty parameter of item I_i. This is equivalent to the assumption that a random variable Z_i exists, which follows the (standard) logistic distribution with expectation θ, and the response to the item corresponds to a single realization of this random variable. The probability of a correct response is then given by

$$P(X_i = 1 \mid \theta) = P(Z_i > \delta_i) .$$

The assumption of local independence is equivalent to the assumption that for all i and j $(i \neq j)$, Z_i and Z_j are independent. For the interpretation, this means that for each observational variable, the student makes an independent attempt to demonstrate his ability. This independence is the reason why the test information function is just the sum of the item information functions.

In a two-facet crossed design n students respond to k items, and each response is graded by r raters. Let X_{vig} be such a rating given by rater R_g to an answer produced by student S_v as a response to item I_i. This answer may be quite complex and exist, for example, of an essay. If the ratings are to be considered as indicators of the writing ability of the student, it is clear that these ratings are not independent, because the student does not produce an independent essay for each of the raters; all raters give a judgment on the same essay, and therefore their judgments will be correlated. If the student has produced an exceptionally good essay, given his writing ability, then all ratings will tend to be relatively high, while in the case of a bad essay all ratings will tend to be relatively low.

In simple applications of the Rasch model with a single measurement facet, the assumption of local independence is usually taken for granted, and no special attention is given to it, for example, in testing the goodness of fit of the model. (A noteworthy exception is Van den Wollenberg, 1982.) If one attempts to generalize this model to designs with more than one measurement facet, it will be necessary to model the above-described dependency in a correct way.

This chapter is meant to serve three purposes. A formal IRT model is developed where the dependency between ratings is taken into account (Section 2). Next, the relation with multilevel analysis is explored (Section 3). The model turns out to be a special case of a nonlinear multilevel regression model so that standard multilevel software can be used for parameter estimation. It is argued, however, that several aspects of the model, which play a central role in an IRT approach, are usually not given much attention in the multilevel approach. In Section 4 it is shown how these aspects can be studied using conditional maximum likelihood. The third purpose of this chapter is to study the consequences of a specification error: what happens if the model is correct, but the dependence between the ratings as specified by the model is ignored? This is the topic of Section 5.

In the sections to follow, reference is made to discrete (binary) and continuous observed variables. The symbol X is used to refer to the binary case, while Y is reserved for continuous variables.

2 The Model

Basic in our approach is the creation of an auxiliary latent variable, which we might label "quality" (of an essay, say). This variable depends stochastically on the latent variable to be measured (writing ability). To fix the ideas, it is assumed that both latent variables (writing ability, denoted θ, and quality, denoted κ_i, where i indexes the item) are continuous variables, and we use a simple model for the conditional distribution of κ_i given θ:

$$\kappa_i \mid \theta \sim \mathcal{N}(\theta - \beta_i^*, \tau^2) \ . \tag{5.1}$$

The model says that the quality of an essay is a random variable with expectation depending on the writing ability θ of the student and the difficulty β_i^* of the item. Furthermore, it is assumed that the random variables κ_i and κ_j $(i \neq j)$ are conditionally independent. It is important to realize that the variance of κ_i, when positive, ensures that there is no one-to-one relationship between the quality of an essay and the writing ability of the student. Of course, we could make the variance depend on the student or the item as well, but we stick to the simplest case where the variance is assumed to be constant across items and students.

As to the notation used in the sequel, realizations of the random variables θ and κ_i are denoted by the same symbols, or as θ_v and κ_{vi} when reference is needed to a particular student S_v.

The next assumption, which seems quite realistic, is that a rater bases his judgment only on the quality of the essay he reads, and additionally, that raters give their ratings independently of each other.

In the case where the judgment is binary, we may use a simple model for the observed rating X_{vig}, giving, for one student–essay combination,

$$P(X_{vig} = 1 \mid \theta_v, \kappa_{vi}) = \frac{\exp(\kappa_{vi} + \psi_g - \beta_i)}{1 + \exp(\kappa_{vi} + \psi_g - \beta_i)} \ , \tag{5.2}$$

where ψ_g is a leniency parameter for rater R_g, and β_i is some general standard for item I_i, summarizing aspects through which the average ratings on the items will differ systematically, and which might be called difficulty.

The relation between the difficulty parameter β_i in (5.2) and the difficulty parameter β_i^* in (5.1) is discussed later.

Notice that the right-hand side of (5.2) does not depend on θ_v, reflecting the assumption that the rater bases his judgment only on the quality of the essay, and not, for example, on some indicators of the writing ability of student S_v, which she might have from other sources. We assume that ratings from different raters are independent, giving, for example, that

$$P(x_{vig}, x_{vih} \mid \theta_v, \kappa_{vi}) = P(x_{vig} \mid \theta_v, \kappa_{vi}) P(x_{vih} \mid \theta_v, \kappa_{vi}) , \qquad g \neq h .$$

If we want to condition only on θ_v, however, we find by combining (5.1) and (5.2), that

$$P(x_{vig} \mid \theta_v) = \int P(x_{vig} \mid \theta_v, \kappa_i) \, g(\kappa_i \mid \theta_v) \, d\kappa_i , \tag{5.3}$$

where $g(\cdot \mid \cdot)$ is the normal probability density function (pdf) corresponding to (5.1). It follows immediately that

$$P(x_{vig}, x_{vih} \mid \theta_v) = \int P(x_{vig} \mid \theta_v, \kappa_i) \, P(x_{vih} \mid \theta_v, \kappa_i) \, g(\kappa_i \mid \theta_v) \, d\kappa_i ,$$

which in general will be unequal to the product of the marginal probabilities, like (5.3), showing that X_{vig} and X_{vih} are not independent given θ_v, unless $\tau^2 = 0$.

If we assume furthermore that a rater gives his ratings on an essay independently of his ratings on other essays from the same or different students, we have by definition

$$P(x_{vig}, x_{vjg} \mid \theta_v) = P(x_{vig} \mid \theta_v) P(x_{vjg} \mid \theta_v) , \tag{5.4}$$

but notice that both factors on the right-hand side of (5.4) are defined by (5.3) and not by (5.2). This leaves us with a quite complicated likelihood function. Denoting the pdf of θ by $h(\theta)$, the marginal likelihood for all ratings pertaining to student S_v is given by

$$L_v = \int \Big[\prod_i \int \prod_g P(x_{vig} \mid \theta, \kappa_i) g(\kappa_i \mid \theta) d\kappa_i \Big] h(\theta) \, d\theta . \tag{5.5}$$

For n students jointly, the likelihood is

$$L = \prod_{v=1}^{n} L_v . \tag{5.6}$$

The intertwining of integrals and products makes marginal maximum likelihood (MML) estimation of the model parameters quite hard. The maximization of (5.6) is not easy as is explained in the next two sections.

3 A Multilevel Interpretation of the Model

In this section it is shown that the model defined by (5.1) and (5.2) is a special case of a nonlinear multilevel model. Before proceeding to a formal derivation, however, it seems useful to elucidate a terminological problem which, in our view, can easily lead to confusion and unjustified use of models and analysis techniques. The keywords of the problem are crossed and nested.

In many introductions to multilevel analysis in educational research, multilevel or hierarchical models are introduced by a tree-like structure, where lower-level entities pertain to precisely one higher level unit. The standard example is the organization of students within classes, classes within schools, and possibly schools within districts. It is then said that students are nested within classes, classes are nested within schools, and schools are nested within districts. This nested structure follows naturally from a hierarchical sampling scheme where the sample of students is not randomly drawn from the population, but where first a sample of units of the highest level (districts in the example) is drawn, and then a sample of the next level (schools) within each sampled unit of the highest level, and so on, until the lowest level. The data collected in such a design can be analyzed adequately by specially developed statistical models, such as an analysis of variance with nested factors, but the confusion may arise from the idea that there is a one-to-one relationship between sampling scheme and model, implying that a hierarchical model should only be applied with a hierarchical sampling scheme.

To see that this is not true, we study an example in a design with two crossed factors. In such a design all levels of the first factor are combined with all levels of the second factor, and for each combination one or more observations are collected. Consider the administration of k items to n students, in a complete design; that is, all students respond to all items. The two factors here are students and items. Suppose that continuous responses Y_{vi} are observed, and the model used is given by

$$Y_{vi} = \theta_v + \eta_i + \varepsilon_{vi} \ , \tag{5.7}$$

where θ_v and η_i are the student and item effects, respectively, and ε_{vi} is the residual. In such a model we might say that the person and item effects are crossed, meaning that the effect of the item is the same for all students and conversely that the effect of the student is the same for all items. We realize that this is not standard terminology, and that the term "additive model" is well suited to describe (5.7), but the term was chosen deliberately to point at the same confusion, namely, that model (5.7) necessarily follows from the crossed data collection design.

Quite a different model for the same data results if we let the value of one effect be dependent on the level of the other effect, as in the model

$$Y_{vi} = \theta_v + \eta_{i(v)} + \varepsilon_{vi} \ , \tag{5.8}$$

where the notation used for the η-parameter is standard in many textbooks to indicate that "items are nested within persons." This model, using nested or hierarchical factors in a crossed design, is the starting point from where the IRT model presented in the preceding section is reconstructed.

Of course, we cannot estimate the parameters in (5.8) if the effects are all considered as fixed, because there are more parameters than observations. But even if we switch to a random effects model, the parameters (the variances of the effects and a single intercept) are not estimable, because the residual ε is confounded completely with the item effects. To disentangle these two effects, we need at least two replications per student–item combination. Using g to index replications, we find as our full model

$$Y_{vig} = \theta_v + \eta_{i(v)} + \varepsilon_{g(vi)} , \tag{5.9}$$

where the residual at replication g can take different values for different combinations of student and item. So, although the data collection design is crossed (all students take all items at each replication) the effects are nested: items are nested within students (i.e., their effects are not constant across students), and the residual (the replication effect) is nested within student–item combinations. Considering all effects as random makes (5.9) a simple multilevel model with three levels.

To come closer to the case with ratings, assume that in model (5.9) a fixed effect of the replications is added, because at each replication the answer of the students is scored by a different judge, for example. In a similar way one could assume that there is also a systematic effect of the items. Then the model becomes

$$Y_{vig} = \theta_v + \eta_{i(v)} + \delta_i + \psi_g + \varepsilon_{g(vi)} . \tag{5.10}$$

To make this model identified and its parameters estimable, several assumptions about the distribution of the random variables involved and their mutual dependence or independence must be added, and several technical restrictions are required. Before discussing these assumptions and restrictions, the IRT model presented in the previous section is reformulated slightly to show its similarity with model (5.10).

Observe in (5.1) that $\kappa_i \mid \theta$ can also be written as

$$\kappa_i = \theta - \beta_i^* + \zeta_i ,$$

where

$$\zeta_i \sim \mathcal{N}(0, \tau^2) ,$$

so that (5.2) can be rewritten as

$$P(X_{vig} = 1 \mid \theta_v, \zeta_{vi}) = \frac{\exp(\theta_v + \zeta_{vi} + \psi_g - \beta_i - \beta_i^*)}{1 + \exp(\theta_v + \zeta_{vi} + \psi_g - \beta_i - \beta_i^*)} . \tag{5.11}$$

From (5.11) it can be seen that the model is not identified since the two parameters β_i and β_i^* are completely confounded; only their sum is identified. Defining

$$\delta_i = -\left(\beta_i + \beta_i^*\right) ,$$

and substituting δ_i into (5.11) gives

$$P(X_{vig} = 1 \mid \theta_v, \zeta_{vi}) = \frac{\exp(\theta_v + \zeta_{vi} + \psi_g + \delta_i)}{1 + \exp(\theta_v + \zeta_{vi} + \psi_g + \delta_i)} . \tag{5.12}$$

The similarity between the two models (5.10) and (5.12) is now clear if we consider some special function of the expected value of the response variable in both models. This special function is called the link function. In the linear model (5.10) the link function is the identity function $I(\cdot)$; in model (5.12) the link function is the logit function defined by

$$\text{logit}(x) = \ln\left(\frac{x}{1-x}\right) , \qquad 0 < x < 1 .$$

Notice that the right-hand side of (5.12) is the expected value of the binary variable X_{vig}. So we find

$$\begin{aligned} I\left[E\left(Y_{vig}\right)\right] &= \theta_v + \eta_{i(v)} + \delta_i + \psi_g , \\ \text{logit}\left[E\left(X_{vig}\right)\right] &= \theta_v + \zeta_{vi} + \delta_i + \psi_g . \end{aligned}$$

Equating $\eta_{i(v)}$ and ζ_{vi} in the latter equations shows that two different transformations (I and logit) of the expected values yield an identical structure, and moreover, that the transformed expected values are linear in the effects. Therefore, both models are examples of the family of generalized linear models (McCullagh & Nelder, 1989).

To define the models completely, one must specify the distribution of all its random variables and impose some identifiability restrictions. For model (5.10) it is usually assumed that the random variables θ_v, $\eta_{i(v)}$, and $\varepsilon_{g(vi)}$ are normally distributed and independent of each other. In model (5.12) θ_v and ζ_{vi} are also considered as independent normal variates, but the observed variable X_{vig}, conditional on θ_v and ζ_{vi}, is Bernoulli distributed with expected value given by (5.12). To make the models identifiable one can fix the expected values of the random variables θ_v, $\eta_{i(v)}$, ζ_{vi}, and $\varepsilon_{g(vi)}$ at an arbitrary constant, usually zero, and impose some normalization on the δ- or ψ-parameters, but not on both (e.g., $\sum \psi_g = 0$), or one could consider $\mu = E(\theta)$ as a parameter to estimate, and impose two normalizations (e.g., $\sum \psi_g = \sum \delta_i = 0$).

Both models, (5.8) and (5.12), can be estimated using current software for multilevel modeling, for example, MLwiN (Goldstein, Rasbash, Plewis, Draper, Browne, Yang, Woodhouse, & Healy, 1998), HLM (Bryk, Raudenbush, & Congdon, 1996), or MIXOR (Hedeker & Gibbons, 1996). Moreover, MLwiN provides facilities to specify the rater effects ψ_g as random. See, for example, Snijders and Bosker (1999, Chap. 11).

4 The IRT Approach

Although the model for binary data, with a (complicated) likelihood function (5.5) and (5.6), can be estimated by existing multilevel software, it may be illuminating to investigate the estimation problem in a more IRT-oriented approach. There are several reasons for doing this.

First, the estimation method mostly used for nonlinear multilevel models is not maximum likelihood (ML), and leans heavily on Taylor approximations to (5.12) (see, e.g., Goldstein, 1995), which in some cases can cause serious biases in the estimates (Rodríguez & Goldman, 1995).

A second and more serious reason not to lean too heavily on the multilevel tradition is the way measurement models are treated and tested in the multilevel literature. It is useful to make a distinction in complicated models between a measurement component and a structural component. The measurement component (i.e., the measurement model) specifies the relation between the observed dependent variable and the latent variables while the structural component describes the distribution of the latent variables and the dependence of this distribution on exogenous variables. Equation (5.12) is a measurement model; the structural component is defined by the distributional assumptions of the θ and ζ variables. Suppose, for example, that the assumption is made that θ is normally distributed in the population, but in reality there is a substantial difference between the average ability of boys and girls. Ignoring this difference constitutes a specification error in the structural model. Quite a different specification error occurs if model (5.12) is valid in the subpopulations of boys and girls taken separately, but not jointly, because for some item I_i the difficulty parameter δ_i takes different values in the two subpopulations; such a specification error is called differential item functioning (DIF), and has received considerable attention in the area of educational testing (see, e.g., Holland & Wainer, 1993). Ignoring DIF is a specification error in the measurement model. Another assumption in the measurement model (5.12) is that all items and raters are considered to discriminate equally well. Violation of this assumption might invalidate (in a complicated and not yet well-understood way) the results of an analysis. In virtually all the multilevel software packages the validity of the measurement is taken for granted. In the IRT literature much more emphasis is placed on tests for model fit (see, e.g., Glas & Verhelst, 1995a).

A third reason is of a didactic nature: estimation of the model parameters turns out to be quite complicated, but at the same time, it is shown that it can be tackled from quite different perspectives which it is hoped will turn out to be complementary. More specifically, it is investigated how the method of conditional maximum likelihood (CML) is applicable to the present model.

Model (5.12) is identified if a proper normalization is chosen; for example,

$$\sum \psi_g = \sum \delta_i = 0 . \tag{5.13}$$

Now define

$$\xi_{vi} = \theta_v + \zeta_{vi} ,$$

and rewrite (5.12) as

$$P(X_{vig} = 1 \mid \xi_{vi}) = \frac{\exp(\xi_{vi} + \psi_g + \delta_i)}{1 + \exp(\xi_{vi} + \psi_g + \delta_i)} . \tag{5.14}$$

Considering the essay–rater combination as an item, (5.14) closely resembles a linear logistic test model (LLTM; see Fischer, 1995a), that is, a Rasch model with a linear structure on the item parameters. It is, however, not a LLTM because the person parameter ξ_{vi} may vary across items. In fact, if we assume that in the population θ is normally distributed with mean μ and variance σ_θ^2, then it can be derived that, for $i \neq j$,

$$\text{Cor}(\xi_i, \xi_j) = \frac{\sigma_\theta^2}{\sigma_\theta^2 + \tau^2} ,$$

which is strictly less than one unless $\tau^2 = 0$. So model (5.14) assumes as many latent variables as there are essays, and from this viewpoint, we can say that it is a linear logistic test model with relaxed assumptions (LLRA; Fischer, 1995b).

It can be shown that

$$S_{vi} = \sum_g X_{vig}$$

is a sufficient statistic for ξ_{vi}, so that it readily follows that the conditional likelihood of the parameters $\boldsymbol{\delta} = (\delta_1, \ldots, \delta_k)'$ and $\boldsymbol{\psi} = (\psi_1, \ldots \psi_r)'$, for a response pattern $\mathbf{x}_{vi} = (x_{vi1}, \ldots, x_{vig}, \ldots, x_{vir})'$ is given by

$$L_c(\boldsymbol{\delta}, \boldsymbol{\psi}; \mathbf{x}_{vi} \mid S_{vi} = s) = \frac{\prod_g \rho_{ig}^{x_{vig}}}{\gamma_s(\boldsymbol{\rho}_i)} , \tag{5.15}$$

where

$$\rho_{ig} = \exp(\psi_g + \delta_i) , \tag{5.16}$$

$\boldsymbol{\rho}_i = (\rho_{i1}, \ldots, \rho_{ig}, \ldots, \rho_{ir})'$, and $\gamma_s(\cdot)$ denotes the basic symmetric function of order s (Fischer, 1974, p. 226).

Although (5.15) looks complicated, it is in fact nothing more than the conditional likelihood function in a LLTM applied in an incomplete test design. The data can be arranged in a matrix with nk rows and rk columns; see Table 5.1 for an example with $k = 2$.

Each row corresponds to a student–essay combination, which can also be called a virtual person. The main point is that the latent variable associated

TABLE 5.1. Incomplete design corresponding to model (5.14).

	(1,1)	(1,2)	$\cdots$	(1,r)	(2,1)	(2,2)	$\cdots$	(2,r)
(1,1)	X_{111}	X_{112}	$\cdots$	X_{11r}	×	×	$\cdots$	×
$\vdots$	$\vdots$	$\vdots$		$\vdots$	$\vdots$	$\vdots$		$\vdots$
(v,1)	X_{v11}	X_{v12}	$\cdots$	X_{v1r}	×	×	$\cdots$	×
$\vdots$	$\vdots$	$\vdots$		$\vdots$	$\vdots$	$\vdots$		$\vdots$
(n,1)	X_{n11}	X_{n12}	$\cdots$	X_{n1r}	×	×	$\cdots$	×
(1,2)	×	×	$\cdots$	×	X_{121}	X_{122}	$\cdots$	X_{12r}
$\vdots$	$\vdots$	$\vdots$		$\vdots$	$\vdots$	$\vdots$		$\vdots$
(v,2)	×	×	$\cdots$	×	X_{v21}	X_{v22}	$\cdots$	X_{v2r}
$\vdots$	$\vdots$	$\vdots$		$\vdots$	$\vdots$	$\vdots$		$\vdots$
(n,2)	×	×	$\cdots$	×	X_{n21}	X_{n22}	$\cdots$	X_{n2r}

Note: × missing data.

with a row is ξ_{vi}, which is assumed to be constant for all the entries in that row. A column of the matrix corresponds to a rater–essay combination and is called a virtual item. Virtual persons are denoted by the pair (v, i) and virtual items by the pair (i, g). Considering ξ_{vi} in (5.14) as the ability of virtual person (v, i), the model is formally equivalent to a LLTM, and Table 5.1 shows that the data are collected in an incomplete design. The only problem that remains is the fact that the design is not linked, meaning that for virtual person (v, i) we have only observations for the virtual items (g, i), $g = 1, \ldots, r$, but none for the virtual items (j, g), $j \neq i$. But this is not prohibitive for using CML-estimation, because of the linear structure of the item parameters. A detailed discussion on the existence and uniqueness of CML-estimates can be found in Fischer (1995a).

It is important to notice that with the CML-approach consistent estimates of the δ- and ψ-parameters are obtained without making assumptions about the distribution of the latent variables or their relation with exogenous variables such as gender. This means that the model assumptions are restricted to the measurement component, and goodness-of-fit tests can be developed that have power against specific violations of the model assumptions. For a general treatment of goodness-of-fit tests when using CML-estimation, see Glas and Verhelst (1995a, 1995b), and Verhelst and Glas (1995); for tests of the LLRA, see Fischer (1995b). Here we give only one example.

Suppose DIF is present for item I_i with respect to the variable gender such that this item is easier for girls than for boys. The null hypothesis of no DIF can be tested by the following procedure. Estimation of the parameters under the null hypothesis is carried out using CML. The value of the conditional loglikelihood function at the CML-estimates is denoted ℓ_0. To estimate under the alternative hypothesis, item I_i is considered as a different item when administered to girls than to boys, having sequence number i when administered to a girl and $k+1$, say, when administered to a boy. Let the maximum of the conditional loglikelihood function in this case be denoted by ℓ_1. The likelihood ratio (LR) statistic

$$\text{LR} = 2(\ell_1 - \ell_0)$$

is asymptotically chi-squared distributed with one degree of freedom given that the null hypothesis is true.

In standard applications of multilevel models, it is far more difficult to devise a test of comparable power against the same alternative hypothesis. A LR test may be built by modeling an interaction between item I_i and the dummy gender indicator, but if this interaction is modeled in the presence of a main effect of gender, it can be expected that the DIF for item I_i will be reflected partly in the main effect: if item I_i is easier for the girls, this will be translated into a main gender effect in favor of the girls, and as a consequence the effect on the interaction term will diminish. Or stated more generally: in the multilevel approach, the structural and the measurement components of the model are strongly intertwined (see the alternation of products and integrals in Equation 5.5), which renders specific model tests much more difficult than in an approach where the measurement component is well separated from the structural component.

It seems good practice therefore to test the measurement component thoroughly, without making assumptions about the structural component before estimating parameters, and making inferences that assume the validity of both components.

5 The Consequences of Ignoring Dependencies

The model given by (5.2), or equivalently by (5.12), takes the dependencies between ratings based on the same student response into account. The CML-estimation procedure, however, is not easily implemented in a computer program. And to our knowledge there is no software to handle the estimation of model (5.12) properly. Adams and Wilson (1996), for example, proposed analyzing data from multiple raters with a simple LLTM, which is defined by dropping the interaction random variable ζ_{vi} from (5.12), and assuming independence between all ratings given to a single student. Of course, this independence is a conditional independence given θ. In this section we try to get an impression of the seriousness of the errors caused by this simplification.

To avoid complicated algebra, due to the nonlinearity of the regression function in IRT models, the derivations in this section are based on a model with a linear regression function that mimics model (5.12) as closely as possible.

Assume a "true" model where the equation for the observations can be written as

$$Y_{vig} = \theta_v + \zeta_{vi} + \delta_i + \psi_g + \varepsilon_{vig} \ , \tag{5.17}$$

where

$$\zeta_{vi} \sim \mathcal{N}(0, \tau^2) \ , \tag{5.18}$$

$$\varepsilon_{vig} \sim \mathcal{N}(0, \sigma^2) \ , \tag{5.19}$$

and where all ζ- and ε-variables are mutually independent. Note that model (5.17) is equivalent to model (5.10).

It may be useful to make a couple of observations at this point. First, the assumptions (5.18) and (5.19) are stricter than necessary. The only crucial assumption in the derivations to follow is that the variances of the ζ- and ε-variables are finite, and of course not to dispense with their mutual independence. The normal distribution is assumed only for didactic reasons here: least squares and ML estimation procedures coincide in the case of normal distributions. So, in what follows, no needless considerations have to be given to the method of estimation. Second, model (5.17) will be easily recognized as a special case of the general model used in two-facet designs in the framework of generalizability theory (e.g., Veldhuijzen, Goldebeld, & Sanders, 1993, pp. 73–79), where the item–rater, the student–rater and the triple interactions vanish. It should be noticed that the present model precisely captures the interaction term ζ_{vi} frequently found to be the most important one in generalizability studies with multiple items and multiple raters (Sanders, 1992, p. 94), giving evidence that the model studied in this chapter is a realistic one.

The model that ignores the dependency is given by

$$Y_{vig} = \theta_v + \psi_g + \delta_i + \varepsilon^*_{vig} \ , \tag{5.20}$$

where $\varepsilon^*_{vig} = \varepsilon_{vig} + \zeta_{vi}$. Let σ^{*2} denote $\text{Var}(\varepsilon^*_{vig})$.

To make both models (5.17) and (5.20) identifiable we impose the normalization (5.13).

First we study some estimators and sampling variances using the true model. Denoting sample averages by replacing subscripts with dots, and using least squares estimates, it is readily found that

$$\hat{\psi}_g = Y_{..g} - Y_{...} \tag{5.21}$$

is an unbiased estimator of ψ_g. It is also the minimum variance estimator. To find its variance, the right-hand side of (5.21) is rewritten as a sum of independent model quantities, giving

$$\hat{\psi}_g = \psi_g + \frac{r-1}{nkr} \sum_v \sum_i \varepsilon_{vig} - \frac{1}{nkr} \sum_v \sum_i \sum_{h \neq g} \varepsilon_{vih} ,$$

whence it follows that

$$\operatorname{Var}(\hat{\psi}_g) = \frac{r-1}{nkr} \sigma^2 . \tag{5.22}$$

The least squares estimator for δ_i is given by

$$\hat{\delta}_i = Y_{.i.} - Y_{...} , \tag{5.23}$$

and by a similar technique as in the preceding case, we find that

$$\operatorname{Var}(\hat{\delta}_i) = \frac{k-1}{nk} \tau^2 + \frac{k-1}{nkr} \sigma^2 . \tag{5.24}$$

Notice that the normalization imposed on the parameters is also valid for the estimators. From (5.21) and (5.23) it follows that

$$\sum_g \hat{\psi}_g = \sum_i \hat{\delta}_i = 0 .$$

If we mimic a procedure applied in many IRT applications, which consists of replacing the true values of the item and rater parameters by their estimates when estimating the person parameter θ_v, we find the least squares estimator

$$\begin{aligned} \hat{\theta}_v &= Y_{v..} - \frac{1}{r} \sum_g \hat{\psi}_g - \frac{1}{k} \sum_i \hat{\delta}_i = Y_{v..} \\ &= \theta_v + \frac{1}{k} \sum_i \zeta_{vi} + \frac{1}{kr} \sum_i \sum_g \varepsilon_{vig} , \end{aligned} \tag{5.25}$$

whence we find that

$$\operatorname{Var}(\hat{\theta}_v) = \frac{\tau^2}{k} + \frac{\sigma^2}{kr} . \tag{5.26}$$

As an aside, we see from (5.25) that the point estimator of θ_v is unaffected by the fact that estimates of the δ- and ψ-parameters are used, instead of the true parameters. This, however, is a consequence of the chosen normalization. This can be seen clearly from the second equality sign in (5.25): the estimation errors in the ψ- and δ-estimates do not influence the final result, because the sum of these estimates is a constant with zero standard error. However, if we base the θ-estimate on $(k-1)$ items, because a particular person did not answer one item, for example, the sum of the $(k-1)$ δ-estimates is a random variable with nonzero standard error, and this will influence the standard error of the θ-estimate. More generally, the

TABLE 5.2. Variance components.

Source	*df*	$E(MS)$
Student–Item (θ, δ)	$(n-1)(k-1)$	$\sigma^2 + r\tau^2$
Student–Rater (θ, ψ)	$(n-1)(r-1)$	σ^2
Item–Rater (δ, ψ)	$(k-1)(r-1)$	σ^2
Residual	$(n-1)(k-1)(r-1)$	σ^2
Total	$nkr - n - r - k + 2$	$\sigma^2 + A\tau^2$

standard error of the estimator depends on the chosen normalization. See Verhelst (1993) for more details.

Of course, when an analysis is carried out on real data, we do not find exactly the right-hand sides of (5.22), (5.24), and (5.26) reported as values of the error variances, because σ^2 and τ^2 are not known, but have to be estimated from the data. If we use the standard methods of variance component analysis, we know, however, that the computed mean squares are unbiased estimators of linear combinations of the variance components. Hence it follows that

$$E(\hat{\tau}^2) = \tau^2 \quad \text{and} \quad E(\hat{\sigma}^2) = \sigma^2 .$$

Therefore we can write

$$E\left[\widehat{\mathrm{Var}}(\hat{\psi}_g)\right] = \frac{r-1}{nkr}\sigma^2 , \tag{5.27}$$

$$E\left[\widehat{\mathrm{Var}}(\hat{\delta}_i)\right] = \frac{k-1}{nk}\tau^2 + \frac{k-1}{nkr}\sigma^2 , \tag{5.28}$$

$$E\left[\widehat{\mathrm{Var}}(\hat{\theta}_v)\right] = \frac{\tau^2}{k} + \frac{\sigma^2}{kr} . \tag{5.29}$$

Next we study what happens if the data are analyzed using an incorrect model, in casu, a model with only main effects and a residual. This means that (5.17) is the true model, but (5.20) is used in the analysis. In Table 5.2 the degrees of freedom and the expected value of the mean squares are given for all sources of variance except for the main effects. The expected values are given using the true model.

The quantity A in the bottom right cell of the table equals

$$A = \frac{(n-1)(k-1)r}{nkr - n - r - k + 2} .$$

In all nontrivial cases, that is, all cases where n, r, $k > 1$, it holds that $0 < A < 1$.

A correct procedure in variance component analysis requires that all sources of variation which are believed to have no systematic variance component are added together in the sum of squares of the residual. If the model with no interactions is used, the sum of squares corresponding to the first four rows in Table 5.2 must be added to give an efficient estimator of the residual variance. The associated degrees of freedom are given by the sum of the degrees of freedom associated with the added rows and displayed in the row labeled Total in the table. The expected value of the sum of squares for each row is just the product of $E(MS)$ and the associated degrees of freedom. Adding the expected sum of squares and dividing by the total number of degrees of freedom gives the bottom right cell of the table. This is precisely what will happen if data, generated by a model with student–item interaction, are analyzed under the assumption of no interactions. So we find that

$$E(\hat{\sigma}^{*2}) = \sigma^2 + A\tau^2 \; . \tag{5.30}$$

Of course, using the wrong model, the same estimators for the parameters ψ, δ, and θ will be used as in the correct model, but the reported error variances will not contain estimates of τ^2, because it is considered to be zero. Therefore we will find

$$\begin{aligned} \widehat{\mathrm{Var}}_M(\hat{\psi}_g) &= \frac{r-1}{nkr}\,\hat{\sigma}^{*2} \, , \\ \widehat{\mathrm{Var}}_M(\hat{\delta}_i) &= \frac{k-1}{nkr}\,\hat{\sigma}^{*2} \, , \\ \widehat{\mathrm{Var}}_M(\hat{\theta}_v) &= \frac{1}{kr}\,\hat{\sigma}^{*2} \, , \end{aligned}$$

where the subscript M has been added to the variance estimator to denote a misspecification of the model. Using (5.30), we find that

$$E\big[\widehat{\mathrm{Var}}_M(\hat{\psi}_g)\big] = \frac{r-1}{nkr}\,\sigma^2 + \frac{A(r-1)}{nkr}\,\tau^2 \, , \tag{5.31}$$

$$E\big[\widehat{\mathrm{Var}}_M(\hat{\delta}_i)\big] = \frac{k-1}{nkr}\,\sigma^2 + \frac{A(k-1)}{nkr}\,\tau^2 \, , \tag{5.32}$$

and

$$E\big[\widehat{\mathrm{Var}}_M(\hat{\theta}_v)\big] = \frac{1}{kr}\,\sigma^2 + \frac{A}{kr}\,\tau^2 \; . \tag{5.33}$$

From the preceding results, the bias can be derived. Defining

$$\zeta^2 = \frac{\tau^2}{\sigma^2} \, ,$$

we find, dividing (5.31) by (5.27), that

$$\frac{E\big[\widehat{\mathrm{Var}}_M(\hat{\psi}_g)\big]}{E\big[\widehat{\mathrm{Var}}(\hat{\psi}_g)\big]} = 1 + A\zeta^2 \, , \tag{5.34}$$

meaning that under the specification error the standard error of ψ_g is biased upwards. For the estimators $\hat{\delta}_i$ and $\hat{\theta}_v$ we find, using (5.28), (5.29), (5.32), and (5.33), that

$$\frac{E\left[\widehat{\mathrm{Var}}_M(\hat{\delta}_i)\right]}{E\left[\widehat{\mathrm{Var}}(\hat{\delta}_i)\right]} = \frac{E\left[\widehat{\mathrm{Var}}_M(\hat{\theta}_v)\right]}{E\left[\widehat{\mathrm{Var}}(\hat{\theta}_v)\right]} = \frac{1 + A\zeta^2}{1 + r\zeta^2}, \tag{5.35}$$

which is less than one if $\tau^2 > 0$, meaning that the standard errors are biased downwards.

In summary, if model (5.12) is replaced by a linear model of the same complexity, and the analysis is done both using the correct model and a model where the student–item interaction is dropped, we find the following.

1. The point estimators of ψ_g, δ_i, and θ_v are exactly the same in both models.
2. The reported standard error of ψ_g, using the incorrect model, is biased upwards.
3. The reported standard errors of δ_i and θ_v, using the incorrect model, are biased downwards with the same factor.
4. The size of the standard errors depends on the chosen normalization.

6 Conclusion

The work reported in this chapter is an attempt to find a satisfactory answer to a long-standing practical problem: is there an easy way to handle multifacet measurement designs in an IRT framework, which yields results comparable to those of generalizability theory? The solution to this problem has become more urgent than a decade ago given the growing interest in performance assessment (see, e.g., Messick, 1994, or Linn, Baker, & Dunbar, 1991) where the use of judges is indispensable. Many discussions within the Dutch psychometric community dealing with the problem of how to model multiple ratings in an IRT approach always focused on the question of local independence, using arguments very similar to the ones used in the introductory section. It is our feeling that the model presented in Section 2 is an original and elegant solution to the problem, nicely fitting into the large family of unidimensional IRT models. The only part of the work left to be done is the development of a computer program to estimate parameters in the model. When we realized, however, the similarities between the model of Section 2 and other approaches, it was decided to explore these first.

The surprising finding that the IRT model is just a special case of a nonlinear multilevel model with three levels opens new perspectives, namely, the possibility of incorporating explanatory variables into the model (such as gender), and making full use of the results of multilevel theory. It is, for example, not difficult to introduce one or more higher levels to model correctly the effects of a hierarchical sampling design. Yet, the differences in perspective between the multilevel tradition and the IRT tradition need to be stressed. These differences can (roughly) be characterized by the relative importance attached to the structural and measurement components of the model. Whereas the main focus in the multilevel approach is modeling the regression model and the variance components, the approach in IRT is to build a valid relation between observed and latent variables. The use of conditional maximum likelihood estimation methods allows one to test the validity of the measurement component of the model without making assumptions about the structural component.

A second finding (which was also surprising) was that the model of Section 2 can be written as a LLRA (see Equation 5.14), allowing CML estimation and powerful tests of the measurement model. Interesting as this finding may be, it should be noted that the use of CML also entails a serious restriction: the rater effects have to be modeled as fixed effects. If these effects are modeled as random, that is, the individual rater effects are not the parameters of interest but only the first two moments, say, of their distribution, the whole construction exemplified by Table 5.1 becomes useless. In situations where large amounts of item responses have to be rated by several judges, as is the case of the Dutch state examinations for Dutch as a second language, quite a large number of raters is used in an incomplete design, making a model with fixed rater effects highly artificial and inefficient. The possibility of modeling the rater effects as crossed random effects in multilevel software certainly is a great advantage not readily available in the IRT tradition.

The third finding is the similarity between the model of Section 2 and a model that fits in generalizability theory. In their original standard work on generalizability theory, Cronbach, Gleser, Nanda, and Rajaratnam (1972, Chap. 2) point out that in a crossed design with two measurement facets, seven different variance components can be estimated: three associated with a main effect (persons, raters, and items), three with first-order interactions, and a residual (second-order interaction and measurement error confounded). A linearization of our model, resulting in model (5.17) is just a special case of the general model presented by Cronbach et al. (1972), where the person–rater and the item–rater interactions vanish. But the important thing to notice is that the person–item interaction term in the model follows directly from the fact that the ratings are based on the same performance. Cronbach et al. (1972, Chap. 6) analyzed test data collected on aphasia patients, and they reported a very substantial person–object variance component (their term "object" refers to specific tasks used in the aphasia research, but it is functionally equivalent with our use of the term item). Their comment is quite remarkable: "The person–object effect implies that some objects ... are much more troublesome for a given patient than their general difficulty and his general impairment would suggest. *To be sure that there is a genuine person–object effect, however, it would be necessary to present the same object more than once.* ... The residual represents discrepancies between scorer ratings of *the same performance.* ... Where the subject performs inconsistently, the scorers agree in their reports of that inconsistency; hence, the variation is assigned to the person–object interaction" (p. 169; italics added). The inconsistency referred to in the quotation should be understood as variation in the performance (quality of the essay, e.g.) given constant ability. A "genuine" person–item effect could result, for example, from unequal discrimination of the items or from a multidimensional ability. The distinction between these two kinds of interactions can only be made by using a different data collection design, where within-person replications of the items are used. How this can be accomplished in the area of performance testing, where tasks are usually rather time consuming, is a hard problem. The model presented in this chapter, and the links it has with other theoretical approaches is to be considered as a modest contribution to the solution of this problem.

Acknowledgments: The authors wish to thank the three anonymous reviewers and the editors for their helpful comments.

References

Adams, R.J., & Wilson, M.R. (1996). Formulating the Rasch model as a mixed coefficients multinomial logit: A generalized approach to fitting Rasch models. In G. Engelhard & M.R. Wilson (Eds.), *Objective measurement III: Theory into practice* (pp. 143–166). Norwood, NJ: Ablex.

Bryk, A.S., Raudenbush, S.W., & Congdon, R.T. (1996). HLM. Hierarchical linear and nonlinear modeling with the HLM/2L and HLM/3L programs [Computer software]. Chicago: Scientific Software.

Cronbach, L.J., Gleser, G.C., Nanda, H., & Rajaratnam, N. (1972). *The dependability of behavioral measurements.* New York: Wiley

Feldt, L.S., & Brennan, R.L. (1989). Reliability. In R.L. Linn (Ed.), *Educational measurement* (3rd ed., pp. 105–146). Washington, DC: American Council on Education.

Fischer G.H. (1974). *Einführung in die Theorie psychologischer Tests.* Bern: Huber.

Fischer G.H. (1995a). The linear logistic test model. In G.H. Fischer & I.W. Molenaar (Eds.), *Rasch models: Foundations, recent developments, and applications* (pp. 131–155). New York: Springer-Verlag.

Fischer G.H. (1995b). Linear logistic models for change. In G.H. Fischer & I.W. Molenaar (Eds.), *Rasch models: Foundations, recent developments, and applications* (pp. 157–180). New York: Springer-Verlag.

Glas, C.A.W., & Verhelst, N.D. (1995a). Testing the Rasch model. In G.H. Fischer & I.W. Molenaar (Eds.), *Rasch models: Foundations, recent developments, and applications* (pp. 69–95). New York: Springer-Verlag.

Glas, C.A.W., & Verhelst, N.D. (1995b). Tests of fit for polytomous Rasch models. In G.H. Fischer & I.W. Molenaar (Eds.), *Rasch models: Foundations, recent developments, and applications* (pp. 326–352). New York: Springer-Verlag.

Goldstein, H. (1995). *Multilevel statistical models* (2nd ed.). London: Arnold.

Goldstein, H., Rasbash, J., Plewis, I., Draper, D., Browne, W., Yang, M., Woodhouse, G., & Healy, M. (1998). A user's guide to MLwiN [Software manual]. London: Multilevel Models Project, Institute of Education, University of London.

Hedeker, D., & Gibbons, R.D. (1996). MIXOR: A computer program for mixed-effects ordinal regression analysis. *Computer Methods and Programs in Biomedicine, 49,* 229–252.

Holland P.W., & Wainer, H. (Eds.). (1993). *Differential item functioning.* Hillsdale, NJ: Erlbaum.

Linn, R.L., Baker, E.L., & Dunbar, S.B. (1991). Complex, performance-based assessment: Expectations and validation criteria. *Educational Researcher, 20 (8),* 15–21.

McCullagh, P., & Nelder, J.A. (1989). *Generalized linear models* (2nd ed.). London: Chapman and Hall.

Messick, S. (1994). The interplay of evidence and consequences in the validation of performance assessments. *Educational Researcher, 23 (2),* 13–23.

Rodríguez, G., & Goldman, N. (1995). An assessment of estimation procedures for multilevel models with binary responses. *Journal of the Royal Statistical Society, Series A, 158,* 73–89.

Sanders, P.F. (1992). *The optimization of decision studies in generalizability theory.* Unpublished doctoral dissertation, University of Amsterdam.

Snijders, T.A.B., & Bosker, R.J. (1999). *Multilevel analysis: An introduction to basic and advanced multilevel modeling.* London: Sage.

Van den Wollenberg, A.L. (1982). Two new test statistics for the Rasch model. *Psychometrika, 47,* 123–140.

Veldhuijzen, N.H., Goldebeld P., & Sanders, P.F. (1993). Klassieke testtheorie en generaliseerbaarheidstheorie [Classical test theory and generalizability theory]. In T.J.H.M. Eggen & P.F. Sanders (Eds.), *Psychometrie in de praktijk* (pp. 33–82). Arnhem: CITO.

Verhelst, N.D. (1993). *On the standard errors of parameter estimates in the Rasch model* (Measurement and Research Department Reports, 93-1). Arnhem: CITO.

Verhelst, N.D., & Glas, C.A.W. (1995). The one parameter logistic model. In G.H. Fischer & I.W. Molenaar (Eds.), *Rasch models: Foundations, recent developments, and applications* (pp. 215–237). New York: Springer-Verlag.

6

Conditional Independence and Differential Item Functioning in the Two-Parameter Logistic Model

Herbert Hoijtink[1]

ABSTRACT The two-parameter logistic model is among other things characterized by conditional independence and the absence of differential item functioning. Two fit statistics are proposed that can be used to investigate conditional independence and differential item functioning globally and at the item level. The statistics are evaluated using their posterior predictive distribution.

1 Introduction

The basic equation of the two-parameter logistic model (2PLM; see, e.g., Mislevy, 1986) is given by

$$P(X_{vi} = x_{vi} \mid \theta_v) = \frac{\exp[\alpha_i(\theta_v - \beta_i)x_{vi}]}{1 + \exp[\alpha_i(\theta_v - \beta_i)]}, \tag{6.1}$$

where X_{vi} is a random variable denoting the response to item I_i. The realizations x_{vi} are dichotomous, where 0 and 1 denote the negative and positive response, respectively. Responses are given by subjects $S_v, v = 1, \ldots, n$, to items $I_i, i = 1, \ldots, k$; θ_v denotes the person location (often called ability in an educational context), and β_i and α_i denote the location (often called difficulty in an educational context) and slope of item I_i, respectively. The assumption of conditional independence (i.e., given θ_v, the item responses of a person are independent) is used to compute the probability of the response vector $\mathbf{X}_v$ of subject S_v,

$$P(\mathbf{x}_v \mid \theta_v) = P(\mathbf{X}_v = \mathbf{x}_v \mid \theta_v) = \prod_{i=1}^{k} P(X_{vi} = x_{vi} \mid \theta_v). \tag{6.2}$$

[1]Department of Methodology and Statistics, University of Utrecht, P.O. Box 80140, 3508 TC Utrecht, The Netherlands; e-mail: *h.hoijtink@fss.uu.nl*

Note that conditional independence is an important assumption. It is the foundation of one of the basic equations of the two-parameter logistic model. There are several situations that will lead to a violation of conditional independence in the two-parameter logistic model. A few examples are given now.

1. The item responses should be explained using a multidimensional instead of a unidimensional item response model. In the context of nonparametric item response models, Stout (1987), Stout, Habing, Douglas, Kim, Roussos, and Zhang (1996), Zhang and Stout (1999a, 1999b), and Stout, Goodwin Froelich, and Gao (2001) use conditional independence as the point of departure to develop an approach that can be used to determine whether a multi- or unidimensional model is more appropriate.

2. The item responses should be explained using a unidimensional unfolding model instead of the 2PLM. In practice it may not always be clear which model should be used; see Post, Van Duijn, and Van Baarsen (2001), and Klinkenberg (2001).

3. The response to an item depends on the responses to the preceding items; see Verhelst and Glas (1993) for a Rasch model in which these dependencies are incorporated.

As far as is known to the author, there is not yet available a goodness-of-fit test for the 2PLM that has power against all the alternatives listed above. In this chapter, such a test is proposed.

Estimation of the parameters of the 2PLM is usually based on a posterior distribution that is proportional to the product of the marginal likelihood function and an informative prior distribution for the item slopes. Mislevy (1986) uses the EM-algorithm (Dempster, Laird, & Rubin, 1977) to obtain maximum a posteriori estimates. In the Appendix of this chapter it is briefly explained how the Gibbs sampler as presented by Zeger and Karim (1991) can be used to obtain the expected a posteriori estimates.

In Section 3 a statistic sensitive to violations of differential item functioning is proposed. This statistic can be used to determine whether the item characteristic curve, (6.1) as a function of θ, differs between two subsamples (e.g., men and women). It may happen that an item is more difficult for women than for men, even if both subsamples have the same ability. This will lead to an underestimation of the ability of the women (the item is biased against women). Removal of such an item from the item set would solve the problem. Since two subsamples have to be distinguished, the marginal likelihood function of the 2PLM used in this chapter contains the distribution of θ for both subsamples. Group membership is indicated by the random variable G, which takes the value 1 if subject S_v belongs to group 1, and the value 2 if S_v is a member of group 2. Thus

$$L(\boldsymbol{\beta}, \boldsymbol{\alpha}, \mu_1, \mu_2, \sigma_1^2, \sigma_2^2 \mid \mathbf{x}_1, \ldots, \mathbf{x}_n, \mathbf{g}) = L(\boldsymbol{\omega} \mid \mathbf{x}_1, \ldots, \mathbf{x}_n, \mathbf{g}) =$$

$$\prod_{v=1}^{n} \int P(\mathbf{x}_v \mid \theta_v)\, h(\theta_v \mid \mu_{g_v}, \sigma_{g_v}^2)\, d\theta \;, \tag{6.3}$$

where $h(\theta \mid \mu_{g_v}, \sigma_{g_v}^2)$ denotes the density of a normal distribution with expectation μ_{g_v} and variance $\sigma_{g_v}^2$, and $\mathbf{g}$ is the $(n \times 1)$ vector indicating group membership of n subjects.

The parameters of the 2PLM are not identified. Addition of the same constant to μ_1, μ_2 and each item location does not change the likelihood. Furthermore, if each person and item location is multiplied by the same constant, and each item slope is divided by that constant, the likelihood will not change. Two restrictions (see also Mislevy, 1986; Mislevy & Bock, 1990) are needed to make the model identifiable; usually, $\mu_1 = 0$ and $\sigma_1^2 = 1$. Even with these restrictions, straightforward estimation of the parameters of the 2PLM is not always possible. It is rather common that the locations and slopes of items located at the extremes of the latent dimension become unrealistically large and small, respectively. Mislevy (1986) showed that these divergence problems can be avoided using a normal prior for the logarithm of the item slope parameters. Multiplication of likelihood and prior renders a function that is proportional to the posterior distribution of the parameters of the 2PLM:

$$\text{Post}(\boldsymbol{\omega} \mid \mathbf{x}_1, \ldots, \mathbf{x}_n, \mathbf{g}) \propto L(\boldsymbol{\omega} \mid \mathbf{x}_1, \ldots, \mathbf{x}_n, \mathbf{g}) \prod_{i=1}^{k} h(\ln \alpha_i) \;, \tag{6.4}$$

where $h(\ln \alpha_i)$ denotes a normal density with an expectation of 0 and a variance of 0.25, evaluated at $\ln \alpha_i$. The chosen values for the expectation and variance (used for all analyses presented in this chapter) imply that values of the item slopes in the range [0.37, 2.72] are a priori considered realistic.

In this chapter global and item level fit statistics with which conditional independence and differential item functioning can be investigated are proposed. These statistics are compared to other statistics that could be used to investigate conditional independence and differential item functioning for the 2PLM. The statistics are evaluated using their posterior predictive distribution (Rubin, 1984; Meng, 1994; Gelman, Meng, & Stern, 1996). A small simulation study and an example are used to illustrate the viability of the proposed methods.

2 A Statistic for Violations of Conditional Independence

As far as the author knows, there is no previous research regarding the detection of violations of conditional independence in the 2PLM. Some authors have investigated conditional independence in the context of the Rasch model (Van den Wollenberg, 1982; Glas & Verhelst, 1995), others in the context of nonparametric item response models (Rosenbaum, 1984; Junker, 1993; Stout, 1987; Hoijtink & Molenaar, 1997; Zhang & Stout, 1999a). Since the 2PLM is more restrictive than a nonparametric item response model, the procedures used to investigate conditional independence for nonparametric item response models could be used in the context of the 2PLM. The approach proposed in this chapter is compared to the procedures developed for nonparametric item response models.

In the case of violations of conditional independence, two courses of action are open to the researcher: remove misfitting items from the item set (which may be appropriate if a few of the items measure a different latent trait, or when there are logical dependencies between a pair of items); or use a model that is better suited to deal with dependence. See, for example, Verhelst and Glas (1993) who present a model where the response to an item depends on the responses given to the preceding test items, or Bock, Gibbons, and Muraki (1988) who discuss a multidimensional extension of the 2PLM.

The implication of (6.2) is that (omitting subscript v) the item responses X_i and X_j are independent conditional on θ. For dichotomously scored items this implies that in the population the conditional covariance $\text{Cov}(X_i, X_j \mid \theta) = 0$. Since θ is unobservable, the assumption of conditional independence cannot be tested using a statistic based on the $\text{Cov}(X_i, X_j \mid \theta)$ observed in the sample.

Let R^{ij} denote a random variable defined as $R^{ij} = \sum_{l \neq i,j} X_l$, that is, a person's restscore. Note that $0 \leq R^{ij} \leq k - 2$. Several authors (Junker, 1993; Stout, Habing, Douglas, Kim, Roussos, & Zhang, 1996; Zhang & Stout, 1999a) have noted that R^{ij} is a useful proxy for θ. If R^{ij} could be transformed to the scale of θ, it would be a consistent but inefficient estimator of θ. The latter implies that $\text{Cov}(X_i, X_j \mid R^{ij}) \approx \text{Cov}(X_i, X_j \mid \theta)$ and thus $\text{Cov}(X_i, X_j \mid R^{ij}) \approx 0$ if conditional independence holds. Several authors (Rosenbaum, 1984; Stout, 1987; Zhang & Stout, 1999a) have noted that the expected value of $\text{Cov}(X_i, X_j \mid R^{ij})$ is larger than or equal to zero if conditional independence holds, and thus, that $\text{Cov}(X_i, X_j \mid R^{ij})$ is a biased estimator of $\text{Cov}(X_i, X_j \mid \theta)$. This phenomenon has to be accounted for whenever tests of conditional independence based on $\text{Cov}(X_i, X_j \mid R^{ij})$ are executed.

Rosenbaum (1984) describes (tests based on) a family of conditional covariances of which $\text{Cov}(X_i, X_j \mid R^{ij})$ is one example. Rosenbaum notes that usually many tests are executed when conditional independence is investigated (e.g., a test of conditional independence for each pair of items). He suggests avoiding capitalization on chance via the use of procedures developed to deal with multiple hypotheses. It is shown that in the context of the 2PLM the conditional covariances can be combined into one statistic. In Section 4 it is shown that a (bias-corrected) posterior predictive p-value can be computed for this combination. The latter is an alternative for Rosenbaum's multiple hypotheses procedure that makes explicit use of the fact that the 2PLM (and not a nonparametric item response model) is being investigated.

In a series of papers (Stout, 1987; Stout et al., 1996, 2001; Zhang & Stout, 1999a, 1999b) Stout and his colleagues develop and investigate procedures that can be used to detect multidimensionality. Stout's DIMTEST procedure is based on a statistic that, like the statistic proposed here, consists of a sum of conditional covariances:

$$\text{DIMTEST} = \sum_{p=0}^{P} \sum_{1 \le i < j \le A} \text{Cov}(X_i^{AT}, X_j^{AT} \mid Z_{PT} = p) , \tag{6.5}$$

where Cov denotes a covariance computed in the sample, P is the number of items in the partitioning test (PT), A is the number of items in the assessment test (AT), and Z_{PT} the sum of the responses obtained for the items in the partitioning test. According to Stout et al. (1996), the assessment test is "the set of items whose dimensionality is compared to the remaining items" and the partitioning test is "the set of items on which the examinees are segregated based on their score" (p. 340). In the papers cited above several methods are described to assign items to the assessment test and the partitioning test such that "the assessment test is as unidimensional as possible and dimensionally distinct from the partitioning test" (Zhang & Stout, 1999a, p. 145). If assessment and partitioning tests are chosen appropriately, (6.5) should have power against multidimensionality. However, if both tests are not chosen appropriately, negative and positive covariances may cancel each other out (note that given the hypothesis of unidimensionality all covariances are approximately zero); that is, the power will decrease.

The item level statistic for conditional independence proposed here is

$$\text{CI}_i = \sum_{j \neq i} \sum_{R^{ij}=0}^{k-2} \sqrt{n_{R^{ij}}}\, \text{Cov}^2(X_i, X_j \mid R^{ij}) , \tag{6.6}$$

where $n_{R^{ij}}$ denotes the number of persons with restscore R^{ij}. The global statistic for conditional independence is

$$\text{CI} = \sum_i \text{CI}_i . \tag{6.7}$$

This statistic does not require the items to be assigned to an assessment and a partitioning test. It uses squared covariances such that positive and negative deviations from (approximately) zero will not cancel each other out. It weights each covariance with the number of persons in the restscore group for which the covariance is computed to ensure that larger groups have more influence on the outcome. It does not only have power against multidimensionality (although probably less than (6.5) if the assessment and partitioning test are chosen appropriately), but also against other violations of conditional independence.

In practice, CI_i and CI are not necessarily evaluated using a sum over all restscore groups. To reduce the influence of the covariances computed in small restscore groups on CI, even beyond the reduction implied by the use of $\sqrt{n_{R^{ij}}}$, adjacent restscore groups will be collapsed if the number of persons in one of them is smaller than 10. Subsequently, the covariances will be computed for the collapsed restscore groups.

If conditional independence holds, CI_i and CI will be positive because the expected value of $\text{Cov}(X_i, X_j \mid R^{ij})$ will be larger than or equal to zero. Stated otherwise, in themselves both statistics do not separate violations of conditional independence from the bias resulting from the fact that R^{ij} is used as a proxy of θ. However, as explained in Section 4, CI_i and CI are evaluated using their posterior predictive distribution. This implies that violations of conditional independence will only be detected if their size is larger than the size that is expected if conditional independence holds.

3 A Statistic for Differential Item Functioning

The technical knowledge with respect to the detection of differential item functioning in the 2PLM is more or less summarized in Muraki, Mislevy, and Bock (1987), Thissen, Steinberg, and Wainer (1988, 1993), Glas (1998), and the references given in these papers. Holland and Wainer (1993) edited a book in which all aspects of differential item functioning are discussed. There appear to be roughly three approaches: Wald tests, likelihood ratio tests, and Lagrange multiplier tests. Each test can, in principle, be applied to a single item, a number of items, or the test as a whole. Furthermore, each test can, in principle, be applied to the detection of items that function differently in two or more samples. See also Glas (2001) who uses a two by two factorial design to find and explain differential item functioning by applying Lagrange multiplier tests.

Common to these approaches is the basic idea that an item does not function differently if its parameters (β and α) are the same in the various samples that are investigated. Although the latter is indeed a very strong indication of the absence of differential item functioning (and implies that these tests are excellent tools to use whenever investigating differential item functioning), there is not a one-to-one correspondence between this basic idea and the usual definition of absence of differential item functioning (omitting subscript v),

$$P(X_i = 1 \mid \theta, G = g) = P(X_i = 1 \mid \theta) , \qquad g = 1, 2 , \tag{6.8}$$

where the binary random variable G denotes subsample membership. It is important to distinguish between the probability of a positive response defined by the 2PLM (6.1) and the probabilities appearing in (6.8). The latter are true probabilities that are not restricted through the use of a model. Absence of differential item functioning implies that the probability of a positive response depends only on a person's location, and not on a person's subsample membership (the item characteristic curves are the same in each subsample of persons), and that the probabilities are equal in each group. In the context of nonparametric item response models Shealy and Stout (1993) use (6.8) as the point of departure for the investigation of differential item functioning. In the context of the Rasch model Glas and Verhelst (1995) also present a procedure based on (6.8). In this section it is shown that a similar approach can also be used in the context of the 2PLM.

Since θ is unobservable, differential item functioning cannot be investigated using a statistic based on the sample probabilities $P(X_i = 1|\theta, G = g)$ for $g = 1, 2$. Again a restscore is used as a proxy of θ. Let R^i denote a random variable defined as $R^i = \sum_{l \neq i} X_l$, that is, a subject's restscore with respect to item I_i. Note that $0 \leq R^i \leq k - 1$. Using this restscore, (6.8) changes to

$$P(X_i = 1 \mid R^i, G = g) \approx P(X_i = 1 \mid R^i) , \qquad g = 1, 2 . \tag{6.9}$$

Note that it is only reasonable to use the restscore as a proxy of θ if it is assumed that the probabilities in (6.8) increase as a function of θ, which implies that the restscore is strongly correlated with θ. In the context of unfolding models the probabilities are a single-peaked function of θ. In that case the restscore is usually uncorrelated with θ and would be a very unreliable proxy of θ.

It is important to realize that, in the population, a strict equality in (6.9) only holds if, in addition to (6.8), $h(\theta \mid \mu_1, \sigma_1^2) = h(\theta \mid \mu_2, \sigma_2^2)$ holds. This implies that differences among the probabilities in (6.9) do not necessarily reflect differential item functioning, but may also reflect differences among the distributions of θ in both subsamples; see Shealy and Stout (1993). Note that in the sample, the probabilities appearing in (6.9) can be directly computed from the $(n_1 + n_2) \times k$ matrix containing the item responses, where n_1 and n_2 denote the sizes of the first and second subsample, respectively, and $n_1 + n_2 = n$. The same holds for all other statistics discussed in this section.

From (6.9) it follows that, for $g = 1, 2$,

$$n(R^i, G = g)P(X_i = 1 \mid R^i, G = g) \approx n(R^i, G = g)P(X_i = 1 \mid R^i) \ , \quad (6.10)$$

where $n(R^i, G = g)$ denotes the observed number of persons in subsample g with restscore R^i with respect to item I_i. Since $P(X_i = 1 \mid R^i, G)$ is approximately equal to $P(X_i = 1 \mid R^i)$ if differential item functioning is absent (cf. Equation 6.9), the left-hand side of (6.10) may be considered as the observed number, and the right-hand side as the expected number of persons with restscore R^i in subsample g and a correct response to item I_i, so that (6.10) can be rewritten as

$$n(X_i = 1 \mid R^i, G = g) \approx m(X_i = 1 \mid R^i, G = g) \ , \qquad g = 1, 2 \ . \quad (6.11)$$

Analogously, the observed and expected number of persons with restscore R^i with respect to item I_i in subsample g with an incorrect response to item I_i, denoted $n(X_i = 0 \mid R^i, G = g)$ and $m(X_i = 0 \mid R^i, G = g)$, respectively, can be calculated. The statistic sensitive to differential item functioning compares the observed with the expected number of correct and incorrect responses as follows,

$$\text{DIF}_i = \sum_{R^i=0}^{k-1} \sum_{g=1}^{2} \sum_{a=0}^{1} \frac{[n(X_i = a \mid R^i, G = g) - m(X_i = a \mid R^i, G = g)]^2}{m(X_i = a \mid R^i, G = g)} \ . \quad (6.12)$$

The item level statistics can be combined into a global statistic that can be used to investigate if the item set as a whole exhibits differential item functioning:

$$\text{DIF} = \sum_i \text{DIF}_i \ . \quad (6.13)$$

Note that the test statistics (6.12) and (6.13) cannot distinguish differential item functioning from differences among the distributions of θ in both subsamples. However, as explained in the next section, these statistics are evaluated using their posterior predictive distribution. This distribution is computed conditional on the differences among $h(\theta \mid \mu_1, \sigma_1^2)$ and $h(\theta \mid \mu_2, \sigma_2^2)$. This implies that DIF and DIF_i will only detect differential item functioning if the differences among corresponding pairs of observed and expected numbers are larger than the differences that are expected to occur given the difference between $h(\theta \mid \mu_1, \sigma_1^2)$ and $h(\theta \mid \mu_2, \sigma_2^2)$. Also note that the same comments regarding the collapsing of restscore groups made at the end of Section 2, apply to the statistics sensitive to differential item functioning.

4 The Posterior Predictive Distribution of Fit Statistics

Let $S(\mathbf{X})$ be the generic notation for any of CI, CI_i, DIF, and DIF_i, where $\mathbf{X}$ denotes the $(n_1 + n_2) \times k$ matrix containing the item responses. Note that the number of persons belonging to both subsamples is the same in the observed data matrix and the replicated data matrices $\mathbf{X}^{rep}$ that are used below; that is, both numbers are ancillary statistics. Rubin (1984), Meng (1994), and Gelman, Meng, and Stern (1996) show that the observed value $S(\mathbf{X})$ of a fit statistic can be evaluated via reference to the posterior predictive distribution of the statistic. The resulting posterior predictive p-value is defined as

$$p = \mathrm{P}[S(\mathbf{X}^{rep}) \geq S(\mathbf{X}) \mid \mathbf{X}] \; . \tag{6.14}$$

This is the posterior probability that $S(\mathbf{X}^{rep})$, the statistic computed for a replication of the data matrix, is equal to or exceeds the value computed for the observed data matrix. This probability is computed over the posterior predictive distribution of the replicated data matrices

$$f(\mathbf{X}^{rep} \mid \mathbf{X}) = \int g(\mathbf{X}^{rep} \mid \boldsymbol{\omega}) \mathrm{Post}(\boldsymbol{\omega} \mid \mathbf{X}) \, d\boldsymbol{\omega} \; , \tag{6.15}$$

where $f(\cdot)$ and $g(\cdot)$ denote the distribution of the replicated data matrices given $\mathbf{X}$ and the parameter vector $\boldsymbol{\omega}$, respectively. Since (6.14) is analytically not tractable, it is computed using a three-step simulation procedure. First, an application of the Gibbs sampler as presented by Zeger and Karim (1991) is used to sample 2000 values of $\boldsymbol{\omega}$ from $\mathrm{Post}(\boldsymbol{\omega} \mid \mathbf{X})$. Some details of the algorithm can be found in the Appendix of this chapter. See Patz and Junker (1999) for a similar algorithm applied to a posterior based on the joint likelihood of the 2PLM. Second, a replicated data matrix is simulated given each sampled $\boldsymbol{\omega}$; that is, a data matrix is sampled from a population defined by $\boldsymbol{\omega}$ with subsample sizes equal to n_1 and n_2. In the third step $S(\cdot)$ is computed for each replicated data matrix and the observed data matrix. The posterior predictive p-value (6.14) is approximated using the proportion of times that $S(\mathbf{X}^{rep})$ is equal to or exceeds $S(\mathbf{X})$.

5 Performance of Fit Statistics: A Small Simulation Study

To obtain an idea about the performance of the fit statistics and the quality of the expected a posteriori (EAP) estimates and the posterior standard deviations (SD), the results of a small simulation study are presented. In Table 6.1 a population that behaves in accordance with the 2PLM (i.e., no

TABLE 6.1. Description of the population used for the simulation study, and the expected a posteriori estimates (EAP) and posterior standard deviations (SD) obtained for the first replication.

I_i	β_i	EAP(β_i)	SD(β_i)	α_i	EAP(α_i)	SD(α_i)
1	−2.5	−2.71	0.63	1.0	0.99	0.21
2	−2.0	−3.04	0.99	0.5	0.33	0.09
3	−1.5	−1.46	0.29	1.0	0.97	0.19
4	−1.0	−0.99	0.16	2.0	1.84	0.38
5	−0.5	−0.55	0.15	1.0	1.24	0.21
6	0.0	−0.15	0.18	0.5	0.72	0.14
7	0.5	0.49	0.14	1.0	1.04	0.17
8	1.0	1.07	0.15	2.0	2.11	0.45
9	1.5	1.92	0.36	1.0	0.82	0.16
10	2.0	1.97	0.42	0.5	0.68	0.14
11	2.5	2.75	0.53	1.0	0.83	0.18
μ_1	0.0	0.00†				
μ_2	0.5	0.40	0.13			
σ_1^2	1.0	1.00†				
σ_2^2	1.0	1.21	0.27			

Note: † fixed parameter.

conditional dependence and no differential item functioning) is described. The item difficulties β_i are in the range −2.5 to 2.5; the item slopes α_i have the values 0.5, 1, or 2. There are two subsamples, the variance of the person locations is 1.0 in each subsample, and the averages are 0 and 0.5, respectively. From this population two samples were drawn of size $n_1 = 200$ and $n_2 = 200$, respectively. Using the Gibbs sampler (see the Appendix of this chapter) for both replications 2000 parameter vectors $\boldsymbol{\omega}$ were sampled from (6.4). For the first replication, the average and standard deviation of the 2000 values sampled are displayed in Table 6.1. As can be seen, most of the population values are within one standard deviation of the corresponding estimates. The quality of the estimates appears to be satisfactory.

For the investigation of conditional independence and differential item functioning, some of the restscore groups were collapsed: 0 and 1, and 9 and 10, with respect to R^i; and 0 and 1, and 8 and 9, with respect to R^{ij}. As we can see in Table 6.2, the posterior predictive p-values for conditional independence are in the range .12 until .97 for both replications. It appears that conditional independence is not violated: the p-values for CI were .54 and .67, respectively. However, looking at the posterior predictive p-values for differential item functioning, it can be seen that in replication one these

TABLE 6.2. Posterior predictive p-values resulting from both replications of the simulation study ($n_1 = 200$, $n_2 = 200$).

I_i	Repl. 1 $p(\text{CI}_i)$	Repl. 2 $p(\text{CI}_i)$	Repl. 1 $p(\text{DIF}_i)$	Repl. 2 $p(\text{DIF}_i)$
1	.18	.75	.77	.07
2	.27	.76	.02	.51
3	.67	.31	.01	.96
4	.12	.40	.94	.42
5	.77	.94	.18	.27
6	.74	.62	.29	.53
7	.30	.32	.58	.28
8	.58	.60	.80	.45
9	.90	.32	.34	.22
10	.81	.97	.52	.40
11	.29	.14	.20	.51
Global	.54	.67	.21	.35

are rather small for items I_2 and I_3. This must be the result of chance since in the population there was no differential item functioning. The posterior predictive p-value of DIF was equal to .21 for replication one, and .35 for replication two. This small simulation study suggests that the fit statistics perform satisfactorily if the assumptions of the 2PLM are not violated.

6 Example: Masculinity and Femininity

The data analyzed in this section were previously discussed by Sanders and Hoijtink (1992). The data consist of the responses of 88 female and 78 male students to 20 items that were constructed for the measurement of femininity and masculinity. The popularity of the items, denoted p_i, is given in Table 6.3. Each student had to indicate whether an item applied to himself or herself on a four-point scale (the responses 1 to 4 indicate "do not apply at all" to "does apply completely"). The data were recoded such that 1 and 2 became 0, and 3 and 4 became 1. The first 10 items in Table 6.3 were considered mainly indicative of femininity and the last 10 items mainly indicative of masculinity.

The main question Sanders and Hoijtink (1992) tried to answer was whether masculinity and femininity as operationalized as in Table 6.3 can be considered to be the same or different latent traits. Their main conclusions, based on analyses with the Rasch model, were that the items Male and Female functioned differently in the female and male subsamples, and

TABLE 6.3. For the Masculine and Feminine items: popularities p_i, posterior predictive p-values for CI and DIF, and item restscore correlations; all 20 items are analyzed together (88 women, 78 men).

I_i	Feminine	p_i	$p(\text{CI}_i)$	$p(\text{DIF}_i)$	ρ_{X_i,R^i}
1	Vulnerable	.75	.10	.36	.22
2	Changeable	.54	.53	.84	.23
3	Dependent	.44	.06	.71	.09
4	Intuitive	.67	.21	.25	.22
5	Emotional	.77	.21	.23	.30
6	Female	.54	.07	.00	.07
7	Oversensitive	.39	.14	.45	.31
8	Touchy	.41	.25	.51	.28
9	Insecure	.47	.01	.80	.10
10	Attentive	.86	.58	.72	.11
I_i	Masculine	p_i	$p(\text{CI}_i)$	$p(\text{DIF}_i)$	ρ_{X_i,R^i}
11	Dominant	.42	.61	.93	.22
12	Self-assured	.72	.19	.25	.06
13	Male	.42	.04	.00	−.05
14	Materialistic	.46	.92	.28	.13
15	Radical	.37	.84	.67	.15
16	Exact	.58	.62	.96	.03
17	Individualistic	.79	.33	.54	.16
18	Independent	.83	.04	.49	−.05
19	Rational	.74	.22	.45	.09
20	Self-conscious	.82	.15	.11	.26

that the correlation between the sum of the responses on the remaining nine Masculine items and the sum of the responses on the remaining nine Feminine items was about zero; that is, both traits are uncorrelated and thus different. Sanders and Hoijtink (1992) did not investigate conditional independence (which should hold if the 20 items are indicative of the same latent trait) but had to "infer" multidimensionality from the "about zero" correlation. In the sequel, using the 2PLM, the procedures proposed in this chapter are applied to these data.

As can be seen in Table 6.3, most of the item restscore correlations ρ_{X_i,R^i} are positive. Furthermore, the reliability of the 20 items is about 0.49. Stated otherwise, if a classical reliability analysis is executed one might conclude that (possibly after the removal of a few items with a negative or very small restscore correlation) the sum of the item responses can be used

TABLE 6.4. For nine Masculine and nine Feminine items: posterior predictive p-values for CI and DIF, and item restscore correlations; Masculine and Feminine items are analyzed separately (88 women, 78 men).

I_i	Feminine	$p(\text{CI}_i)$	$p(\text{DIF}_i)$	ρ_{X_i,R^i}
1	Vulnerable	.15	.20	.39
2	Changeable	.03	.52	.26
3	Dependent	.12	.40	.34
4	Intuitive	.33	.14	.22
5	Emotional	.02	.10	.36
7	Oversensitive	.41	.45	.45
8	Touchy	.45	.79	.40
9	Insecure	.07	.99	.36
10	Attentive	.39	.62	.01
I_i	Masculine	$p(\text{CI}_i)$	$p(\text{DIF}_i)$	ρ_{X_i,R^i}
11	Dominant	.42	.43	.20
12	Self-assured	.03	.46	.31
14	Materialistic	.64	.55	.09
15	Radical	.52	.30	.17
16	Exact	.25	.99	.18
17	Individualistic	.35	.55	.29
18	Independent	.60	.56	.23
19	Rational	.21	.78	.21
20	Self-conscious	.54	.24	.43

as a moderately reliable measurement of an unspecified latent trait. Given the about zero correlation discussed in the previous paragraph, this conclusion seems to be invalid. In the fourth and fifth columns of Table 6.3, the posterior predictive p-values of item level tests for conditional independence (6.6) and no differential item functioning (6.12) are presented. As can be seen item I_{13} Male and item I_6 Female function differently for the male and the female subsamples. Since the latter was also confirmed in separate analyses of the Masculine and Feminine items, both items were deleted from the item set. As can also be seen, the p-values for the tests of conditional independence are rather small, for 6 of the 20 items smaller than .10. The p-value of the global test for conditional independence (6.7) was .14. These results indicate that it is questionable whether conditional independence holds for the whole set of 20 items. Note that 6 out of 20 p-values smaller than .10 seems to be a rather strong indication of violations of conditional independence. This is not as strongly reflected by the

p-value of the global test for conditional independence. Maybe one of the other ideas to construct a test statistic that are given in the discussion of this chapter will render a more powerful global test. Note that to compute the test for conditional independence restscore groups 0 to 7 and 14 to 18 were collapsed. To compute the tests for differential item functioning restscore groups 0 to 8 and 15 to 19 were collapsed.

In Table 6.4 the results of separate analyses of the remaining nine Feminine and Masculine items are presented. It appears that the Masculine items are conditionally independent (a posterior predictive p-value for the global test of .27), do not show differential item functioning, and have reasonable item restscore correlations. Three of the nine Feminine items have p-values for the test of conditional independence smaller than .07 (the p-value of the global test was .09). It appears that conditional independence is questionable for the remaining nine items. However, the Feminine items do not show differential item functioning and have (possibly with one exception) satisfactory item restscore correlations. The reliabilities of the scales formed by the nine Feminine and Masculine items were 0.63 and 0.52, respectively. To compute the test for differential item functioning, for both sets of items restscore groups 0 to 2 and 7 to 8 were collapsed. To compute the test for conditional independence, restscore groups 0 to 1 and 6 to 7 were collapsed for the Feminine items, and restscore groups 0 to 2 and 6 to 7 for the Masculine items.

The correlations between the responses to the Feminine items Changeable and Emotional for each of the (collapsed) restscore groups, were, subsequently, .82, .33, −.40, −.03, .07, and −.19. As can be seen in Table 6.4, both items have rather small p-values with respect to the test for conditional independence. This is nicely reflected in the conditional correlations: .82 is rather large (i.e., far away from zero); and −.40 and −.19 are rather negative (i.e., not positive as is expected under conditional independence). The correlations between the Masculine items Dominant and Materialistic were −.41, −.07, .23, .12, and −.02. On face value these correlations look more in accordance with the assumption of conditional independence than the correlations for both Feminine items; the latter is confirmed by the p-values of both items, .42 and .64, respectively.

We conclude that the remaining set of nine Masculine items does not seem to violate the assumptions of conditional independence or show differential item functioning. As can be seen in Table 6.5, with respect to masculinity the average location of the men (0.52) is higher than the average location of the women (0.00); that is, men are more masculine than women (and, as an aside, the women are on the average more feminine than the men). The expected a posteriori estimates and the corresponding standard deviations of the location and slope of each item are presented in Table 6.5 (computational details can be found in the Appendix of this chapter). We also conclude that conditional independence is questionable for the Feminine items.

TABLE 6.5. Expected a posteriori estimates and posterior standard deviations for separate analyses of the Masculine and the Feminine items (88 women, 78 men).

I_i	Feminine	EAP(β_i)	SD(β_i)	EAP(α_i)	SD(α_i)
1	Vulnerable	−1.63	0.30	1.46	0.42
2	Changeable	−0.78	0.33	0.64	0.18
3	Dependent	−0.09	0.26	0.80	0.21
4	Intuitive	−1.79	0.47	0.65	0.17
5	Emotional	−1.92	0.38	1.19	0.32
7	Oversensitive	−0.01	0.19	1.61	0.44
8	Touchy	−0.03	0.24	0.95	0.24
9	Insecure	−0.28	0.24	0.91	0.24
10	Attentive	−5.54	1.87	0.41	0.13
	Subsample	EAP(μ)	EAP(σ^2)		
	Women	0.00†	1.00†		
	Men	−1.01	1.37		
I_i	Masculine	EAP(β_i)	SD(β_i)	EAP(α_i)	SD(α_i)
11	Dominant	0.80	.35	0.69	0.22
12	Self-assured	−0.85	.33	1.16	0.35
14	Materialistic	0.60	.44	0.47	0.15
15	Radical	1.06	.38	0.79	0.26
16	Exact	−0.45	.41	0.57	0.18
17	Individualistic	−1.46	.50	0.99	0.29
18	Independent	−1.99	.76	0.88	0.28
19	Rational	−1.66	.71	0.67	0.23
20	Self-conscious	−1.00	.27	2.05	0.65
	Subsample	EAP(μ)	EAP(σ^2)		
	Women	0.00†	1.00†		
	Men	0.52	0.84		

Note: † fixed parameter.

The analyses described so far in this section provide evidence that the Masculine and the Feminine items are not measuring the same latent trait, which is a nice result since the sums of the responses on both sets of items correlate about zero. However, the results of a final analysis (a joint analysis of the 9 Female and the 9 Male items listed in Table 6.5) were a little disappointing: there appeared to be instances of differential item functioning (a global p-value of .09), while clear violations of conditional independence (a global p-value of .26) were not detected. More knowledge about the fit statistics proposed in this chapter and their posterior predictive p-values is needed, before this phenomenon can be explained.

7 Discussion

The methods to investigate conditional independence and differential item functioning proposed in this chapter are based on recent developments in statistical computing (the Gibbs sampler) and statistical theory (posterior predictive checks). The scope of the simulation study and the example presented in this chapter are rather limited. Although the results are encouraging, more research is needed. An interesting topic could be the frequency properties of the proposed procedures. To investigate these properties a more elaborate simulation study should be executed to determine the degree to which the posterior predictive p-values are biased, and to examine the power against a number of specific model violations. Also interesting would be a comparison of the procedures proposed with other procedures that exist for the 2PLM and with procedures that have been developed in the context of nonparametric item response models; see the references given in Sections 2 and 3.

Since (6.6) and (6.7) were chosen because of their simplicity, it is likely that there are other combinations of conditional covariances that have more power against violations of conditional independence. Some ideas follow.

1. Since $\text{Cov}(X_i, X_j \mid R^{ij}) \geq 0$ if conditional independence holds, the power of (6.6) and (6.7) may be increased when negative conditional covariances receive a larger weight than positive conditional covariances.

2. In Section 4 the posterior predictive distribution of (replicated) data matrices was introduced. Replicated data matrices are sampled from populations for which the 2PLM holds. This implies that the assumption of conditional independence is valid for each replicated data matrix. The conditional covariances can be computed for each of these replicated data matrices. Let $\mathbf{c}^{rep}$ denote a vector in which the conditional covariances computed from one of the replicated data matrices are collected. Let $\mathbf{c}$ denote a vector in which the conditional covariances for the observed data matrix are collected. Since $\mathbf{c}^{rep}$ can be computed for each replicated data matrix, it is rather easy to compute $\Sigma_{\mathbf{c}^{rep}}$, an estimate of the covariance matrix of $\mathbf{c}^{rep}$. This matrix can be used to construct a statistic sensitive to violations of conditional independence in which the variances of and covariances among the conditional covariances are accounted for, namely, $\mathbf{c}'\Sigma_{\mathbf{c}^{rep}}^{-1}\mathbf{c}$.

3. In this chapter only statistics have been considered. Meng (1994) and Gelman, Meng, and Stern (1996) also discuss discrepancy measures,

which are functions of the data and the model parameters. Discrepancy measures can be evaluated using posterior predictive p-values in a way that explicitly accounts for the fact that these measures depend on unknown parameter values. An interesting discrepancy measure could be $[\mathbf{c} - E(\mathbf{c} \mid \boldsymbol{\omega})]'\Sigma^{-1}_{\mathbf{c} \mid \boldsymbol{\omega}}[\mathbf{c} - E(\mathbf{c} \mid \boldsymbol{\omega})]$, that is, a comparison of the conditional covariances with the expected values weighted by the covariance matrix of the conditional covariances (given that conditional independence is appropriate).

4. Note finally, that the DIMTEST statistic (6.5) can also be evaluated using its posterior predictive distribution.

It would be interesting to include a few of these ideas when the properties of different methods to investigate conditional independence are evaluated.

8 Appendix: Implementation of the Gibbs Sampler

The Gibbs sampler as described by Zeger and Karim (1991) was used to obtain a sample from (6.4) and (6.15). For all analyses presented in this chapter, the Gibbs sampler was run across five steps for $d = 1, \ldots, 110,000$ iterations. The first 10,000 iterations served as a "burn-in" period and were discarded. Subsequently every 50th iteration was saved, leaving a sample of 2000 parameter vectors $\boldsymbol{\omega}$ from (6.4). Since, in this chapter, the Gibbs sampler is used to calculate posterior predictive p-values, the investigation of convergence and dependence among subsequent iterations of the Gibbs sampler does not receive much attention. The reader is referred to Hoijtink and Molenaar (1997) and Hoijtink (1998). Their methods to investigate convergence and to deal with dependence were also used in this chapter. The five steps of our application of the Gibbs sampler in iteration d are as follows.

1. For $v = 1, \ldots, n_1, n_1 + 1, \ldots, n$ sample $\theta_{v,d}$ from its conditional posterior distribution

$$\text{Post}(\theta_{v,d} | \mathbf{x}_v, g_v, \boldsymbol{\omega}_{d-1}) \propto P(\mathbf{x}_v | \theta_{v,d})\, h(\theta_{v,d} | \mu_{g_v,d-1}, \sigma^2_{g_v,d-1}) \,. \tag{6.16}$$

   Since the conditional posterior (6.16) is not proportional to a standard distribution, a sample is obtained using rejection sampling (see below).

2. Sample $\mu_{2,d}$ from its conditional posterior distribution

$$\text{Post}(\mu_{2,d} \mid \sigma^2_{2,d-1}, \theta_{n_1+1,d}, \ldots, \theta_{n,d}) =$$

$$\mathcal{N}\Big(\sum_{v=n_1+1}^{n} \frac{\theta_{v,d}}{n_2}, \sigma^2_{2,d-1}\Big) \, . \tag{6.17}$$

3. Sample $\sigma^2_{2,d}$ from $\text{Post}(\sigma^2_{2,d} \mid \mu_{2,d}, \theta_{n_1+1,d}, \ldots, \theta_{n,d})$. This is a scaled inverse chi-square distribution with scale parameter

$$\sum_{v=n_1+1}^{n} \frac{(\theta_{v,d} - \mu_{2,d})^2}{n_2 - 2} \tag{6.18}$$

   and $n_2 - 2$ degrees of freedom. See Gelman, Carlin, Stern, and Rubin (1995, p. 47, p. 480), for a definition of the scaled inverse chi-square distribution and for a procedure to sample from this distribution.

4. For $i = 1, \ldots, k$ sample $\beta_{i,d}$ from its conditional posterior distribution

$$\text{Post}(\beta_{i,d} \mid x_{1i}, \ldots, x_{ni}; \boldsymbol{\theta}_d, \alpha_{i,d-1}) \propto \prod_{v=1}^{n} P(\mathbf{x}_v \mid \theta_{v,d}) \, . \tag{6.19}$$

   Since (6.19) is not proportional to a standard distribution, a sample is obtained using rejection sampling (see below).

5. For $i = 1, \ldots, k$ sample $\ln \alpha_{i,d}$ from its conditional posterior distribution

$$\text{Post}(\ln \alpha_{i,d} | x_{1i}, \ldots, x_{ni}; \boldsymbol{\theta}_d, \beta_{i,d}) \propto \prod_{v=1}^{n} P(\mathbf{x}_v | \theta_{v,d}) \, h(\ln \alpha_{i,d}) \, , \tag{6.20}$$

   and compute $\alpha_{i,d} = \exp(\ln \alpha_{i,d})$. Note that the restriction $\alpha_{i,d} > 0$ is always satisfied if $\ln \alpha_{i,d}$ is sampled and subsequently transformed. Note furthermore, that (6.1) can be written as a function of $\ln \alpha_i$ if α_i is replaced by $\exp(\ln \alpha_i)$. Since (6.20) is not proportional to a standard distribution, a sample is obtained using rejection sampling.

The sampling in Steps 1, 4 and 5 is done using rejection sampling (see, e.g., Gelman et al., 1995, pp. 303–305). The first step is the choice of a distribution that is approximately proportional to the distribution from which the sample has to be obtained. The conditional posterior distributions of β_i and α_i were approximated using a normal distribution with expectation equal to the mode m of the conditional posterior, and variance equal to twice the inverse of the observed Fisher information. Initial experiences indicated that a normal distribution is a rather poor approximation of (6.16). Instead a Student t-distribution with expectation equal to the mode of the conditional posterior, variance equal to three times the inverse of the observed Fisher information, and degrees of freedom equal to four, was used. See Gelman et al. (1995, p. 476 and p. 481) for the formula of this distribution and a procedure to sample from it.

In the second step of rejection sampling, a number ξ is sampled from the approximating distribution Prox($\cdot$). This number is accepted as a draw from the conditional posterior at hand with probability

$$\frac{\text{Cond}(\xi)}{\text{Prox}(\xi)\frac{\text{Cond}(m)}{\text{Prox}(m)}}, \tag{6.21}$$

where Cond($\cdot$) denotes the right-hand sides of (6.16), (6.19), and (6.20), respectively, that is, the unnormalized conditional posterior. Sampling is repeated until a number ξ is accepted. The factor Cond(m)/Prox(m) appearing in (6.21) is added to ensure that the maximum value of the conditional distribution is equal to the maximum value of the approximating distribution. Note that the occurrence of values of (6.21) larger than one implies that the approximating distribution does not envelop the corresponding conditional posterior as required. This occurred less than 100 times during the many millions of times rejection sampling was applied in our study. It appears that the rejection sampling procedure performed almost perfectly. The reader is referred to Chib and Greenberg (1995) for a description of the Metropolis–Hastings algorithm, which is an alternative for rejection sampling.

For each parameter, the average and standard deviation of the 2000 sampled values can be computed. These quantities are close approximations of the expected a posteriori estimate of each parameter and the corresponding posterior standard deviation.

References

Bock, R.D., Gibbons, R., & Muraki, E., (1988). Full information item factor analysis. *Applied Psychological Measurement, 12*, 261–280.

Chib, S., & Greenberg, E. (1995). Understanding the Metropolis–Hastings algorithm. *The American Statistician, 49*, 327–335.

Dempster, A.P., Laird, N.M., & Rubin, D.B. (1977). Maximum likelihood from incomplete data via the EM algorithm. *Journal of the Royal Statistical Society, Series B, 39*, 1–38.

Gelman, A., Carlin, J.B., Stern, H.S., & Rubin, D.B. (1995). *Bayesian data analysis.* London: Chapman and Hall.

Gelman, A., Meng, X., & Stern, H.S. (1996). Posterior predictive assessment of model fitness via realized discrepancies (with discussion). *Statistica Sinica, 6*, 733–807.

Glas, C.A.W. (1998). Detection of differential item functioning using Lagrange multiplier tests. *Statistica Sinica, 8*, 647–667.

Glas, C.A.W. (2001). Differential item functioning depending on general covariates. In A. Boomsma, M.A.J. van Duijn, & T.A.B. Snijders (Eds.), *Essays on item response theory* (pp. 131–148). New York: Springer-Verlag.

Glas, C.A.W., & Verhelst, N.D. (1995). Testing the Rasch model. In G.H. Fischer & I.W. Molenaar (Eds.), *Rasch models: Foundations, recent developments, and applications* (pp. 69–96). New York: Springer-Verlag.

Hoijtink, H. (1998). Constrained latent class analysis using the Gibbs sampler and posterior predictive p-values: Application to educational testing. *Statistica Sinica, 8*, 691–711.

Hoijtink, H., & Molenaar, I.W. (1997). A multidimensional item response model: Constrained latent class analysis using the Gibbs sampler and posterior predictive checks. *Psychometrika, 62*, 171–189.

Holland, P.W., & Wainer. H. (1993). *Differential item functioning.* Hillsdale, NJ: Erlbaum.

Junker, B.W. (1993). Conditional association, essential independence and monotone unidimensional item response models. *The Annals of Statistics, 3*, 1359–1378.

Klinkenberg, E.L. (2001). A logistic IRT model for decreasing and increasing item characteristic curves. In A. Boomsma, M.A.J. van Duijn, & T.A.B. Snijders (Eds.), *Essays on item response theory* (pp. 173–192). New York: Springer-Verlag.

Meng, X. (1994). Posterior predictive p-values. *The Annals of Statistics, 22*, 1142–1160.

Mislevy, R.J. (1986). Bayes modal estimation in item response models. *Psychometrika, 51*, 177–195.

Mislevy, R.J., & Bock, R.D. (1990). BILOG 3: Item analysis and test scoring with binary logistic models [Software manual]. Mooresville, IN: Scientific Software.

Muraki, E., Mislevy, R.J., & Bock, R.D. (1987). BIMAIN: A program for item pool maintenance in the presence of item parameter drift and item bias [Software manual]. Mooresville, IN: Scientific Software.

Patz, R.J., & Junker, B.W. (1999). A straightforward approach to Markov chain Monte Carlo methods for item response models. *Journal of Educational and Behavioral Statistics, 24*, 146–177.

Post, W.J., Van Duijn, M.A.J., & Van Baarsen, B. (2001). Single-peaked or monotone tracelines? On the choice of an IRT model for scaling data. In A. Boomsma, M.A.J. van Duijn, & T.A.B. Snijders (Eds.), *Essays on item response theory* (pp. 391–414). New York: Springer-Verlag.

Rosenbaum, P.R. (1984). Testing the conditional independence and monotonicity assumptions of item response theory. *Psychometrika, 49*, 425–435.

Rubin, D.R. (1984). Bayesianly justifiable and relevant frequency calculations for the applied statistician. *The Annals of Statistics, 12*, 1151–1172.

Sanders, K., & Hoijtink, H. (1992). Androgynie bestaat [Androgyny exists]. *Nederlands Tijdschrift voor de Psychologie, 47*, 123–133.

Shealy, R., & Stout, W. (1993). A model based standardization approach that separates true bias/DIF from group ability differences and detects test bias/DTF as well as item bias/DIF. *Psychometrika, 58,* 159–194.

Stout, W. (1987). A nonparametric approach for assessing latent trait dimensionality. *Psychometrika, 52*, 589–618.

Stout, W., Goodwin Froelich, A., & Gao, F. (2001). Using resampling methods to produce an improved DIMTEST procedure. In A. Boomsma, M.A.J. van Duijn, & T.A.B. Snijders (Eds.), *Essays on item response theory* (pp. 357–375). New York: Springer-Verlag.

Stout, W., Habing, B., Douglas, J., Kim, H.R., Roussos, L., & Zhang, J. (1996). Conditional covariance-based nonparametric multidimensionality assessment. *Applied Psychological Measurement, 20*, 331–354.

Thissen, D., Steinberg, L., & Wainer, H. (1988). Use of item response theory in the study of group differences in trace lines. In H. Wainer & H.I. Braun (Eds.), *Test validity* (pp. 147–169). Hillsdale, NJ: Erlbaum.

Thissen, D., Steinberg, L., & Wainer, H. (1993). Detection of differential item functioning using the parameters of item response models. In P.W. Holland & H. Wainer (Eds.), *Differential item functioning* (pp. 67–113). Hillsdale, NJ: Erlbaum.

Van den Wollenberg, A.L. (1982). Two new test statistics for the Rasch model. *Psychometrika, 47*, 123–139.

Verhelst, N.D., & Glas, C.A.W. (1993). A dynamic generalization of the Rasch model. *Psychometrika, 58*, 395–416.

Zeger, S.L., & Karim, M.R. (1991). Generalized linear models with random effects: A Gibbs sampling approach. *Journal of the American Statistical Association, 86*, 79–86.

Zhang, J., & Stout, W. (1999a). Conditional covariance structure for generalized compensatory multidimensional items. *Psychometrika, 64*, 129–152.

Zhang, J., & Stout, W. (1999b). The theoretical DETECT index of dimensionality and its application to approximate simple structure. *Psychometrika, 64*, 213–250.

7

Differential Item Functioning Depending on General Covariates

Cees A.W. Glas[1]

ABSTRACT Item response theory (IRT) is a powerful tool for the detection of differential item functioning (DIF). It is shown that the class of IRT models with manifest predictors is a comprehensive framework for the detection of DIF. These models also support the investigation of the causes of DIF. In principle, the responses to every item in a test can be subject to DIF, and traditional IRT-based detection methods require one or more estimation runs for every single item. Therefore, Glas (1998) proposed an alternative procedure that can be performed using only a single estimate of the item parameters. This procedure is based on the Lagrange multiplier test or the equivalent Rao efficient score test. In this chapter, the procedure is generalized in various directions, the most important one being the possibility of conditioning on general covariates. A small simulation study is presented to give an impression of the power of the test. In an example using real data it is shown how the method can be applied to the identification of main and interaction effects in DIF.

1 Introduction

Differential item functioning (DIF) is a difference in item responses between equally proficient members of two or more groups. For instance, a dichotomous item is subject to DIF if, conditionally on the proficiency level, the probability of a correct response differs between groups. One might think of a test of foreign language comprehension, where girls are impeded by items referring to a football setting. The poor performance of the girls on the football-related items must not be attributed to their poor level of comprehension of the foreign language but to their lack of knowledge of football. Since DIF is highly undesirable in fair testing, methods for the detection of DIF are extensively studied. Overviews are provided by the books by Holland and Wainer (1993) and by Camilli and Shepard (1994).

[1]Department of Educational Measurement and Data Analysis, University of Twente, P.O. Box 217, 7500 AE Enschede, The Netherlands; e-mail: *glas@edte.utwente.nl*

Several techniques for detection of DIF have been proposed. Most of them are based on evaluation of differences in response probabilities between groups, conditional on some measure of ability (in this chapter proficiency and ability are used interchangeably). The most generally used technique is based on the Mantel–Haenszel statistic (Holland & Thayer, 1988), others are based on log-linear models (Kok, Mellenbergh, & Van der Flier, 1985; Swaminathan & Rogers, 1990), and on IRT models (Hambleton & Rogers, 1989; Kelderman, 1989; Muraki, & Bock, 1991; Glas & Verhelst, 1995; Glas, 1998).

In the Mantel–Haenszel (MH) approach, the respondent's unweighted sum score is used as a proxy for proficiency, and DIF is evaluated by testing whether the response probabilities, given the unweighted sum scores, differ among the groups. Though the MH test works quite well in practice, Fischer (1993, 1995) points out that its application is based on the assumption that the Rasch model holds, and in applying the MH test in other cases, such as with data following the two- or three-parameter logistic model (2PLM or 3PLM), several theoretical limitations arise. In the log-linear approach the unweighted sum score is taken as a proxy for proficiency too, and the issues raised by Fischer (1993, 1995; see also Meredith & Millsap, 1992) apply to this approach as well.

In an IRT model, proficiency is represented by a latent variable θ, and an obvious solution to the problem mentioned is to apply the MH and the log-linear approach using subgroups that are homogeneous with respect to an estimate $\hat{\theta}$ of θ. This, however, introduces the problem that $\hat{\theta}$ is subject to estimation error, which is difficult to take into account when forming the subgroups. The obvious solution is to view DIF as a special case of misfit of an IRT model and to use the machinery for IRT model-fit evaluation to explore DIF. An overview of this approach was given by Thissen, Steinberg, and Wainer (1993). In that overview, evaluation of item parameter invariance over subgroups using likelihood ratio and Wald statistics was presented as the main statistical tool for detection of DIF. However, Glas (1998) argued that the likelihood ratio and Wald statistics approach is not very efficient, and he presented an application of the Lagrange multiplier (LM) test or the equivalent Rao efficient score test for evaluating DIF in the context of the 2PLM and the nominal response model. In the present chapter, this approach is generalized in two directions: the approach is generalized to the 3PLM; and more important, the method is generalized in such a way that DIF can be systematically related to multiple background variables of the respondents.

This chapter is organized as follows. In Section 2 it is shown how multiple background variables can be introduced to the IRT model. Then, the procedure for evaluation of DIF using the LM test is described. In the following sections, this approach is implemented in a marginal maximum likelihood (MML) framework, a power study is presented, and, using a real data example, the outcome of the LM approach is compared with the results of the

MH procedure. Finally, in Section 7, some suggestions for further research are formulated.

2 Modeling Differential Item Functioning

In the 3PLM, the probability of a correct response of a subject S_v to an item I_i, denoted $X_{vi} = 1$, is given by

$$P_i(\theta_v) = P(X_{vi} = 1 \mid \theta_v, \alpha_i, \beta_i, \gamma_i) = \gamma_i + (1 - \gamma_i)\psi_i(\theta_v) , \tag{7.1}$$

with

$$\psi_i(\theta_v) = \frac{\exp(\alpha_i\theta_v - \beta_i)}{1 + \exp(\alpha_i\theta_v - \beta_i)} , \tag{7.2}$$

where the subject parameter θ_v is a proficiency parameter and the item parameters α_i, β_i, and γ_i are a discrimination, difficulty, and guessing parameter, respectively (see, for instance, Lord, 1980). In the present approach, DIF is modeled by introducing a more general alternative to the 3PLM; that is,

$$P_i(\theta_v, \mathbf{y}_v) = P(X_{vi} = 1 \mid \theta_v, \boldsymbol{\alpha}_i, \boldsymbol{\beta}_i, \boldsymbol{\gamma}_i, \mathbf{y}_v) = \boldsymbol{\gamma}_i'\mathbf{y}_v + (1 - \boldsymbol{\gamma}_i'\mathbf{y}_v)\psi_i(\theta_v, \mathbf{y}_v) ,$$

with

$$\psi_i(\theta_v, \mathbf{y}_v) = \frac{\exp(\boldsymbol{\alpha}_i'\mathbf{y}_v\theta_v - \boldsymbol{\beta}_i'\mathbf{y}_v)}{1 + \exp(\boldsymbol{\alpha}_i'\mathbf{y}_v\theta_v - \boldsymbol{\beta}_i'\mathbf{y}_v)} . \tag{7.3}$$

Below, this model is referred to as the 3PL-DIF model. In the 3PL-DIF model, $\mathbf{y}_v$ is a vector of background variables with elements y_{vg}, $g = 0, \ldots, m$, where m is the number of background variables considered. Furthermore, $\boldsymbol{\alpha}_i'$, $\boldsymbol{\beta}_i'$, and $\boldsymbol{\gamma}_i'$ are row vectors $(\alpha_{i0}, \ldots, \alpha_{im})$, $(\beta_{i0}, \ldots, \beta_{im})$, and $(\gamma_{i0}, \ldots, \gamma_{im})$, respectively. It is assumed that $y_{v0} = 1$ for all v, so item parameters indexed with $i0$ can be interpreted as the parameters of a 3PLM which will serve as a baseline no-DIF model. The other item parameters gauge changes in the item response curves of this baseline model as a function of background variables. These background variables comprise a general class of covariates which might be either categorical dummy coded covariates, as, for instance, gender or race, or discrete or continuous background variables, such as age, social-economic status, intelligence, or educational achievement. Rogers, Swaminathan, and Egan (1999) point out that entering background variables is a useful approach to modeling and explaining DIF. An example of this approach is evaluating whether differences in item response behavior between two race groups vanish when differences in social-economic status are taken into account.

The application in this chapter, however, deals with categorical dummy coded covariates, that is, dichotomous background variables defined as

$$y_{vg} = \begin{cases} 1\,, & \text{if the respondent } v \text{ belongs to group } g, \\ 0\,, & \text{otherwise,} \end{cases}$$

for $g = 1, \ldots, m$. Below, it is assumed that a respondent can belong to more than one group. For instance, consider the case where one is interested in distinguishing between the effects of gender-related DIF and DIF related to cultural background, say, the effects of being either nonnative or native. These effects are modeled by introducing the variables

$$y_{v1} = \begin{cases} 1\,, & \text{if } v \text{ is female,} \\ 0\,, & \text{if } v \text{ is male,} \end{cases} \tag{7.4}$$

and

$$y_{v2} = \begin{cases} 1\,, & \text{if } v \text{ is nonnative,} \\ 0\,, & \text{if } v \text{ is native.} \end{cases} \tag{7.5}$$

In studies of DIF, one usually speaks of a reference population, say, the majority group or an advantaged group, and a focal population, say, a minority or a disadvantaged group. Notice that the parameterization induced by (7.4) and (7.5) implies that the native males are taken as a baseline; that is, their response behavior is modeled by the parameters α_{i0}, β_{i0}, and γ_{i0}. The main effects of being female are modeled by α_{i1}, β_{i1}, and γ_{i1}, and the main effects of being nonnative are modeled by α_{i2}, β_{i2}, and γ_{i2}. In the same manner, by introducing

$$y_{v3} = \begin{cases} 1\,, & \text{if } y_{v1} = 1 \text{ and } y_{v2} = 1, \\ 0\,, & \text{otherwise,} \end{cases} \tag{7.6}$$

it is also possible to model the interaction between being female and nonnative by α_{i3}, β_{i3}, and γ_{i3}. So the effects of belonging to a focal group are modeled by parameters that reflect the difference between the item response curves of the focal and the reference group.

The model discussed here is closely related to several other models. In the framework of the partial credit model (PCM), a comparable model for the analysis of DIF using dichotomous covariates was proposed by Wang (in press). In the general formulation with general covariates, the model presented in this chapter is analogous to the class of response models with manifest predictors considered by Zwinderman (1991, 1997). One could go even further and notice that the model is a special case of the generalized linear model (see, for instance, McCullagh & Nelder, 1989) where latent variables, such as ability, and manifest covariates, such as gender, are entered into the model simultaneously. But although the correspondence between the models is important from a theoretical point of view, the practical implications should not be overstressed. Cressie and Holland (1983) point out that an IRT model must be viewed as a multinomial model for the set of all possible response patterns. Notice that with k dichotomous items, there are 2^k different response patterns. As was also pointed out by Zwinderman (1991, 1997), this limits the utility of general-purpose software for logistic regression as a tool for IRT analyses to the case of very short tests. So in practice, and at this point in time, dedicated IRT software is needed to perform the computations introduced below.

3 Evaluation of Differential Item Functioning

There are various approaches to trying to identify DIF using the 3PL-DIF model. One could concurrently estimate the complete 3PL-DIF model with all possible item effects included, or one could estimate a model with DIF parameters for one item only, and repeat the procedure for all items separately. The first procedure involves many parameters, several of which might be poorly conditioned, and estimates might be hard to obtain. The second approach needs a lot of separate parameter estimation runs. Therefore, an alternative approach is suggested where only the estimates of the parameters of the 3PLM are needed, and tests for DIF are performed for one item at a time. This approach is based on the Lagrange multiplier (LM) test by Aitchison and Silvey (1958), or on the equivalent efficient score test by Rao (1947). Details of these tests can be found in Rao (1973, pp. 417–418), or Cox and Hinkley (1974, p. 315 and p. 355).

The principle can be described as follows. Consider a general model with parameters $\boldsymbol{\eta}$. For the case discussed here, the special model is derived from the general model by setting one or more parameters equal to zero. So if the parameter vector $\boldsymbol{\eta}$ is partitioned $\boldsymbol{\eta} = (\boldsymbol{\eta}_1', \boldsymbol{\eta}_2')'$, the null hypothesis entails $\boldsymbol{\eta}_2 = \mathbf{0}$. Let $\mathbf{h}(\boldsymbol{\eta}) = (\mathbf{h}_1(\boldsymbol{\eta})', \mathbf{h}_2(\boldsymbol{\eta})')'$ be the partial derivatives of the log-likelihood of the general model, so $h_p(\boldsymbol{\eta}) = \partial \ln L(\boldsymbol{\eta})/\partial \boldsymbol{\eta}_p$, for $p = 1, 2$. Notice that the vector of partial derivatives gauges the change of the log-likelihood as a function of local changes in $\boldsymbol{\eta}$. The test is based on the statistic

$$\mathrm{LM} = \mathbf{h}_2(\boldsymbol{\eta})'\boldsymbol{\Sigma}^{-1}\mathbf{h}_2(\boldsymbol{\eta}) \;, \tag{7.7}$$

where

$$\boldsymbol{\Sigma} = \boldsymbol{\Sigma}_{22} - \boldsymbol{\Sigma}_{21}\boldsymbol{\Sigma}_{11}^{-1}\boldsymbol{\Sigma}_{12} \;,$$

and

$$\boldsymbol{\Sigma}_{pq} = -\frac{\partial^2 \ln L(\boldsymbol{\eta})}{\partial \boldsymbol{\eta}_p \partial \boldsymbol{\eta}_q'} \;,$$

for $p = 1, 2$ and $q = 1, 2$. LM is evaluated using the maximum likelihood estimates of the parameters of the special model of the null hypothesis. The LM statistic has an asymptotic χ^2-distribution with degrees of freedom equal to the number of parameters in $\boldsymbol{\eta}_2$ (Aitchison & Silvey, 1958; Rao,

1947). The importance of a significant result can be evaluated by computing an estimator of $\boldsymbol{\eta}_2$, say, $\hat{\boldsymbol{\eta}}_2$, based on one Newton–Raphson (ONR) step, that is,

$$\hat{\boldsymbol{\eta}}_2 = \boldsymbol{\Sigma}^{-1}\mathbf{h}_2(\boldsymbol{\eta}) \ . \tag{7.8}$$

In the sequel, $\hat{\boldsymbol{\eta}}_2$ is called the ONR estimator. In the next section, it is shown how this can be implemented to evaluate the fit of IRT models.

4 The MML Framework

Marginal maximum likelihood (MML) estimation is probably the most used technique for the estimation of parameters of IRT models. This framework is also used to implement the LM tests for DIF. For the 1PLM, 2PLM, and 3PLM, MML estimation was developed by Bock and Aitkin (1981; see also Bock & Zimowski, 1997; Rigdon & Tsutakawa, 1983; Mislevy, 1984, 1986). The computations can be made using the software package BILOG-MG (Zimowski, Muraki, Mislevy, & Bock, 1996). In the MML framework adopted here, it is assumed that a number of groups of respondents can be distinguished, and that the ability parameters of the respondents within a group have a normal distribution indexed by a group-specific mean and variance parameter. The density of θ_v is denoted $g(\theta_v; \boldsymbol{\lambda}, \mathbf{y}_v)$, where $\boldsymbol{\lambda}$ is a vector of the means and variances of the groups and $\mathbf{y}_v$ $(m \times 1)$ is a vector that defines the m groups to which subject S_v does or does not belong. Usually, to identify the model, the mean and the variance of one of the groups are set to zero and unity, respectively. The item parameters are stacked in a vector $\boldsymbol{\xi}$, so this vector contains all discrimination, difficulty, and guessing parameters. Finally, $\boldsymbol{\eta}$ is the vector of all item parameters $\boldsymbol{\xi}$ and the parameters $\boldsymbol{\lambda}$ of the ability distributions. The first- and second-order derivatives of the marginal likelihood of $\boldsymbol{\eta}$, needed for the LM or efficient score test, are derived as follows. The loglikelihood function of $\boldsymbol{\eta}$ can be written as

$$\ln L(\boldsymbol{\eta} \mid \mathbf{X}, \mathbf{Y}) = \sum_{v=1}^{n} \ln \int p(\mathbf{x}_v \mid \theta_v, \boldsymbol{\xi}, y_v) g(\theta_v; \boldsymbol{\lambda}, y_v)\, d\theta_v \ , \tag{7.9}$$

with

$$p(\mathbf{x}_v \mid \theta_v, \boldsymbol{\xi}, \mathbf{y}_v) = \prod_{i=1}^{k} P_i(\theta_v, \mathbf{y}_v)^{x_{vi}} [1 - P_i(\theta_v, \mathbf{y}_v)]^{(1-x_{vi})} \ , \tag{7.10}$$

where $\mathbf{x}_v$ is the response pattern of subject S_v, $v = 1, \ldots, n$, $\mathbf{X}$ is an $n \times k$ matrix of all item responses, and $\mathbf{Y}$ is an $n \times m$ matrix containing the group membership variables.

The first-order derivatives of (7.9) with respect to $\boldsymbol{\eta}$ can be derived using Fisher's identity (Efron, 1977; Louis 1982):

$$\mathbf{h}(\boldsymbol{\eta}) = \frac{\partial}{\partial \boldsymbol{\eta}} \ln L(\boldsymbol{\eta} \mid \mathbf{X}) = \sum_v E[\boldsymbol{\omega}_v(\boldsymbol{\eta}) \mid \mathbf{x}_v, \mathbf{y}_v; \boldsymbol{\eta}] \;, \tag{7.11}$$

with

$$\boldsymbol{\omega}_v(\boldsymbol{\eta}) = \frac{\partial}{\partial \boldsymbol{\eta}} \ln p(\mathbf{x}_v, \theta_v \mid \mathbf{y}_v; \boldsymbol{\eta}) \;. \tag{7.12}$$

The expectation in (7.11) is taken with respect to the posterior distribution $p(\theta_v \mid \mathbf{x}_v, \mathbf{y}_v; \boldsymbol{\eta})$.

Application of this framework to derive the first- and second-order derivatives of the loglikelihood of the 3PL-DIF model with respect to the item DIF parameters proceeds as follows. Let P_i and ψ_i be a shorthand notation for $P_i(\theta_v, \mathbf{y}_v)$ and $\psi_i(\theta_v, \mathbf{y}_v)$, respectively. By taking first-order derivatives of the logarithm of (7.10), the expressions for the elements of $\boldsymbol{\omega}_v(\boldsymbol{\eta})$ are found:

$$\omega_v(\alpha_{ig}) = y_{vg} \frac{(x_{vi} - P_i)(1-\gamma_i)\psi_i(1-\psi_i)\theta_v}{P_i(1-P_i)} \;,$$

$$\omega_v(\beta_{ig}) = y_{vg} \frac{(P_i - x_{vi})(1-\gamma_i)\psi_i(1-\psi_i)}{P_i(1-P_i)} \;,$$

$$\omega_v(\gamma_{ig}) = y_{vg} \frac{(x_{vi} - P_i)(1-\psi_i)}{P_i(1-P_i)} \;.$$

Inserting these expressions into (7.11) gives the required first-order derivatives. Glas (1999) points out that the matrix of second-order derivatives can be approximated as

$$\mathbf{H}(\boldsymbol{\eta}, \boldsymbol{\eta}') \approx \sum_v E[\boldsymbol{\omega}_v(\boldsymbol{\eta}) \mid \mathbf{x}_v, \mathbf{y}_v; \boldsymbol{\eta}] \; E[\boldsymbol{\omega}_v(\boldsymbol{\eta}) \mid \mathbf{x}_v, \mathbf{y}_v; \boldsymbol{\eta}]' \;. \tag{7.13}$$

In practice, the MML approach to parameter estimation is often extended to Bayes modal estimation. This is motivated by the fact that item parameter estimates in the 3PLM are sometimes hard to obtain, because the parameters are poorly determined by the available data, in the sense that in the region of the ability scale where the respondents are located, the item characteristic curves can be appropriately described by many item parameter values $(\alpha_i, \beta_i, \gamma_i)$. As a result, the estimates of the three item parameters in the 3PLM are often highly correlated. To obtain "reasonable" and finite estimates, Mislevy (1986) considers a number of Bayesian approaches. These entail the introduction of prior distributions on item parameters. Parameter estimates are then computed by maximizing the posterior density of $\boldsymbol{\eta}$, which is proportional to $\ln L(\boldsymbol{\eta}\,; \mathbf{X}) +$

$\ln p(\boldsymbol{\eta} \mid \boldsymbol{\zeta}) + \ln p(\boldsymbol{\zeta})$, where $p(\boldsymbol{\eta} \mid \boldsymbol{\zeta})$ is the prior density of the $\boldsymbol{\eta}$, characterized by parameters $\boldsymbol{\zeta}$, which in turn follow a density $p(\boldsymbol{\zeta})$. In one approach the prior distribution is fixed; in another approach, often called empirical Bayes, the parameters of the prior distribution are estimated along with the other parameters. In the first case, the likelihood equations (7.9) change to $\partial \ln L(\boldsymbol{\eta}; \mathbf{X})/\partial\boldsymbol{\eta} + \partial \ln p(\boldsymbol{\eta} \mid \boldsymbol{\zeta})/\partial\boldsymbol{\eta} = \mathbf{0}$. In the second case, in addition to these modified likelihood equations, the additional equations $\partial \ln p(\boldsymbol{\zeta})/\partial\boldsymbol{\zeta} = \mathbf{0}$ must also be solved. See Mislevy (1986) for details. Glas (1999) showed how the LM test can be modified to adapt to this approach. His method boils down to adding terms to the matrix of second-order derivatives related to the added parameters and the modified posterior.

5 A Power Study

To give an idea of the output and the power of the procedure, some simulated examples are presented. All simulations concern a test of 10 items. For the first simulations, the test had no DIF, and the item parameter values are given in the first columns of the first panel of Table 7.1.

It was assumed that the test was administered to four groups: native men, nonnative men, native women, and nonnative women. The four groups had

TABLE 7.1. Simulation values and data summary ($n = 4000$).

Item	α_i	β_i	γ_i	p_i	ρ_{it}	Score	Frequency
						0	5
1	1.0	0.0	0.2	0.64	0.47	1	42
2	1.0	0.0	0.2	0.66	0.45	2	125
3	1.0	0.5	0.2	0.57	0.44	3	258
4	1.0	1.0	0.2	0.49	0.45	4	419
5	1.5	−0.5	0.2	0.74	0.48	5	554
6	1.5	0.0	0.2	0.66	0.51	6	627
7	1.5	0.5	0.2	0.60	0.51	7	710
8	0.5	−0.5	0.2	0.72	0.35	8	648
9	0.5	0.0	0.2	0.63	0.36	9	419
10	0.5	0.5	0.2	0.53	0.34	10	193
Group	μ_θ	σ_θ	**Mean**	**SD**	**Alpha**		
1	0.00	1.0	5.6	2.0	0.49		
2	0.25	1.0	6.1	2.1	0.55		
3	0.50	1.0	6.5	2.0	0.50		
4	0.75	1.0	6.8	1.9	0.51		

TABLE 7.2. Estimated parameter values and standard errors ($n = 4000$).

Item	$\hat{\alpha}_i$	SE($\hat{\alpha}_i$)	$\hat{\beta}_i$	SE($\hat{\beta}_i$)	$\hat{\gamma}_i$	SE($\hat{\gamma}_i$)
1	1.04	0.14	0.15	0.24	0.18	0.07
2	0.89	0.11	0.01	0.23	0.19	0.07
3	0.92	0.16	0.68	0.34	0.23	0.07
4	0.87	0.13	0.81	0.25	0.13	0.05
5	1.41	0.19	−0.28	0.26	0.24	0.08
6	1.50	0.21	0.15	0.26	0.19	0.06
7	1.40	0.21	0.56	0.29	0.19	0.06
8	0.49	0.06	−0.47	0.19	0.20	0.09
9	0.48	0.07	0.04	0.24	0.20	0.08
10	0.50	0.12	0.84	0.41	0.26	0.08
Group	$\hat{\mu}_\theta$	SE($\hat{\mu}_\theta$)	$\hat{\sigma}_\theta$	SE($\hat{\sigma}_\theta$)		
1	0.00†	–	1.00†	–		
2	0.33	0.06	1.06	0.07		
3	0.63	0.06	0.98	0.07		
4	0.81	0.07	1.00	0.07		

Note: † fixed parameter.

different normal ability distributions with means and standard deviations as displayed in the first columns of the second panel of Table 7.1. Every group consisted of 1000 subjects. The columns of the first panel contain the proportion of correct item responses p_i, item-test correlations ρ_{it}, and the distribution of the respondents' sum scores, all aggregated over the four groups. The columns of the second panel contain the mean and the standard deviation of the score distributions per group, and Cronbach's alpha as an indication of the classical test theory reliability, also per group.

The parameter estimates and their standard errors as obtained using BILOG-MG are given in Table 7.2. Notice that the model was identified by setting the mean and the standard deviation of the ability distribution of the first group equal to zero and one, respectively.

The power of the LM test (7.7) and the usefulness of the effect sizes computed with the ONR estimator (7.8) were investigated by introducing DIF to the first two items. Four conditions were used: a condition with a main DIF effect of gender (cf. Equation 7.4), a condition with a main DIF effect of cultural background (cf. Equation 7.5), a condition where both main effects were present, and a condition where an interaction effect was added to the two main effects (cf. Equation 7.6). DIF was only imposed on the item difficulties. These conditions are denoted Data Model B1, Data Model B2, Data Model B1+B2, and Data Model B1+B2+B3, respectively. For a given effect size δ, this was implemented as follows.

TABLE 7.3. Overview of test statistics.

Statistic	General Model	*df*	Symbol
$\text{LM}(\beta_{i1})$	Main Effect β_{i1}	1	·····▲·····
$\text{LM}(\beta_{i2})$	Main Effect β_{i2}	1	-·-·-■-·-·-
$\text{LM}(\beta_{i1}, \beta_{i2})$	Main Effects β_{i1} and β_{i2}	2	——●——
$\text{LM}(\beta_{i1}, \beta_{i2}, \beta_{i3})$	Main Effects β_{i1}, β_{i2}, and Interaction Effect β_{i3}	3	-----◆-----
$\text{LM}(\alpha_{i*}, \beta_{i*}, \gamma_{i*})$	Main and Interaction Effects $\alpha_{i1}, \alpha_{i2}, \alpha_{i3}, \beta_{i1}, \beta_{i2}, \beta_{i3}$, γ_{i1}, γ_{i2}, and γ_{i3}	9	———▼———

- Data Model B1. Gender-related DIF: $\beta_{11} = \delta$ and $\beta_{21} = -\delta$.
- Data Model B2. Race-related DIF: $\beta_{12} = \delta$ and $\beta_{22} = -\delta$.
- Data Model B1+B2. Main effects of gender- and race-related DIF: $\beta_{11} = \beta_{12} = \delta$ and $\beta_{21} = \beta_{22} = -\delta$.
- Data Model B1+B2+B3. Main effects of gender- and race-related DIF with an additional interaction effect: $\beta_{11} = \beta_{12} = \beta_{13} = \delta$ and $\beta_{21} = \beta_{22} = \beta_{23} = -\delta$.

To investigate the power of the LM tests, data were generated under the no-DIF model ($\delta = 0$) and the four data models with effect sizes $\delta = 0.10$, $\delta = 0.25$, and $\delta = 0.50$. For every condition, 100 replications were made, and for every item, the proportion of tests significant at a level of 10% was computed. Five LM statistics were studied, which are denoted $\text{LM}(\beta_{i1})$, $\text{LM}(\beta_{i2})$, $\text{LM}(\beta_{i1}, \beta_{i2})$, $\text{LM}(\beta_{i1}, \beta_{i2}, \beta_{i3})$, and $\text{LM}(\alpha_{i*}, \beta_{i*}, \gamma_{i*})$. The statistics are characterized in Table 7.3.

Notice that the first four statistics are focused at model violations that were actually imposed in the simulations. The last statistic has a more general focus; that is, the general model on which the statistic is based contains discrimination and guessing parameters that were actually equal over groups. The symbols in the last column refer to the results displayed in Figure 7.1, showing the power curves aggregated over the two items with DIF.

In the first panel of Figure 7.1, the panel pertaining to Data Model B1, it can be seen that the power of the $\text{LM}(\beta_{i1})$ test was highest, while the power of the $\text{LM}(\beta_{i2})$ test was virtually equal to the significance level. So the test aimed at the model violation actually imposed was highest, while the power of the test aimed at an absent model violation was equal to the significance level. The order of the power curves of the $\text{LM}(\beta_{i1}, \beta_{i2})$, $\text{LM}(\beta_{i1}, \beta_{i2}, \beta_{i3})$,

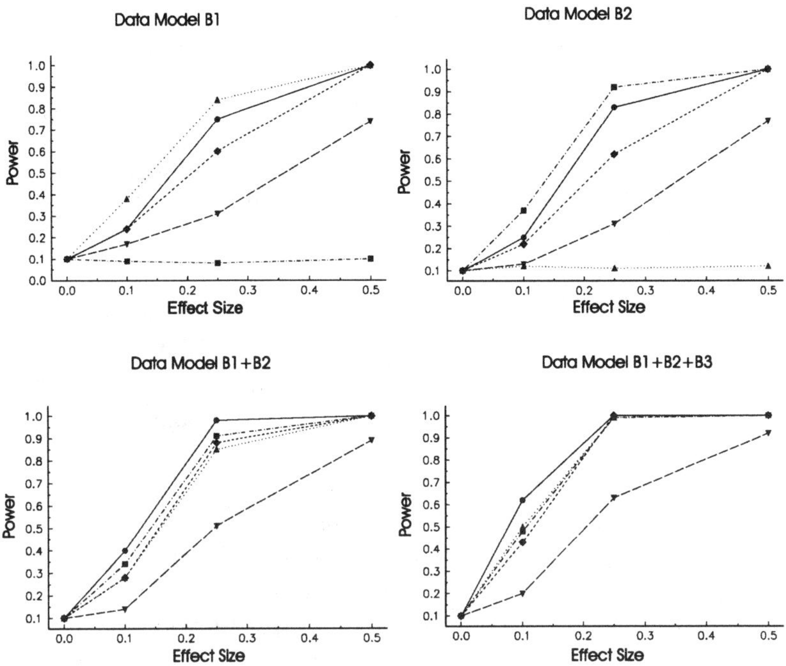

FIGURE 7.1. Power of LM tests: Percentage of hits.

and $\mathrm{LM}(\alpha_{i*}, \beta_{i*}, \gamma_{i*})$ tests was consistent with the degree to which the test was focused on the model violation imposed: the $\mathrm{LM}(\beta_{i1}, \beta_{i2})$ test focuses on two DIF parameters, while the $\mathrm{LM}(\alpha_{i*}, \beta_{i*}, \gamma_{i*})$ test focuses on nine DIF parameters, and with only one DIF parameter present, the power of the former was higher than the power of the latter. The second panel, associated with Data Model B2, shows that imposing a DIF effect β_{i2} resulted in an analogous result: the $\mathrm{LM}(\beta_{i2})$ test had the highest power and the power of the $\mathrm{LM}(\beta_{i1})$ test was approximately 0.10. In the third panel, it can be seen that for the Data Model B1+B2 the $\mathrm{LM}(\beta_{i1}, \beta_{i2})$ test had the highest power. Finally, in the fourth panel, it can be seen that the power of all except the $\mathrm{LM}(\alpha_{i*}, \beta_{i*}, \gamma_{i*})$ test already had a power equal to one for an effect size $\delta = 0.25$. In all analyses, the proportion of false alarms aggregated over the eight items without DIF never exceeded 0.20, and was usually very close to the significance probability.

The decision that an interaction effect might be present cannot be based on the significance of the LM tests alone: significance of $\mathrm{LM}(\beta_{i1}, \beta_{i2}, \beta_{i3})$ is an indication that one of the DIF parameters β_{i1}, β_{i2}, or β_{i3} is not equal to zero, but it cannot be concluded which of the three parameters caused the significant effect. This can be further investigated through computation of effect sizes with the ONR estimator defined by (7.8). To explore the

TABLE 7.4. The means and standard deviations of ONR estimates of absolute values of DIF effects aggregated over two DIF items and 100 replications.

	Effect Size $\delta = 0.25$		Effect Size $\delta = 0.50$	
	Data Model		Data Model	
Parameter	B1+B2	B1+B2+B3	B1+B2	B1+B2+B3
$\mid \beta_{i1} \mid$	0.28	0.28	0.55	0.59
SD($\mid \beta_{i1} \mid$)	0.14	0.13	0.19	0.93
$\mid \beta_{i2} \mid$	0.29	0.27	0.54	0.55
SD($\mid \beta_{i2} \mid$)	0.14	0.12	0.19	0.86
$\mid \beta_{i3} \mid$	0.00	0.22	0.07	0.55
SD($\mid \beta_{i3} \mid$)	0.19	0.17	0.19	0.49

feasibility of this approach, in every replication of the simulation studies with Data Model B1+B2 and Data Model B1+B2+B3, and with effect sizes $\delta = 0.25$ and $\delta = 0.50$, ONR estimates of β_{i1}, β_{i2}, and β_{i3} were computed. The means and standard deviations of these estimates over the 100 replications are displayed in Table 7.4. The ONR estimates in Table 7.4 are aggregated over the two DIF items. Since δ was applied as a positive effect to the first item and as a negative effect to the second item, the absolute values of the ONR estimates were aggregated. The results support the usefulness of the ONR estimates: the estimates of $\mid \beta_{i3} \mid$, for instance, were close to zero for Data Model B1+B2, and close to δ for Data Model B1+B2+B3.

6 A Comparison of the MH and LM Approach

To discard items, or flag items for review, several testing companies have adopted a system developed by the Educational Testing Service (ETS) for categorizing the severity of DIF. This system is both based on the statistical significance and on the magnitude of a DIF index known as MHD (see Zieky, 1993). The index is defined as MHD $= -2.35 \ln(\hat{\alpha}_{MH})$, where $\hat{\alpha}_{MH}$ is an estimate of the Mantel–Haenszel (1959) conditional odds-ratio. Items receive a categorization C, which represents moderate to large DIF, when the absolute value of MHD is at least 1.5 and significantly greater than 1 (at $\alpha = 0.05$). The computations for this procedure are very simple, and the procedure has become a "standard of industry" in all large-scale testing programs. Usually, the items receiving a categorization C are investigated further using more complicated IRT-based methods. The aim of this example is to investigate the relation between the results of the MHD and LM procedures.

TABLE 7.5. Data summary of the mathematics test.

Gender	Race	n	p	Alpha	$\hat{\mu}_\theta$	$\hat{\sigma}_\theta$
Men	White	2290	0.62	0.87	0.00†	1.00†
					(–)	(–)
	Black	403	0.51	0.81	–0.76	0.89
					(0.05)	(0.04)
Women	White	1521	0.60	0.81	–0.15	0.79
					(0.03)	(0.02)
	Black	322	0.51	0.77	–0.76	0.79
					(0.05)	(0.04)

Note: † fixed parameter.

The example is a mathematics test for students from 14 to 18 years old, published by CTB/McGraw-Hill (see also Meijer & Van Krimpen-Stoop, 2001). The test consisted of 50 multiple choice items. Responses of 4536 students were available, together with information on the background variables, gender and race. Some summary statistics of the data are given in Table 7.5. The column labeled p gives the mean proportion of correct responses in the four groups of students. The column labeled Alpha gives Cronbach's alpha as an estimate of the reliability of the test in the four groups. The estimates of the means and standard deviations of the proficiency distributions issued by BILOG-MG are given in the last two columns. Standard errors of these estimates are within parentheses.

Given the total sample size of 4536 students, it is to be expected that the power of both the LM and the MHD statistic is quite high, because an IRT model is always a simplification of actual response behavior, and the number of small but significant model deviations increases with the sample size. In simulation studies, such as the studies reported above, it is usually assumed that only one or two items violate the model while the other items conform to the model perfectly. Therefore, results from simulated data and results from real data are often quite different. Because of this power problem, in a large sample of real data, the primary interest is not in the significance of the tests, but in the magnitude of the effect sizes. Below, a two-step approach to identify DIF is adopted: first, items with significant DIF tests are identified, and next, from this set of items effect sizes are used to evaluate DIF.

Two MH analyses were made: with gender as a background variable and with race as a background variable. For the gender analysis, 26 out of 50 MH tests were significant at a 5% significance level; for the analysis with race as a background variable, 21 out of 50 MH tests were significant. Using the ETS classification system, 12 items received a C classification in the gender-related DIF analysis and 6 in the race-related-DIF analysis. Next,

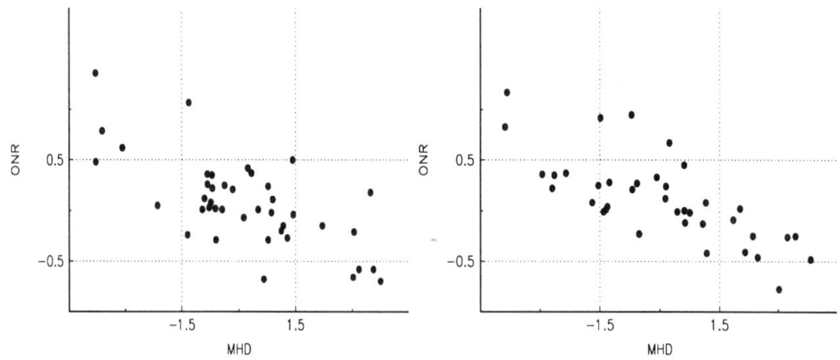

FIGURE 7.2. Plot of MHD statistics against ONR estimates.

an LM analysis was carried out where gender and race were entered as background variables simultaneously. Again using a 5% significance level, this resulted in 19 significant LM($\beta_{i1}, \beta_{i2}, \beta_{i3}$) tests, 21 significant LM(β_{i1}) tests, and 14 significant LM(β_{i2}) tests. With a rather arbitrarily chosen ONR estimate $|\hat{\beta}_{i1}| > 0.50$ as critical region, 9 items were flagged as items with gender DIF. Using an ONR estimate $|\hat{\beta}_{i2}| > 0.50$ as a critical region for race-related-DIF, 6 items were flagged.

To investigate the agreement between the LM and MH procedures in terms of analogously flagged and nonflagged items, the values of the MHD statistics and the ONR estimates obtained in the two analyses were plotted. The result is shown in Figure 7.2. A positive value of the ONR estimate and a negative value of the MHD statistic indicate that, conditional on the ability, the item is more difficult for the focal group than for the reference group. Therefore, the upper left-hand, middle, and lower right-hand sections in Figure 7.2 contain items analogously classified by the two procedures. The agreement between the two procedures in terms of analogously classified items was 88% for the gender-related DIF analysis, and 70% for the race-related DIF analysis. It can be seen that, in general, the results of the two approaches are comparable, but not equivalent.

In both panels of Figure 7.2, one of the 50 items was deleted because its values on MHD and ONR were much more extreme than the analogous values of the other items, and this outlier would have obscured the pictures. Further analysis of this item showed that separate MH analyses with gender and race as background variables, and LM analyses with both background variables entered simultaneously, can certainly lead to quite different conclusions. With the covariates dummy-coded as in (7.4) through (7.6), the ONR estimates of the DIF parameters were $\hat{\beta}_{i1} = -4.68$, $\hat{\beta}_{i2} = -4.22$, and $\hat{\beta}_{i3} = 4.12$. The interpretation of the estimates is that, on this item, the difference in performance between white men and the other three groups is about equal (where the white men perform less well). The estimate $\hat{\beta}_{i3} = 4.12$ signifies that there is no additional effect of both being female and black; the advantage of this group on this item is the same as the advantage of white women and black men. In the separate MH analyses, however, the item was only flagged as a DIF item in the gender-related DIF analysis. So in the culture-related DIF analysis, gender may have acted as a suppressor variable.

7 Discussion

LM tests provide a powerful tool for evaluation of fit to IRT models. Apart from the applications presented above, Jansen and Glas (2001) apply the LM test to evaluate differential test functioning in Rasch's model for speed tests. In the present chapter, the approach is applied to the 3PLM. Both the simulation study and the real data example show that the LM tests for DIF have good power characteristics. In fact, the real data example shows that the power of the test becomes very high with large sample sizes, and even very small model violations result in significant model tests. Therefore, the effect size of the model violation must also be taken into account. It is shown that one Newton–Raphson step estimates of the DIF parameters provide information about the degree to which the item response curve of a focal group is removed from the item response curve of the reference group. In many situations, differences between response curves may be informative to gauge DIF; in many other situations, one may desire information closer to the application considered. For instance, one may want an index for the impact of DIF on ability estimates, or on the proportion of misclassifications in a mastery test. Such indices are an important point for further study.

References

Aitchison, J., & Silvey, S.D. (1958). Maximum likelihood estimation of parameters subject to restraints. *Annals of Mathematical Statistics 29*, 813–828.

Bock, R.D., & Aitkin, M. (1981). Marginal maximum likelihood estimation of item parameters: An application of an EM-algorithm. *Psychometrika, 46*, 443–459.

Bock, R.D., & Zimowski, M.F. (1997). Multiple group IRT. In W.J. van der Linden & R.K. Hambleton (Eds.), *Handbook of modern item response theory* (pp. 433–448). New York: Springer-Verlag.

Camilli, G., & Shepard, L.A. (1994). *Methods for identifying biased test items.* Thousand Oaks, CA: Sage.

Cox, D.R., & Hinkley, D.V. (1974). *Theoretical statistics.* London: Chapman and Hall.

Cressie, N., & Holland, P.W. (1983). Characterizing the manifest probabilities of latent trait models. *Psychometrika, 48*, 129–141.

Efron, B. (1977). Discussion on maximum likelihood from incomplete data via the EM algorithm (by A. Dempster, N. Laird, and D. Rubin). *Journal of the Royal Statistical Society, Series B, 39,* 1–38.

Fischer, G.H. (1993). Notes on the Mantel–Haenszel procedure and another chi-square test for the assessment of DIF. *Methodika, 7,* 88–100.

Fischer, G.H. (1995). Some neglected problems in IRT. *Psychometrika, 60,* 459–487.

Glas, C.A.W. (1998). Detection of differential item functioning using Lagrange multiplier tests. *Statistica Sinica, 8*, 647–667.

Glas, C.A.W. (1999). Modification indices for the 2-PL and the nominal response model. *Psychometrika, 64*, 273–294.

Glas, C.A.W., & Verhelst, N.D. (1995). Tests of fit for polytomous Rasch models. In G.H. Fischer & I.W. Molenaar (Eds.), *Rasch models: Foundations, recent developments, and applications* (pp. 325–352). New York: Springer-Verlag.

Hambleton, R.K., & Rogers , H.J. (1989). Detecting potentially biased test items: Comparison of IRT area and Mantel–Haenszel methods. *Applied Measurement in Education, 2,* 313–334.

Holland, P.W., & Thayer, D.T. (1988). Differential item functioning and the Mantel–Haenszel procedure. In H. Wainer & H.I. Braun (Eds.), *Test validity* (pp. 129–145). Hillsdale, NJ: Erlbaum.

Holland, P.W., & Wainer, H. (1993). *Differential item functioning.* Hillsdale, NJ: Erlbaum.

Jansen, M.G.H., & Glas, C.A.W. (2001). Statistical tests for differential test functioning in Rasch's model for speed tests. In A. Boomsma, M.A.J. van Duijn, & T.A.B. Snijders (Eds.), *Essays on item response theory* (pp. 149–162). New York: Springer-Verlag.

Kelderman, H. (1989). Item bias detection using loglinear IRT. *Psychometrika,54,* 681–697.

Kok, F.G., Mellenbergh, G.J., & van der Flier, H. (1985). Detecting experimentally induced item bias using the iterative logit method. *Journal of Educational Measurement, 22,* 295–303.

Lord, F.M. (1980). *Applications of item response theory to practical testing problems.* Hillsdale, NJ: Erlbaum.

Louis, T.A. (1982). Finding the observed information matrix when using the EM algorithm. *Journal of the Royal Statistical Society, Series B, 44*, 226–233.

Mantel, N., & Haenszel, W. (1959). Statistical aspects of the analysis of data from retrospective studies of disease. *Journal of the National Cancer Institute*, 22, 719–748.

McCullagh, P., & Nelder, J. (1989). *Generalized linear models* (2nd ed.). London: Chapman and Hall.

Meijer, R.R., & Van Krimpen-Stoop, E.M.L.A. (2001). Person fit across subgroups: An achievement testing example. In A. Boomsma, M.A.J. van Duijn, & T.A.B. Snijders (Eds.), *Essays on item response theory* (pp. 377–390). New York: Springer-Verlag.

Meredith, W., & Millsap, R.E. (1992). On the misuse of manifest variables in the detection of measurement bias. *Psychometrika, 57,* 289–311.

Mislevy, R.J. (1984). Estimating latent distributions. *Psychometrika, 49,* 359–381.

Mislevy, R.J. (1986). Bayes modal estimation in item response models. *Psychometrika, 51,* 177–195.

Muraki, E., & Bock, R.D. (1991). PARSCALE: Parameter scaling of rating data [Computer software]. Chicago: Scientific Software.

Rao, C.R. (1947). Large sample tests of statistical hypotheses concerning several parameters with applications to problems of estimation. *Proceedings of the Cambridge Philosophical Society, 44,* 50–57.

Rao, C. R. (1973). *Linear statistical inference and its applications.* New York: Wiley.

Rigdon, S.E., & Tsutakawa , R.K. (1983). Parameter estimation in latent trait models. *Psychometrika, 48,* 567–574.

Rogers, H.J., Swaminathan, H., & Egan, K. (1999, April). *A multi-level approach for investigating differential item functioning.* Paper presented at the Annual Meeting of the NCME, Montreal.

Swaminathan, H., & Rogers , H.J. (1990). Detecting differential item functioning using logistic regression procedures. *Journal of Educational Measurement, 27,* 361–370.

Thissen, D., Steinberg, L., & Wainer, H. (1993). Detection of differential item functioning using the parameters of IRT models. In P.W. Holland & H. Wainer (Eds.), *Differential item functioning* (pp. 67–113). Hillsdale, NJ: Erlbaum.

Wang, W.-C. (in press). Modeling effects of differential item functioning in polytomous items. *Journal of Outcome Measurement.*

Zieky, M. (1993). Practical questions in the use of DIF statistics in test development. In P.W. Holland & H. Wainer (Eds.), *Differential item functioning* (pp. 337–347). Hillsdale, NJ: Erlbaum.

Zimowski, M.F., Muraki, E., Mislevy, R.J., & Bock, R.D. (1996). BILOG-MG: Multiple-group IRT analysis and test maintenance for binary items [Computer software]. Chicago: Scientific Software.

Zwinderman, A.H. (1991). A generalized Rasch model for manifest predictors. *Psychometrika, 56,* 589–600.

Zwinderman, A.H. (1997). Response models with manifest predictors. In W.J. van der Linden & R.K. Hambleton (Eds.), *Handbook of modern item response theory* (pp. 433–448). New York: Springer-Verlag.

8

Statistical Tests for Differential Test Functioning in Rasch's Model for Speed Tests

Margo G.H. Jansen[1] and Cees A.W. Glas[2]

ABSTRACT We consider a latent trait model developed by Rasch for the response time on a set of pure speed tests, which is based on the assumption that the test response times are approximately gamma distributed with known index parameters and scale parameters depending on subject ability and test difficulty parameters. In this chapter, the principle of Lagrange multiplier tests is used to evaluate differential test functioning and subgroup invariance of the test parameters. Two numerical illustrations are given.

1 Introduction

In the area of measurement of educational achievement we usually focus on how well a subject can perform a task, and we neglect the time it takes to complete it. A notable exception can be found in the area of measurement of reading development and, in general, in the area of reading research. Children learning to read are supposed to do so correctly and fluently. Tests developed to measure reading performance, for example, single-word reading tests, tend to have scoring rules where both aspects are scored separately or are combined in some way. Several studies have shown that speed and accuracy in reading are related to speed and accuracy in word retrieval, as demonstrated in various naming tests (see, for instance, Denckla & Rudel, 1974, 1976; Wolf, Bally, & Morris, 1986; Kail & Hall, 1994; Meyer, Wood, Hart, & Felton, 1998). In particular, naming speed deficits are connected with dyslexia. Naming speed tests make use of a variety of visual stimuli, such as letters, numerals, colors, and objects, and

[1]Department of Education, University of Groningen, Grote Rozenstraat 38, 9712 TJ Groningen, The Netherlands; e-mail: *g.g.h.jansen@ppsw.rug.nl*

[2]Department of Educational Measurement and Data Analysis, University of Twente, P.O. Box 217, 7500 AE Enschede, The Netherlands; e-mail: *glas@edte.utwente.nl*

are given in different formats. In the continuous-naming-test the subjects are asked to name, as rapidly as possible, a set of serially presented stimuli and the total time it takes to name all items is recorded. In the discrete stimuli format, the stimuli are presented and timed individually, usually using the computer.

Several of the tests mentioned above can be viewed as speed tests. The definition of a pure speed test used here goes back to Gulliksen (1950; see also Lord & Novick, 1968; Meredith, 1971). By an itemized test of pure speed, we mean a set of easy items administered under strict time-limit conditions. First, the test items are supposed to be of the same type and to depend on a single latent trait. Easy items are items that any member of the subject population should be able to answer correctly given sufficient time. Second, the time limit will be such that no subject will be able to complete all items. Subjects differ in speed, and will therefore complete different numbers of items in any fixed time period. Speed tests can also be administered without imposing a time limit, instructing the subjects to complete the entire test as rapidly as possible. The response variable to record under this condition is the total response time.

Considering the abundance of item response theory (IRT) models for response accuracy, models for response speed are rare and less well developed; see, for instance, Thissen (1983) or Maris (1993). However, the applications of IRT, such as optimal test construction, adaptive testing, test equating, evaluation of differential item functioning, and the many possibilities opened up by the combination of IRT and incomplete item administration designs, can also play an important role in the field of measurement of speed; see, for instance, Visser (1997). But as with IRT models for response accuracy, IRT models for speed test applications are only valid as far as the model holds. Therefore, evaluation of model fit must always be an integral part of IRT analyses, and this is the main topic of this chapter.

More specifically, a statistical test of model fit is presented for evaluation of differential test functioning (DTF) and subgroup invariance of test parameters. The concepts of differential test functioning and subgroup invariance of test parameters in the framework of IRT models for speed tests are the analogues of differential item functioning and subgroup invariance of item parameters in the framework of IRT models for response accuracy. The difference in naming is related to the difference in data format: in measurement of accuracy, responses to individual items are the relevant outcome variable, while in speed measurement the number of items completed or the response time on a number of speed tests is the relevant outcome variable. DTF is a difference of test performance between groups that cannot be attributed to differences in proficiency. It is an important issue because DTF often discriminates against a disadvantaged group on grounds unrelated to the proficiency that should be measured. A stereotype example is a text about football in a test of language comprehension. Items related to this text will probably discriminate against girls. Subgroup invariance is an essential assumption of IRT, entailing that response behavior is described by the same item or test parameters in all subgroups. It can be evaluated by forming subgroups using some background variable. If the background variable is no part of the speed test itself, this procedure is equivalent to assessing DTF. An alternative procedure is to form homogeneous score groups of test takers, but this approach is beyond the scope of this chapter. In the sequel, the terminology related to evaluation of DTF is adopted, but the approach applies to general evaluation of subgroup invariance as well.

This chapter is organized as follows. In the next section, an IRT model for speed tests as developed by Georg Rasch is described. In Sections 3 and 4, a marginal maximum likelihood framework for parameter estimation and a procedure for evaluation of model fit are discussed. Then, in Section 5, the principle of Lagrange multiplier tests is described, which is used to evaluate DTF and subgroup invariance. Finally, in Sections 6 and 7, some numerical illustrations are given.

2 An IRT Model for Response Times

The model considered here is an extension of the Rasch model (Rasch, 1960) for response times on pure speed tests, developed in the context of measuring oral reading speed. Adopting the assumptions of a homogeneous Poisson process, Rasch derived a latent trait model for simple time-limit tests. It is assumed that the probability p, of a subject completing an item in any small time interval, depends on the length of the interval and that, as the interval length goes to zero, the probability of more than one completion occurring in the interval is negligible, compared to either none or one. The probability of completing y items in k time periods can then be approximated by a Poisson distribution. Notice that kp is the expected number of items completed in the total time period. The probability p of an item being completed within a time period need not be the same for all items in the test, as was pointed out by Lord and Novick (1968, p. 490). The really important assumption for the approximation to hold is that p is sufficiently small on average. Lindgren (1968, p. 169), for instance, gives an example which shows that the Poisson approximation of a binomial distribution with $k = 20$ and $p = 0.05$ is quite close. By writing kp as λt in the limit, with λ the average number of items completed per time unit, the Poisson probability function for the number of items y completed in total test time t is written as

$$p(y \mid t) = \frac{(\lambda t)^{y}}{y!} \exp(-\lambda t) \; . \tag{8.1}$$

Speed tests can also be administered without imposing a time limit by instructing the subjects to complete the entire test as rapidly as possible.

The response variable to record under this condition is the total response time. The probability model for the response rate can be obtained directly from (8.1). If the Poisson distribution holds for the number of items completed, then the response times are exponentially distributed with scale parameter λ. The time t_v it takes subject S_v to complete k items is gamma distributed with a known index parameter k and scale or rate parameter λ. Thus, the probability density of the total response time of subject S_v on the k items of the test follows a gamma distribution, given by

$$p(t_v \mid \lambda) = \frac{\lambda^k t_v^{k-1}}{\Gamma(k)} \exp(-\lambda t_v) \; . \tag{8.2}$$

Rasch (1960) assumed that the scale or rate parameter depends on two other parameters, namely, a subject parameter, which might be labeled speed, pace, or ability, and a test easiness parameter. Although we have assumed that items in a speed test are so easy that no member of the target population should fail them, subjects will make mistakes. This might be related to working under pressure. It also seems reasonable to assume that response speed is, at least to a certain degree, chosen voluntary, and some kind of trade-off between speed and accuracy is to be expected. In the following, we consider the Rasch speed model for a set of pure speed tests measuring the same trait, and possibly varying in length and difficulty. For the rate parameters λ, we use Rasch's multiplicative decomposition, which is given by

$$\lambda_{vj} = \theta_v \varepsilon_j \; , \tag{8.3}$$

where θ_v refers to the ability or speed of subject S_v and ε_j to the easiness of test j with k_j items, $j = 1, \ldots, m$. The larger θ_v and ε_j, the shorter is the expected test response time.

3 Estimation

By treating the ability parameters as gamma distributed random variables, Jansen (1997a, 1997b) derived marginal maximum likelihood (MML) estimators for the test parameters and the parameters of the ability distribution. It is assumed that θ has a gamma distribution with mean μ and index parameter τ, which implies a variance μ^2/τ. The density of θ is denoted $g(\theta; \mu, \tau)$, and μ and τ are called population parameters. The assumption of one population distribution for θ can be generalized to the assumption of multiple populations, each having a unique gamma distribution for θ. Rasch's gamma response speed model with the additional assumption that the ability parameters are also gamma distributed is known as the multiplicative gamma response speed model.

To facilitate the description of the Lagrange multiplier test for differential test functioning, the derivation of the MML estimation procedure by Jansen

(1997a, 1997b) is rephrased in an alternative form. Let $\boldsymbol{\xi}$ be the vector of test and population parameters; that is, $\boldsymbol{\xi}$ is the concatenation of μ, τ, and $\boldsymbol{\varepsilon}$. Response times are observed on m tests; the m-dimensional vector $\mathbf{t}_v$ stands for the vector of response times of subject $S_v, v = 1, \ldots, n$. The loglikelihood can be written as

$$\ln L(\boldsymbol{\xi} \mid \mathbf{T}) = \sum_v \ln p(\mathbf{t}_v \mid \boldsymbol{\xi}) \ , \tag{8.4}$$

where $\mathbf{T}$ is the $n \times m$ data matrix with rows $\mathbf{t}'_v$.

To obtain the MML estimation equations, the first-order derivatives of (8.4) with respect to $\boldsymbol{\xi}$ have to be obtained. By adopting an identity due to Louis (1982), which also provides the basis for the EM algorithm for the computation of the MML estimates, these first-order derivatives can be written as

$$\frac{\partial}{\partial \boldsymbol{\xi}} \ln L(\boldsymbol{\xi} \mid \mathbf{T}) = \sum_v E(\boldsymbol{\omega}_v(\boldsymbol{\xi}) \mid \boldsymbol{\xi}, \mathbf{t}_v) \ , \tag{8.5}$$

with

$$\boldsymbol{\omega}_v(\boldsymbol{\xi}) = \frac{\partial}{\partial \boldsymbol{\xi}} \ln p(\mathbf{t}_v, \theta_v \mid \boldsymbol{\xi}) = \frac{\partial}{\partial \boldsymbol{\xi}} [\ln p(\mathbf{t}_v \mid \theta_v, \boldsymbol{\varepsilon}) + \ln g(\theta_v; \mu, \tau)] \ . \tag{8.6}$$

Notice that (8.6) defines $\boldsymbol{\omega}_v(\boldsymbol{\xi})$ as a vector of the derivatives of the simultaneous probability function of the response times $\mathbf{t}_v$ and the ability parameter θ_v with respect to $\boldsymbol{\xi}$.

By taking the first-order derivatives, the elements of $\boldsymbol{\omega}_v(\boldsymbol{\xi})$ are obtained as

$$\omega_v(\mu) = -\frac{\tau}{\mu} + \frac{\tau}{\mu^2} \theta_v \ , \tag{8.7}$$

$$\omega_v(\tau) = 1 - \psi(\tau) + \ln\left(\frac{\tau}{\mu}\right) + \ln \theta_v - \frac{\theta_v}{\mu} \ , \tag{8.8}$$

and

$$\omega_v(\varepsilon_j) = \frac{k_j}{\varepsilon_j} - \theta_v t_{vj} \ , \qquad j = 1, \ldots, m \ , \tag{8.9}$$

where $\psi(\tau)$ indicates the first-order derivative of the gamma function $\Gamma(\tau)$, and k_j is the number of items in test j.

Inserting these three expressions in (8.5), the likelihood equations are obtained as

$$\mu = \frac{1}{n} \sum_v E(\theta_v \mid \mathbf{t}_v, \boldsymbol{\xi}) \ , \tag{8.10}$$

$$1 - \psi(\tau) + \ln\left(\frac{\tau}{\mu}\right) = \frac{1}{n}\Big[\sum_v E\Big(\frac{\theta_v}{\mu} \mid \mathbf{t}_v, \boldsymbol{\xi}\Big) - \sum_v E(\ln \theta_v \mid \mathbf{t}_v, \boldsymbol{\xi})\Big] \ , \tag{8.11}$$

and

$$\frac{k_j}{\varepsilon_j} = \frac{1}{n} \sum_v t_{vj} E(\theta_v \mid \mathbf{t}_v, \boldsymbol{\xi}) \ , \qquad j = 1, \ldots, m \ . \tag{8.12}$$

For identifiability we need one suitable restriction on either the test difficulties or the parameters of the subject distribution. In this chapter, the mean of the first ability distribution is set equal to 1.0.

4 Model Tests

The multiplicative gamma response-speed model allows for a simple and elegant analysis of test-response times. Moreover, it can be used for solving some of the practical problems we encounter in achievement testing, such as test calibration in the context of incomplete designs (Jansen, 1997a). However, the model is based on strong assumptions, and if these are not fulfilled, the results of the analysis may not be valid. From this point of view, the availability of suitable model tests is of prime importance. Unfortunately, methods for testing the fit of the model are, as yet, not very well developed. Likelihood ratio tests and an informal procedure for inspecting the pattern of correlations between weighted test scores were given by Jansen (1997b).

A potential source of misfit can be found in subgroup differences in large heterogeneous target populations. Ideally, the test parameters should not be affected by, for example, cultural or gender differences. Differences in test parameters between groups of respondents, that is, DTF forms a serious threat to the generality of the inferences made on the basis of the parameter estimates. A possible global type of model control is the comparison of test parameter estimates from subsamples of subjects, defined on the basis of some background variable, by means of a likelihood ratio test.

The likelihood ratio statistic for testing the null hypothesis of identical test parameters for all subjects, against the alternative hypothesis of different test parameters for subpopulations, can be calculated as follows. Under the null model, the samples are combined and analyzed simultaneously for the test difficulty estimation, but allowing for differences in the subject distribution parameters of the subpopulations. Under the alternative model, the subgroups are analyzed separately and their loglikelihoods are added. In the case of two subpopulations, the test statistic is

$$\mathrm{LR} = -2(\ln L - \ln L^{(1)} - \ln L^{(2)}) , \qquad (8.13)$$

where L is the likelihood of the null model, and $L^{(g)}$ the likelihood in subgroup $g, g = 1, 2$, under the alternative model. Under the null hypothesis, this statistic is asymptotically χ^2-distributed with degrees of freedom equal to the difference between the number of parameters of the alternative model and the number of parameters of the null model. In the next section, we develop an alternative method for detecting differential test functioning on the level of the individual tests, which does not need estimation of the parameters of alternative models.

5 A Lagrange Multiplier Test for DTF

The principle of Lagrange multiplier (LM) tests (Aitchison & Silvey, 1958) or the equivalent score test (Rao, 1947) has proved to be very useful in the framework of the derivation of tests for violations of IRT models, such as the one-, two-, and three-parameter logistic model, the nominal response model, and the generalized partial credit model (Glas & Verhelst, 1995; Glas, 1998, 1999). The basic idea of the procedure is to test a special model, parameterized by $\boldsymbol{\eta}_1$, by adding parameters $\boldsymbol{\eta}_2$ to build a more general model parameterized by $\boldsymbol{\eta} = (\boldsymbol{\eta}_1', \boldsymbol{\eta}_2')'$ where essential assumptions of the special model are relaxed. The special model can be derived from the general model by setting $\boldsymbol{\eta}_2$ equal to zero. For the LM test, only the parameters of the special model $\boldsymbol{\eta}_1$ have to be estimated, denoted $\hat{\boldsymbol{\eta}}_1$. Let $\mathbf{h}_2(\boldsymbol{\eta})$ denote the partial derivatives with respect to $\boldsymbol{\eta}_2$ of the loglikelihood of the general model. The test is then based on the LM statistic

$$\mathrm{LM} = \mathbf{h}_2(\boldsymbol{\eta})'\boldsymbol{\Sigma}^{-1}\mathbf{h}_2(\boldsymbol{\eta}) , \tag{8.14}$$

to be evaluated at $\boldsymbol{\eta} = (\hat{\boldsymbol{\eta}}_1', \mathbf{0}')'$, where $\boldsymbol{\Sigma}$ denotes the asymptotic covariance matrix of the parameter estimates of the general model (cf. Equation 7.7 in Glas, 2001, and see also Glas, 1999, for details).

Asymptotically, the LM statistic has a χ^2-distribution with degrees of freedom equal to the number of parameters in $\boldsymbol{\eta}_2$ (Aitchison & Silvey, 1958). If the absolute values of the first-order derivatives are large, the fixed parameters are bound to change when they are unconstrained, and the test is likely to be significant; that is, the special model is rejected. If the absolute values of these derivatives are small, the fixed parameters will probably show little change should they be set free; that is, the values at which these parameters are fixed in the special model are adequate and the test is likely not to be significant.

The framework of LM tests is also well suited for developing a statistical test for differential test functioning in the framework of the speed test model. In the following, we derive a LM test for DTF in the context of MML estimation. Consider the case of two populations, say, Population 1 and Population 2.

We define the background variable,

$$x_v = \begin{cases} 0 \, , & \text{if subject } S_v \text{ belongs to Population 1,} \\ 1 \, , & \text{if subject } S_v \text{ belongs to Population 2.} \end{cases} \tag{8.15}$$

Under the null hypothesis, DTF is supposed to be absent; that is, the respondents of different populations of equal ability θ_v have the same expected test scores. This does not imply that the expected scores in the two subpopulations are the same, since the ability distributions of the two populations will usually be different; that is, μ and τ have different values for the two populations. Under the alternative hypothesis, DTF entails that the test parameters differ over populations. This is modeled by introducing a test parameter $\varepsilon_j + x_v\delta_j$. So in Population 1, where $x_v = 0$, the test parameter is equal to the original test easiness parameter ε_j, while in Population 2, where $x_v = 1$, the test parameter becomes $\varepsilon_j + \delta_j$. The additional location parameter δ_j thus denotes the difference between the easiness parameters of test j in Population 1 and Population 2. The probability density of test response t_{vj} under the alternative model is then given by

$$p(t_{vj} \mid x_v, \theta_v, \varepsilon_j, \delta_j) = \frac{[\theta_v(\varepsilon_j + x_v\delta_j)]^{m_j}\, t^{m_j-1}}{\Gamma(m_j)}\, e^{-\theta_v(\varepsilon_j + x_v\delta_j)t_{vj}}\,, \tag{8.16}$$

which can be viewed as a generalization of (8.2).

Under the null hypothesis that no DTF is present, the additional location parameters δ_j are fixed at zero. The corresponding LM statistic is given by (8.14), with $\boldsymbol{\eta}_2 = \delta_j$ and $\boldsymbol{\eta}_1$ a vector of population parameters μ_g and τ_g, $g = 1, 2$, and test parameters $(\varepsilon_1, \ldots, \varepsilon_k)$. The mean of Population 2 is fixed to an arbitrary value of 1.0 to ensure identifiability of the model.

To compute the LM statistic, we need the first-order derivatives (8.6) with respect to δ_j. The derivative $\omega_v(\delta_j)$ is given by

$$\omega_v(\delta_j) = x_v \frac{k_j}{\varepsilon_j + x_v\delta} - x_v\theta_v t_{vj}\,. \tag{8.17}$$

Substituting $\delta_j = 0$, and taking expectations results in

$$h_v(\delta) = x_v\left[\frac{m_j}{\varepsilon_j} - t_{vj}E(\theta_v \mid \mathbf{t}_v, \boldsymbol{\xi})\right]. \tag{8.18}$$

The estimate of the asymptotic covariance matrix of $\mathbf{h}(\boldsymbol{\eta}_2)$, $\boldsymbol{\Sigma}$, can also be based on a function of the first-order derivatives (8.17); for details see Glas (1999). Inserting (8.18) in (8.14) results in a LM statistic with an asymptotic χ^2-distribution with one degree of freedom.

6 A Simulation Study

A small simulation study was carried out to illustrate the approach presented in this chapter. Data were generated conforming to the model described by (8.2) and (8.16). The test lengths were assumed to be known and set to 25 for each test. In each of the simulations, we generated scores for 200 subjects, 100 for a sample from Population 1 and 100 for a sample from Population 2. The small sample size was chosen to resemble the practical applications of the model in reading speed research. Scores were generated for five tests of varying difficulty, the last being the least difficult. A value of $\delta = 0$ was used for a first series of analyses to provide a benchmark. Test bias was introduced by specifying a nonzero value for a δ for the third

TABLE 8.1. Average estimates of the model parameters and their standard errors, over 500 replications, for the case where $\delta = 0.125$ for the third test.

	True Values		Estimates	
Test	ε	δ	$\hat{\varepsilon}$	SE($\hat{\varepsilon}$)
1	0.640	–	0.635	0.029
2	0.800	–	0.792	0.036
3	1.000	0.125	1.050	0.048
4	1.250	–	1.240	0.057
5	1.563	–	1.550	0.071
Population	μ		$\hat{\mu}$	SE($\hat{\mu}$)
1	1.00		$1.00^{\dagger}$	–
2	1.20		1.22	0.077
Population	τ		$\hat{\tau}$	SE($\hat{\tau}$)
1	5.00		5.15	0.745
2	5.00		5.25	0.762

Note: † fixed parameter.

test in Population 2. Two different nonzero values for δ were used, namely, 0.0625 and 0.125. The size of the nonzero bias of the third test was equal to once or twice the estimated standard errors of the difficulty parameter estimates for the given sample size. In all simulations, the subjects were sampled from gamma distributions with different means and equal index parameters. The true parameter values are given in Table 8.1. Note that Population 2 is more able than Population 1, and the biased test is easier in Population 2 than in Population 1.

In the columns under the label Estimates, we find the average parameter estimates and their standard errors based on 500 replications for the case where the third test has an additional location parameter $\delta = 0.125$. Notwithstanding the presence of DTF, in the column labeled $\hat{\varepsilon}$ it can be seen that the parameters of all tests are well recovered, in the sense that all parameters are well within 90% confidence intervals that can be inferred from the given standard errors based on a normal approximation. Furthermore, as expected, the estimate of the parameter of the third test seems to be a compromise between the values in the two populations, which are 1.000 and 1.125, respectively.

In Table 8.2 the results of applying the LM test for DTF can be found. In the second and third columns, labeled LM and Reject %, respectively, we find the values of the LM test statistic averaged over 500 replications, and the percentage of cases in which the null hypothesis of no DTF had to

TABLE 8.2. Results of the LM test for differential test functioning, averaged over 500 replications.

	$\delta = 0.00$		$\delta = 0.0625$		$\delta = 0.125$	
Test	LM	Reject %	LM	Reject %	LM	Reject %
1	1.119	6.2	1.279	8.2	2.008	17.2
2	0.977	5.2	1.272	8.2	1.907	15.4
3	1.010	6.8	4.571	45.8	14.924	94.0
4	0.960	4.2	1.358	9.8	1.777	14.6
5	0.877	3.2	1.230	8.6	1.872	15.2

be rejected using a significance level of 0.05. From the results we may infer that the LM test statistic performs adequately in the case where no bias is present. The rejection rate varied between 3.2 and 6.8%. The subsequent columns contain the same results for the intermediate and the large values of δ. The bias in the third test was detected in 46% of the cases for the intermediate case, and in 94% of the cases for the large case. The rejection rate for the tests for which the null hypothesis holds was also inflated. The rejection rate went up to approximately 9% in the case of an intermediate amount of bias, and to at least 14.6% in the case of a relatively large amount of bias. This inflated rejection rate is due to the fact that the LM statistics for the nonbiased tests depend on the estimates of all parameters, including the estimate of the parameter of the biased test. In practice, however, the inflated rejection rate for the tests without DTF is not very important. This is because the best procedure for detection of DTF is to identify the test with the largest significant value on the LM statistic, remove this test or model DTF by adding a δ parameter to the test parameter, and to run a new LM analysis to evaluate DTF in the remaining items. This process can be iterated until no further DTF is identified.

7 An Empirical Example

The data in the following example are derived from a larger study on the development of reading skills of Dutch primary school pupils (Van den Bos & Lutje Spelberg, 1997). The test material consisted of two tests for single word reading; the first makes use of real and the second of pseudo words. The subjects were instructed to read as fast as possible without making errors (reaction times on individual test items were not available). Response speed was measured by timing the first 50 test items on each test.

TABLE 8.3. Descriptive statistics of two tests of single word reading.

	Girls			Boys		
Test	Mean	SD	n	Mean	SD	n
Real Words	77.6	63.05	47	98.5	84.42	49
Pseudo Words	130.6	85.02	47	155.9	110.08	49

	Normal			Dyslectic		
Test	Mean	SD	n	Mean	SD	n
Real Words	59.1	30.33	71	169.5	98.21	26
Pseudo Words	123.1	144.50	71	241.0	121.52	26

Data on several background variables were also collected. To illustrate the method for detection of test bias presented above, two of these background variables are used. The first background variable is sex, where we expect differences in proficiency but no DTF. The second is a classification in poor readers (dyslectics) and normal readers. Defining the normal readers as Population 1, and the dyslectic pupils as Population 2, we might find indications for DTF. The reason is that the good readers are expected to be faster on both tests, but the real-word test may even be easier for the normal readers, because they have the advantage of recognizing the real words.

The sample consisted of 100, partially incomplete, cases. Table 8.3 contains a number of descriptive statistics of the observed response times of the complete cases, such as means and standard deviations of the reading times per test. It can be seen that the mean time to complete the first 50 items of the reading tests is lower for girls than for boys. Furthermore, it can be seen that the real words have a lower mean response time than the pseudo words, and that normal readers have lower mean response times than dyslectics. Note that these differences are not significant due to the large standard deviations.

Table 8.4 contains estimates of the test easiness parameters, estimates of the parameters of the subject parameter distributions for both groups, and results from the computation of the LM statistics. The estimated standard errors were computed using the observed information matrix. With sex as a background variable, the LM statistic is not significant. When the two populations are defined on the basis of reading skills, the LM statistic becomes significant, and it can be concluded that DTF plays a role here. Notice that, due to the fact that only two tests are used, bias in one test implies bias in the other. So in the present case, DTF cannot be attributed

TABLE 8.4. Results of the estimation and testing procedure.

Test	$\hat{\varepsilon}$	SE($\hat{\varepsilon}$)	Sex Group	$\hat{\mu}$	SE($\hat{\mu}$)	$\hat{\tau}$	SE($\hat{\tau}$)
Real Words	0.897	0.020	Girls	1.00†	–	3.81	0.793
Pseudo Words	0.511	0.011	Boys	0.88	0.076	3.06	0.617
LM = 0.85, $p = 0.356$							

Test	$\hat{\varepsilon}$	SE($\hat{\varepsilon}$)	Dyslexia Group	$\hat{\mu}$	SE($\hat{\mu}$)	$\hat{\tau}$	SE($\hat{\tau}$)
Real Words	0.988	0.022	Normal	1.00†	–	5.46	0.942
Pseudo Words	0.554	0.012	Dyslectic	0.46	0.057	3.16	0.884
LM = 17.74, $p = 0.000$							

Note: † fixed parameter.

to a specific test, and the only conclusion that can be drawn is that the two tests have different measurement properties for the two groups.

8 Discussion

Until recently, a statistical test to assess the fit of the Rasch model for speed tests was not readily available. In this chapter, the Lagrange multiplier test was applied to the multiplicative gamma model. The test statistic described here is aimed at detection of differential test functioning, a type of model violation analogous to DIF in IRT. In a small preliminary simulation study the LM test for DTF was found to be performing adequately. Its performance with data from a larger number of tests and larger sample sizes has not been studied yet.

Essentially, the test presented here is used for the evaluation of subgroup invariance of item parameters. This assumption is tested by forming subgroups using some background variable. In the present case, the background variable is not part of the speed test itself. An alternative procedure is to form homogeneous score groups of test takers. An application of that approach to the two-parameter logistic model was given by Glas (1999) and to the three-parameter logistic model by Glas (2001). An analogous test for the multiplicative gamma model is a point for further research.

Acknowledgments: We wish to thank K.P. van den Bos who gave us permission to use his data for the empirical example.

References

Aitchison, J., & Silvey, S.D. (1958). Maximum likelihood estimation of parameters subject to restraints. *Annals of Mathematical Statistics, 29*, 813–828.

Denckla, M.B., & Rudel, R.G. (1974). Rapid automatized naming of picture objects, colors, letters and numbers by normal children. *Cortex, 10,* 186–202.

Denckla, M.B., & Rudel, R.G. (1976). Rapid automatized naming (RAN): Dyslexia differentiated from other learning abilities. *Neuropsychologica, 14*, 471–479.

Glas, C.A.W. (1998). Detection of differential item functioning using Lagrange multiplier tests. *Statistica Sinica, 8*, 647–667.

Glas, C.A.W. (1999). Modification indices for the 2PL and the nominal response model. *Psychometrika, 64,* 273–294.

Glas, C.A.W. (2001). Differential item functioning depending on general covariates. In A. Boomsma, M.A.J. van Duijn, & T.A.B. Snijders (Eds.), *Essays on item response theory* (pp. 131–148). New York: Springer-Verlag.

Glas, C.A.W., & Verhelst, N.D. (1995). Testing the Rasch model. In G.H. Fischer & I.W. Molenaar (Eds.), *Rasch models: Foundations, recent developments, and applications* (pp. 69–96). New York: Springer-Verlag.

Gulliksen, H. (1950). The reliability of speeded tests. *Psychometrika, 15*, 259–269.

Jansen, M.G.H. (1997a). The Rasch model for speed tests and some extensions with applications to incomplete designs. *Journal of Educational and Behavioral Statistics, 22*, 125–140.

Jansen, M.G.H. (1997b). Rasch's model for reading speed with manifest exploratory variables. *Psychometrika, 62*, 393–409.

Kail, R., & Hall, L.K. (1994). Processing speed, naming speed, and reading. *Developmental Psychology, 30*, 949–954.

Lindgren, B.W. (1968). *Statistical theory* (2nd ed.). New York: MacMillan.

Louis, T.A. (1982). Finding the observed information matrix when using the EM algorithm. *Journal of the Royal Statistical Society, Series B, 44*, 226–233.

Lord, F.M., & Novick, M.R. (1968). *Statistical theories of mental test scores.* Reading, MA: Addison-Wesley.

Maris, E. (1993). Additive and multiplicative models for gamma distributed random variables, and their applications as psychometric models for response times. *Psychometrika, 58*, 445–469.

Meredith, W. (1971). Poisson distributions of error in mental test theory. *British Journal of Mathematical and Statistical Psychology, 24*, 49–82.

Meyer, M.S., Wood, F.B., Hart, L.A., & Felton, R.H. (1998). Selective predictive value of rapid automatized naming in poor readers. *Journal of Learning Disabilities, 31*, 106–117.

Rao, C.R. (1947). Large sample tests of statistical hypothesis concerning several parameters with applications to problems of estimation. *Proceedings of the Cambridge Philosophical Society, 44*, 50–57.

Rasch, G. (1960). *Probabilistic models for some intelligence and attainment tests.* Copenhagen: The Danish Institute of Educational Research. (Expanded edition, 1980. Chicago: The University of Chicago Press.)

Thissen, D. (1983). Timed testing: An approach using item response theory. In D.J. Weiss (Ed.), *New horizons in testing: Latent trait theory and computerized adaptive testing* (pp. 179–203). New York: Academic Press.

Van den Bos, K.P., & Lutje Spelberg, H.C.L. (1997). Measuring word identification skills and related variables. In C.K. Leong & L.M. Joshi (Eds.), *Cross-language studies of learning to read and spell* (pp. 271–281). Boston, MA: Kluwer.

Visser, J.J.C.M. (1997). *Lezen op AVI-niveau. De schaling, normering en validering van de nieuwe AVI-toetskaarten* [Reading on AVI-level: Scaling, norming and validating the AVI-test]. 's-Hertogenbosch: KPC.

Wolf, M., Bally, H., & Morris, R. (1986). Automaticity, retrieval processes, and reading: A longitudinal study in average and impaired children. *Child Development, 57*, 988–1000.

9

Expected Response Functions

Charles Lewis[1]

ABSTRACT Item response theory makes use of what are sometimes referred to as item response functions. Such a function is actually a probability density function (pdf) for the response of an individual to a test item, given the values of certain parameters, classified as item parameters and person parameters (or abilities). In testing, there is typically a calibration phase, in which item parameters are estimated and abilities are ignored. This is followed by an application phase, in which abilities are estimated while conditioning on the estimated values of the item parameters. A Bayesian alternative to treating the item parameters as known quantities involves replacing item response functions with another class of pdfs, referred to as expected response functions (ERFs). The latter take the uncertainty regarding item parameters into account for purposes of estimating abilities. This chapter provides a formal description of ERFs and briefly illustrates their application to the Rasch model for binary item responses.

1 Introduction

In practical applications of item response theory (IRT), estimates of item parameters are obtained in calibration samples and then used as though they were the true values when making inferences about abilities for individuals; see, for instance, Hoijtink and Boomsma (1995). The consequences of ignoring the uncertainty regarding item parameters have been studied in various contexts, and various approaches to addressing the problem have been proposed (for instance, Albert, 1992; Tsutakawa & Johnson, 1990; Tsutakawa & Soltys, 1988). One of these approaches—the use of expected response functions—was described in a talk given by the author at a meeting of the Psychometric Society (Lewis, 1985), and subsequently applied in various contexts by Mislevy and his colleagues (Mislevy, Sheehan, & Wingersky, 1993; Mislevy, Wingersky, & Sheehan, 1994). It is the goal of this chapter to provide a formal description of expected response functions and briefly illustrate their application to the Rasch model (RM).

[1]Educational Testing Service, Rosedale Road, Mailstop 03-T, Princeton, NJ 08541, USA; e-mail: *clewis@ets.org*

Taking a Bayesian point of view regarding the inferences to be made allows incorporation of the uncertainty about item parameters by modifying the item response functions (IRFs) from their original form, as specified by the IRT model being used. These modified curves are called expected response functions (ERFs) because they are the expectations of the original IRFs with respect to the posterior distributions of the item parameters. The general effect of taking expectations is to "flatten" the resulting functions in comparison to the IRFs based on point estimates of the item parameters. Consequently, posterior distributions for abilities will generally show increased variability when ERFs are used in place of the usual IRFs.

2 Theoretical Development of ERFs

In general, let $\boldsymbol{\theta}$ denote a vector of abilities for an individual to whom a set of items is administered. Let $\boldsymbol{\beta}_i$ denote a vector of parameters for the ith item in the set ($i = 1, \ldots, k$). Let $\mathbf{x}_i$ denote the vector response of the individual to the ith item, and let $p(\mathbf{x}_i \mid \boldsymbol{\theta}, \boldsymbol{\beta}_i)$ denote the item response function for the item response model being used. It provides the probability (or density) for response $\mathbf{x}_i$ given $\boldsymbol{\theta}$ and $\boldsymbol{\beta}_i$. If conditional independence is a part of the item response model, then the likelihood function for $\boldsymbol{\theta}$ and the $\boldsymbol{\beta}_i$ may be written as

$$L(\boldsymbol{\theta}, \boldsymbol{\beta}_1, \ldots, \boldsymbol{\beta}_k) = \prod_i p(\mathbf{x}_i \mid \boldsymbol{\theta}, \boldsymbol{\beta}_i) . \tag{9.1}$$

In order to continue with a Bayesian approach to inference, a prior distribution for $\boldsymbol{\theta}$ and the $\boldsymbol{\beta}_i$ is needed. It is natural to assume (prior) independence between the person and item parameters, giving

$$p(\boldsymbol{\theta}, \boldsymbol{\beta}_1, \ldots, \boldsymbol{\beta}_k) = p(\boldsymbol{\theta}) p(\boldsymbol{\beta}_1, \ldots, \boldsymbol{\beta}_k) . \tag{9.2}$$

It is less natural, but required for the development of ERFs, to assume joint independence among the item parameters:

$$p(\boldsymbol{\beta}_1, \ldots, \boldsymbol{\beta}_k) = \prod_i p(\boldsymbol{\beta}_i) . \tag{9.3}$$

It may be noted that, if the person parameters for the individuals in the sample(s) used to calibrate the items were known, then use of a prior distribution satisfying (9.3), combined with conditional independence of the item responses used for calibration, would yield a posterior distribution for the $\boldsymbol{\beta}_i$ that still satisfied (9.3). In practice, only marginal information (point estimates and standard errors) regarding item parameters is routinely available from calibrations, so working with (9.3) would seem to be almost a necessity.

Combining (9.1) through (9.3) using Bayes' Theorem provides the joint posterior distribution for $\boldsymbol{\theta}$ and $\boldsymbol{\beta}_1, \ldots, \boldsymbol{\beta}_k$:

$$p(\boldsymbol{\theta}, \boldsymbol{\beta}_1, \ldots, \boldsymbol{\beta}_k \mid \mathbf{x}_1, \ldots, \mathbf{x}_k) \propto p(\boldsymbol{\theta}) \prod_i p(\mathbf{x}_i \mid \boldsymbol{\theta}, \boldsymbol{\beta}_i)\, p(\boldsymbol{\beta}_i) \; . \tag{9.4}$$

To obtain the marginal posterior distribution for $\boldsymbol{\theta}$, this joint posterior must be integrated with respect to the $\boldsymbol{\beta}_i$. Because of the product structure of (9.4), the result may be written as

$$p\,(\boldsymbol{\theta} \mid \mathbf{x}_i, \ldots, \mathbf{x}_k) \propto p(\boldsymbol{\theta}) \prod_i \int p\,(\mathbf{x}_i \mid \boldsymbol{\theta}, \boldsymbol{\beta}_i)\, p(\boldsymbol{\beta}_i)\, d\boldsymbol{\beta}_i \; . \tag{9.5}$$

Now consider the term inside the product in (9.5), defined as

$$p_i^*\,(\mathbf{x}_i \mid \boldsymbol{\theta}) = \int p\,(\mathbf{x}_i \mid \boldsymbol{\theta}, \boldsymbol{\beta}_i)\, p\,(\boldsymbol{\beta}_i)\; d\boldsymbol{\beta}_i \; . \tag{9.6}$$

This distribution may be interpreted as the expectation of the IRF $p\,(\mathbf{x}_i \mid \boldsymbol{\theta}, \boldsymbol{\beta}_i)$ with respect to the marginal prior distribution of $\boldsymbol{\beta}_i$. In fact, (9.6) is the formal definition of the expected response function (or ERF) for the ith item. With this notation, (9.5) may be written as

$$p\,(\boldsymbol{\theta} \mid \mathbf{x}_1, \ldots, \mathbf{x}_k) \propto p(\boldsymbol{\theta}) \prod_i p_i^*\,(\mathbf{x}_i \mid \boldsymbol{\theta}) \; . \tag{9.7}$$

Compare this expression with

$$p(\boldsymbol{\theta} \mid \mathbf{x}_1, \ldots, \mathbf{x}_k, \hat{\boldsymbol{\beta}}_1, \ldots, \hat{\boldsymbol{\beta}}_k) \propto p(\boldsymbol{\theta}) \prod_i p(\mathbf{x}_i \mid \boldsymbol{\theta}, \hat{\boldsymbol{\beta}}_i) \; , \tag{9.8}$$

where estimated item parameters $\hat{\boldsymbol{\beta}}_i$ have been substituted for the true values in the IRFs. As a practical matter, the only difference between (9.7) and (9.8) is that IRFs in the latter have been replaced by ERFs in the former. In other words, to make inferences about $\boldsymbol{\theta}$ taking the uncertainty in the item parameters into account, the standard (Bayesian) approach may be used, but with a new set of item response functions, namely, ERFs.

3 ERFs for the Rasch Model

In the foregoing, a general form of the IRF, $p(\mathbf{x}_i \mid \boldsymbol{\theta}, \boldsymbol{\beta}_i)$, was used. The main reason for this choice was to emphasize the range of applicability of the ERF concept. For further discussion, however, attention is focused on a very special case, namely, IRFs specified by the logistic form of the Rasch model for binary data. In this case, x_i is a scalar random variable taking on the values 0 and 1, and θ and β_i are also scalars, each with prior densities defined on $(-\infty, \infty)$. Defining

$$P(\theta \mid \beta_i) = \frac{\exp(\theta - \beta_i)}{1 + \exp(\theta - \beta_i)} ,$$

and

$$Q(\theta \mid \beta_i) = 1 - P(\theta \mid \beta_i) ,$$

the Rasch IRF may be written as

$$p(x_i = 1 \mid \theta, \beta_i) = P(\theta \mid \beta_i) ,$$

and

$$p(x_i = 0 \mid \theta, \beta_i) = Q(\theta \mid \beta_i) ,$$

or generally

$$p(x_i \mid \theta, \beta_i) = P(\theta \mid \beta_i)^{x_i} \, Q(\theta \mid \beta_i)^{(1-x_i)} . \tag{9.9}$$

Substituting (9.9) into (9.8) with $\hat{\beta}_i$ in place of β_i gives the usual approximation to the posterior distribution of θ:

$$p(\theta \mid x_1, \ldots, x_k, \hat{\beta}_1, \ldots, \hat{\beta}_k) \propto p(\theta) \prod_i P(\theta \mid \hat{\beta}_i)^{x_i} \, Q(\theta \mid \hat{\beta}_i)^{(1-x_i)} . \tag{9.10}$$

Now define

$$P_i^*(\theta) = \int P(\theta \mid \beta_i) \, p(\beta_i) \, d\beta_i , \tag{9.11}$$

and

$$Q_i^*(\theta) = 1 - P_i^*(\theta) .$$

Then the ERF for the Rasch model may be written as

$$p_i^*(x_i \mid \theta) = P_i^*(\theta)^{x_i} \, Q_i^*(\theta)^{(1-x_i)} , \tag{9.12}$$

and the posterior distribution of θ becomes

$$p(\theta \mid x_1, \ldots, x_k) \propto p(\theta) \prod_i P_i^*(\theta)^{x_i} \, Q_i^*(\theta)^{(1-x_i)} . \tag{9.13}$$

In this expression, the expected response function (for $x_i = 1$), namely, $P_i^*(\theta)$, plays the same role as the standard item response function $P(\theta \mid \beta_i)$, but without dependence on the unknown item parameter β_i.

It may be instructive to study an example of the behavior of expected response functions for the Rasch model. Suppose that the prior distribution for the item parameter β_i is $\mathcal{N}(0, \sigma_\beta^2)$. Values of $P_i^*(\theta)$ may be obtained via

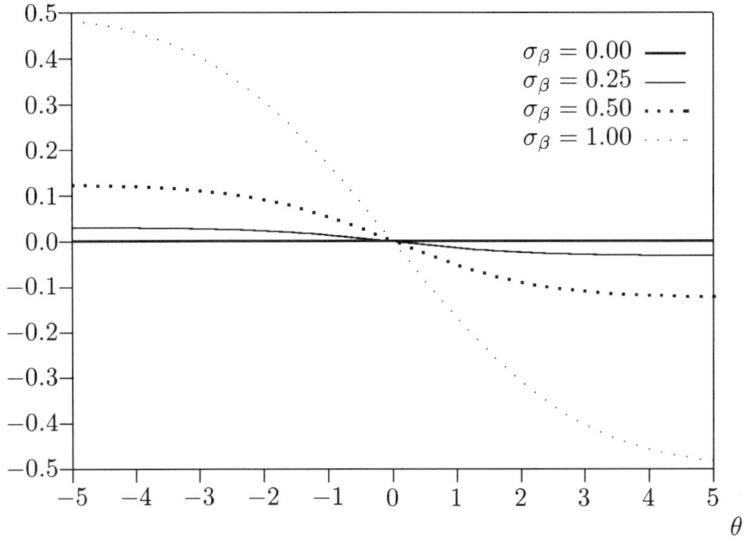

FIGURE 9.1. Logit$[P_i^*(\theta)]$ − logit$[P(\theta \mid 0)]$ based on Normal priors for β_i with $\mu_\beta = 0$ and different values of σ_β.

numerical evaluation of the integral in (9.11). Figure 9.1 compares $P_i^*(\theta)$ based on different values of σ_β to the Rasch $P(\theta \mid \beta_i)$ with $\beta_i = 0$. Specifically, it plots logit$[P_i^*(\theta)]$ − logit$[P(\theta \mid 0)]$ as a function of θ.

All the curves pass through the origin because the $P_i^*(\theta)$ all take the value of 0.5 when $\theta = 0$. The difference curve for $P_i^*(\theta)$ based on $\sigma_\beta = 0.25$ differs from 0 by less that 0.05 logits throughout the range of the figure, suggesting that there will be only a minimal difference between inferences about θ based on the Rasch IRF and those based on this ERF. The fact that all the curves in Figure 9.1 have negative slopes, with a more negative slope for larger values of σ_β indicates that the $P_i^*(\theta)$ are flatter functions than the Rasch $P(\theta \mid 0)$, with the flatness increasing as σ_β increases.

If the ERFs represented in Figure 9.1 could be adequately fit by two-parameter logistic IRFs, the curves in the figure would all be approximately straight lines. In fact, they only appear to be more or less linear for values of θ within about two logits of $\beta_i = 0$. It may be worth noting that, starting with a one-parameter Normal ogive IRF and integrating with respect to a Normal prior for the item difficulty parameter, an ERF is obtained that is exactly a two-parameter Normal ogive IRF with a difficulty parameter of μ_β and a discrimination parameter equal to $(1 + \sigma_\beta^2)^{-1/2}$. As Figure 9.1 illustrates, such a simple result does not hold in the case of the Rasch model.

Examination of $P_i^*(\theta)$, while informative, does not directly address the question of the impact of uncertainty about item parameters on inferences about θ. For this purpose, it is appropriate to directly evaluate posterior

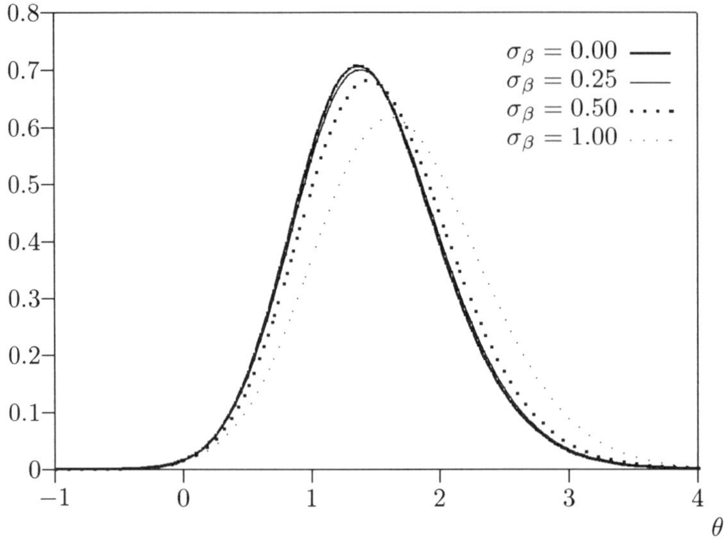

FIGURE 9.2. Posterior densities for θ using different ERFs for a test with $k = 20$ equivalent items, $r = 16$ correct responses ($p_r = 0.8$), and a Normal prior for θ with $\mu_\theta = 0$ and $\sigma_\theta = 5$.

distributions for θ based on ERFs, as given in (9.13). Consider a very simple case, where the k items in a test all have the same prior density for their item parameter. Specifically, suppose that the β_i are i.i.d. $\mathcal{N}(0, \sigma_\beta^2)$. This implies that the $P_i^*(\theta)$ are identical, so that the test items are effectively equivalent. As a result, the posterior density (9.13) may be simplified to

$$p(\theta \mid x_i, \ldots, x_k) \propto p(\theta) P_i^*(\theta)^r Q_i^*(\theta)^{k-r} , \tag{9.14}$$

where $r = \sum_i x_i$ denotes the number of correct responses.

Figure 9.2 plots four examples of the density given in (9.14), using the same set of ERFs that formed the basis for Figure 9.1. The prior used for θ is $\mathcal{N}(0, 25)$, chosen to have a minimal impact on the posterior, so that the effects of different ERFs may be seen more clearly. The test on which the posteriors are based is taken to have $k = 20$ items, of which $r = 16$ have been answered correctly (for a proportion of correct responses given by $p_r = r/k = 0.8$). The density corresponding to $\sigma_\beta = 0.00$ is that arising from using the Rasch IRF. When $\sigma_\beta = 0.25$, the resulting posterior for θ is hardly distinguishable from the Rasch result, thus confirming the impression gained from Figure 9.1 regarding the impact of this degree of uncertainty about the item parameters. As σ_β increases, two effects on the posterior density for θ are evident: both its mean and its variance increase. With more uncertainty about the item parameters, it is perhaps intuitively reasonable that there should be more posterior uncertainty about θ, hence

TABLE 9.1. Posterior means and standard deviations (*in italics*) for θ as functions of σ_β, p_r, and k, with $\mu_\beta = 0.0$, prior mean $\mu_\theta = 0.0$, and prior standard deviation $\sigma_\theta = 5.0$.

		σ_β			
p_r	k	0.00	0.25	0.50	1.00
0.6	10	.441	.446	.460	.511
		.674	*.682*	*.706*	*.791*
	20	.423	.429	.444	.501
		.467	*.473*	*.491*	*.554*
	40	.414	.420	.437	.495
		.326	*.331*	*.344*	*.390*
0.8	10	1.539	1.553	1.595	1.749
		.852	*.858*	*.876*	*.949*
	20	1.463	1.480	1.527	1.700
		.582	*.587*	*.602*	*.659*
	40	1.425	1.442	1.493	1.675
		.404	*.407*	*.418*	*.461*

a greater posterior variance. The effect of that uncertainty on the posterior mean may be less clear. In the simple case leading to (9.14), and disregarding the effect of the prior for θ (which was chosen to be relatively uniform in the range of interest), the posterior mode is given by the solution to the equation

$$P_i^*(\theta) = p_r \; . \tag{9.15}$$

As $P_i^*(\theta)$ becomes flatter with p_r fixed (and not equal to 0.5), the value of θ that satisfies (9.15) becomes more extreme, relative to the threshold value for $P_i^*(\theta)$. For $p_r > 0.5$, this means that the posterior mode will increase.

A summary of the effects of varying σ_β, k, and p_r on the posterior mean and standard deviation for θ is provided in Table 9.1. The row of the table corresponding to $p_r = 0.8$ and $k = 20$ gives the values for the densities appearing in Figure 9.2. The pattern of increasing values of the mean and standard deviation as a function of σ_β, seen in the figure and in this row of the table, is repeated in the remaining rows. In addition, the usual effect of more extreme scores, namely, more extreme means and larger standard deviations, is clearly present in all comparisons between the upper and lower panels of Table 9.1. Within each panel, the effect of increasing test length k is to reduce the standard deviation and produce a slightly less extreme mean. This latter result does not occur with the posterior mode, as can be seen from an inspection of (9.15), and is associated with a slight decrease in the skewness of the posterior density as k increases. Note that the greater test lengths, at least in the range studied, do not appear to reduce the relative effects of increasing uncertainty about the item parameters. The fact that increasing k has no effect on the posterior mode is clear from an inspection of (9.15). The reasons for the relative stability of the ratios of posterior standard deviations for θ based on ERFs to those based on IRFs are not as clear and require further study.

To summarize the most important result of this case study, the effects of uncertainty regarding item parameters in the Rasch model on the posterior distribution of θ are a more extreme mean and a larger standard deviation. In this example, the effects are negligible for $\sigma_\beta = 0.25$, noticeable for $\sigma_\beta = 0.50$, and substantial for $\sigma_\beta = 1.00$. It should be noted that with calibration sample sizes greater than 100, posterior standard deviations for Rasch item parameters are typically less than 0.25. In other words, concern about the effect of item parameter uncertainty on inferences about θ in the case of the Rasch model should probably be confined to cases where the items were calibrated using responses from fewer than 100 individuals. This conclusion supports the prediction made by Hoijtink and Boomsma (1995):

> It may be that taking uncertainty about the item parameter estimates into account is not all that important in the RM (p. 59).

Acknowledgments: The author expresses his sincere appreciation to Liz Brophy for her skill in preparing the manuscript for this chapter. He also acknowledges the many valuable suggestions and patience of the editors, as well as the very helpful comments provided by anonymous reviewers of an earlier version of the text. Without the contributions of all these individuals, a published version of the author's thoughts on expected response functions would still be a fantasy, rather than the reality it has become.

References

Albert, J.H. (1992). Bayesian estimation of normal ogive item response curves using Gibbs sampling. *Journal of Educational Statistics, 17*, 251–269.

Hoijtink, H., & Boomsma, A. (1995). On person parameter estimation in the dichotomous Rasch model. In G.H. Fischer & I.W. Molenaar (Eds.), *Rasch models: Foundations, recent developments, and applications* (pp. 67–84). New York: Springer-Verlag.

Lewis, C. (1985). *Estimating individual abilities with imperfectly known item response functions.* Paper presented at the Annual Meeting of the Psychometric Society, Nashville, TN.

Mislevy, R.J., Sheehan, K.M., & Wingersky, M.S. (1993). How to equate tests with little or no data. *Journal of Educational Measurement, 30,* 55–78.

Mislevy, R.J., Wingersky, M.S., & Sheehan, K.M. (1994). *Dealing with uncertainty about item parameters: Expected response functions* (Research Report, 94–28–ONR). Princeton, NJ: Educational Testing Service.

Tsutakawa, R.K., & Johnson, J. (1990). The effect of uncertainty of parameter estimation on ability estimates. *Psychometrika, 55,* 371–390.

Tsutakawa, R.K., & Soltys, M.J. (1988). Approximation for Bayesian ability estimation. *Journal of Educational Statistics, 13,* 117–130.

10

A Logistic IRT Model for Decreasing and Increasing Item Characteristic Curves

Edwin L. Klinkenberg[1]

ABSTRACT In item response theory, dominance relations are modeled by cumulative models, and proximity relations by unfolding models. Usually cumulative models are used for the measurement of latent abilities and unfolding models for the measurement of latent attitudes and preferences. The distinction between both types of measurement models is best represented by the shape of the item characteristic curve which is monotone increasing in the cumulative model and single-peaked in the unfolding model. A boundary case is a situation in which some items are monotone decreasing and others are monotone increasing. In this chapter, an extension of the one-parameter logistic model is proposed for that situation. It is shown that the parameters of the proposed model can be estimated by a simple data transformation. The model is illustrated using an empirical data set.

1 Introduction

In *A Law of Comparative Judgement* Thurstone (1927) introduced two basic mechanisms for subjects responding to items. He made a distinction between what is presently known in item response theory (IRT) as the dominance relation and the proximity relation. Basically, this distinction has led to the development of two classes of models that are nowadays widely applied in psychometric research.

The first class of models is the dominance-relation-based cumulative models that are known from their application in ability measurement. In probabilistic unidimensional versions of this model like the Rasch model (see, e.g., Molenaar, 1995), the response of a subject to an item is modeled by a monotone increasing item characteristic curve (ICC). Given the loca-

[1]Department of Psychiatry and Neuropsychology, Maastricht University, P.O. Box 616, 6200 MD Maastricht, The Netherlands; e-mail: *e.klinkenberg@np.unimaas.nl*

tion of an item on the latent trait (i.e., the item difficulty), the increasing ICC denotes the increasing probability of mastering the item with increasing ability. By the use of several items that differ in their difficulty, the grid of a measurement instrument is constituted by which measurement of some cumulative latent trait such as ability becomes possible.

The class of models that postulates a proximity relation contains the unfolding models which are applied in the measurement of attitudes and preferences. The key assumption of these models is that a subject's response to an item is a function of the *absolute* location difference. In the probabilistic unidimensional PARELLA model (Hoijtink, 1990) this relation is modeled by a single-peaked ICC, representing the decreasing probability of a positive response given the increasing distance between subject location and item location. That is, the larger the distance between subject and item, the less likely it is that a subject will agree with the item. As for the cumulative models, the use of several items that differ in location on the attitude dimension forms a measurement grid.

Although the theoretical distinction between both classes of models is generally associated with a distinction in their object of measurement, it is by no means exhaustive. For the case of attitude measurement, Jansen (1983) stated that "attitude theory is a definitorial jungle" (p. 45), and demonstrated how the Rasch model can be used for the measurement of latent attitudes. Another example is the use of items that are negatively or positively formulated towards an attitude in some controversial issue. Although generally accepted, this type of item construction affects the interpretation of the response data and the choice of a model. In this chapter a theoretical investigation into this type of item construction is presented and an extension of a cumulative model that can be used for these data is proposed. Finally, the proposed model is applied to empirical data and compared to the analysis with an unfolding model.

2 Attitudes on Issues

The definition of a latent trait and the construction of indicative items are the most important steps in the construction of a measurement instrument. Besides constructing indicative items, a researcher has to make sure that items are constructed that represent different levels of the latent trait to achieve discrimination among subjects. Contrary to ability measurement, where the construction of increasingly difficult items is usually straightforward, the construction of items that are indicative of different levels of some attitude is often complicated by its social and psychological connotation. Item construction is especially complicated when an attitude concerns a controversial issue that appeals to strong social and political beliefs, such as capital punishment or environmental conservation. The bipolar nature of such issues makes it easy to formulate items that represent either an extreme position in favor of or against the issue. However, it is generally much more difficult (and in case of capital punishment maybe even impossible) to formulate items that represent more moderate positions on such issues.

Theoretically, the absence or lack of moderately formulated items has important implications for the properties of a measurement instrument. In terms of item locations, these items constitute the area of the measurement grid between items that are more extremely formulated in favor of and against the issue. Depending on the strength of the bipolar issue, the lack of moderately located items will therefore constitute an increasingly large gap in the measurement grid. This gap seriously affects the measurement properties of the instrument when a majority of subjects is inclined to hold more or less neutral positions on the issue at stake, that is, moderate locations on the latent trait. In that case, a considerable part of the distribution of subject locations is not discriminated by item locations, hence, measurement between several moderate levels of the attitude is not achieved.

The use of more or less extremely located items has theoretical implications for the interpretation of the latent trait. When assuming single-peaked ICCs, where the probability of a positive response decreases with the absolute difference between subject and item location, a negative response only indicates that the subject is too far away from the item. The subject may be located on either side of the item, also on the more extreme—implausible—side. Consider, for instance, an item that expresses a strong position in favor of capital punishment. Disagreement with this item, for reasons that it is not extreme enough, postulates a meaningful interpretation of the latent trait beyond this extreme item. In many applications, the assumption of nondecreasingness of the ICCs at the extreme sides of the scale is sufficient; that is, it is not necessary to assume single-peakedness too. In our example, this implies that an increasing ICC can be assumed. In general, increasing ICCs are assumed for the right side of the scale (indicative of an attitude in favor of the issue measured by the latent trait), whereas decreasing ICCs are assumed for the left side of the scale (indicative of an attitude against the issue measured by the latent trait).

Recent research on the behavior of scaling data indicates that the choice for single-peaked or monotone ICCs should be regarded with caution to avoid wrongful representations of the underlying response mechanisms. An example of such a representation is provided by Van Schuur (1998), who demonstrated that any two arbitrary cumulative scales can be represented by one unfolding scale. The opposite situation is described by Van Schuur and Kiers (1994) who showed that factor analysis of unfolding data resulted in two dimensions. Post, Van Duijn, and Van Baarsen (2001) confirm these findings in an application, comparing the results of analyses of IRT models with single-peaked tracelines with those from analyses of IRT models with monotone ICCs.

In this chapter a unidimensional model is developed that allows decreasing and increasing ICCs for data that do not conform to an unfolding structure.

3 The Signed One-Parameter Logistic Model

A cumulative model can be extended to a model in which an ICC is allowed to be either monotone decreasing or monotone increasing. This extension is made by allowing the slope parameter of a parametric cumulative model to take on negative values. Consider the dichotomous one-parameter logistic model (OPLM) by Verhelst and Glas (1995), defining increasing ICCs:

$$P(X_{vi} = 1 \mid \theta_v, \beta_i, \alpha_i) = \frac{\exp[\alpha_i(\theta_v - \beta_i)]}{1 + \exp[\alpha_i(\theta_v - \beta_i)]}\,, \tag{10.1}$$

where X_{vi} is the response of subject S_v, $v = 1, \ldots, n$, to item I_i, $i = 1, \ldots, k$; θ_v is the location of subject S_v, β_i is the difficulty of item I_i, and α_i is the slope of item I_i which expresses the amount of discrimination that is achieved by the item. Both location parameters, θ_v and β_i, are estimated from the data; the slope α_i, however, is a *pseudo parameter* to which, prior to an analysis, the researcher has to assign *positive* integer values. Verhelst and Glas (1995) provided a data-based method to obtain these slope values, and they developed goodness-of-fit tests to evaluate their adequacy.

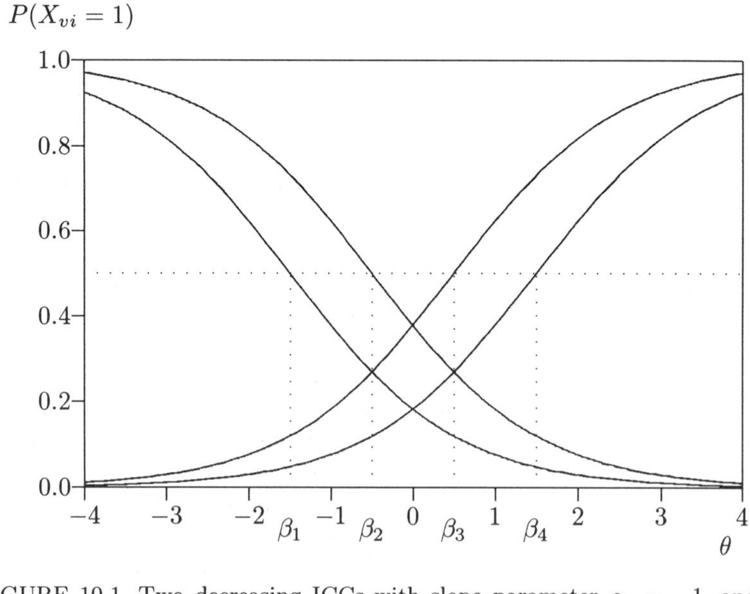

FIGURE 10.1. Two decreasing ICCs with slope parameter $\alpha_i = -1$, and two increasing ICCs with slope parameter $\alpha_i = 1$.

TABLE 10.1. A data matrix with the perfect responses of five subjects to two decreasing and two increasing items; 1/0 denoting agreement/disagreement.

	Item			
Subject	1	2	3	4
1	1	1	0	0
2	0	1	0	0
3	0	0	0	0
4	0	0	1	0
5	0	0	1	1

Model (10.1) can be extended by allowing α_i to take on negative values as well. In this way decreasing ICCs can be created. See Figure 10.1, in which two decreasing ICCs ($\alpha_i = -1$) and two increasing ICCs ($\alpha_i = 1$) are presented; the parameters β_i, $i = 1, \ldots, 4$, denote the four item locations in this figure. These ICCs either express the decreasing or increasing probability of agreement to items with increasing values of θ_v. Thus, the model can be used for attitude measurement.

As an example of this situation, a data matrix is given in Table 10.1, showing perfect response patterns of five subjects to the four ordered items in Figure 10.1; 1/0 denoting agreement/disagreement. The ordering of the subjects is determined by two factors: the number of 1s in the response patterns and the slopes of the items for which there is agreement. In terms of a bipolar issue, disagreement with all items (Subject 3) indicates a neutral or indifferent position, reflected by a location somewhere in the middle of the latent trait (between Items 2 and 3). Agreement with any item indicates a position that deviates, to the left or right, from neutral, depending on the number of items for which there is agreement. Agreement with Item 2 (Subject 2) denotes a moderate position on the theoretical left side of the issue, reflected by a location somewhere to the left of Item 2 and to the right of Item 1. However, agreement with both Items 1 and 2 (Subject 1) denotes a stronger position on the left side of the issue, reflected by a location to the left of Item 1. The opposite reasoning holds for a moderate and strong position on the theoretical right side of the issue (Subjects 4 and 5). Note that the response patterns 0110 and 1111 are not included in Table 10.1; they are perfect in an unfolding model but not in the signed OPLM.

An estimation procedure for the proposed signed OPLM is obtained by reversing the decreasing ICCs into increasing ICCs. To achieve this, the 0s in the first two columns of Table 10.1 are recoded into 1s, and the 1s into 0s. For Table 10.1 this item reversal results in a scalogram structure that governs the ICCs given in Figure 10.2. Note that the proposed data

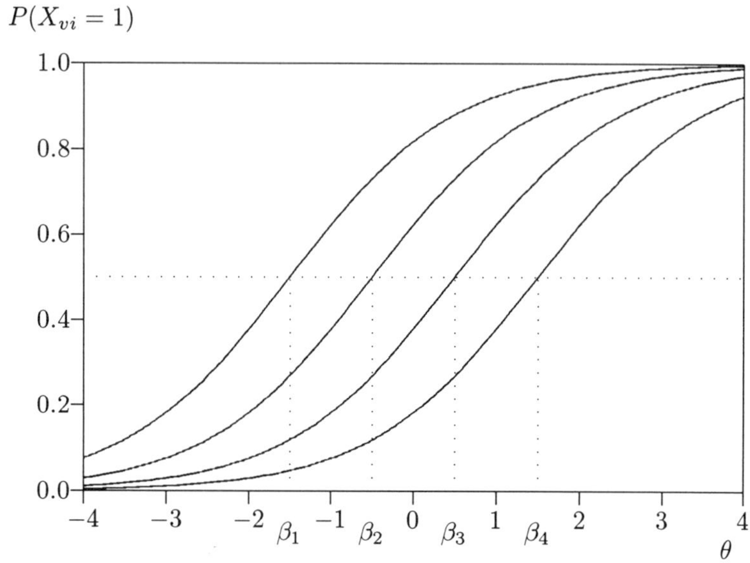

FIGURE 10.2. Four increasing ICCs with slope parameter $\alpha_i = 1$.

transformation does not change the location of Items 1 and 2, because instead of modeling $P(X_{vi} = 1|\theta_v, \beta_i, \alpha_i)$ the complementary ICC, $P(X_{vi} = 0 \mid \theta_v, \beta_i, \alpha_i)$, is modeled, which intersects the former ICC at the point that determines its location. This data transformation makes the proposed extension of the OPLM equivalent to the OPLM introduced by Verhelst and Glas (1995). A proof of the equivalence between both models is given in the Appendix (see Section 7), showing that the estimation equations of the signed OPLM can be rewritten as the estimation equations of the OPLM. Because of the equivalence of the estimation equations the OPLM software (Verhelst, Glas, & Verstralen, 1995) can be used for parameter estimation, and for the goodness-of-fit tests of the model and the ICCs.

Note that the proposed reversal of item scores is only permitted if the signs of the slopes are known prior to an analysis, as may be obtained from the definition of the latent trait and the formulation of the items. Under this restriction the content of the items is not reversed and the interpretation of the latent trait is unaffected. Reversal of item scores without prior knowledge about the signs of the slopes will change the content of the items when the reversal is not connotatively consistent with the unreversed item score (Chang, 1995). If the shape of the ICC is not monotone increasing or decreasing, the reversed item score will lead to different results of the scale analysis, as was demonstrated by Post et al. (2001) and Chang (1995). Moreover, in that case, item reversal will affect the reliability of the measurement instrument and possibly render it invalid (Schriesheim, Eisenbach, & Hill, 1991; Schriesheim & Hill, 1981).

4 The PARELLA Model

As indicated in the introduction, the PARELLA model (Hoijtink, 1990) is a parametric unfolding model that can be applied to dichotomously scored items for which a proximity relation is assumed. A single-peaked ICC represents the decreasing probability of a subject's agreement with an item with an increasing distance between subject location and item location:

$$P(X_{vi} = 1 \mid \theta_v, \beta_i, \gamma) = \frac{1}{1 + |\theta_v - \beta_i|^{2\gamma}}, \tag{10.2}$$

where the interpretation of X_{vi}, θ_v, and β_i is the same as in the OPLM (10.1). In addition to the parameters for subject and item locations, the PARELLA model incorporates an overall parameter γ that affects the shape of the single-peaked ICCs. This parameter, referred to as the *power* parameter, determines the steepness of the ICC. Note that this interpretation of the power parameter is equivalent to the one given for the slope parameter in the OPLM. PARELLA, however, distinguishes only one power parameter, whereas the (signed) OPLM features as many slope parameters as there are items. The steepness of the ICCs in PARELLA gives an indication of the degree of randomness of a subject's choice. The larger γ, the more the response is determined by the distance between subject location and item location, and thus the more deterministic the subject's choice is. For large γ, the ICCs of the PARELLA model approach the ICCs of Coombs' deterministic parallelogram model (Coombs, 1964). For a smaller value of γ, however, the response is determined to a lesser degree by the subject-item distance and is more "random." A common interpretation of this randomness is that other variables such as subject characteristics are needed to explain the subject's item response better.

For estimation of the parameters of the PARELLA model, Hoijtink (1990) proposed the use of a marginal maximum likelihood (MML) procedure (Bock & Aitkin, 1981), as is incorporated in the PARELLA program (Hoijtink, 1993a). In this procedure a discrete distribution $g(\theta)$, approximated by a step function with known nodes B_q and weights π_q, $q = 1, \ldots, Q$, is estimated in such a way that the number of parameters does not increase with the number of subjects n. In this procedure the subject parameters are replaced by the nodes B_q, which changes the interpretation of (10.2) to the probability that a subject located at node B_q will agree with item I_i. Simulation results (Hoijtink, 1990) indicate that the estimates of the PARELLA model are consistent and unbiased. Experience with this estimation procedure using empirical data (Hoijtink, 1990, 1991) shows that it functions adequately and that a limited number of nodes ($Q = 10$) is usually sufficient for approximating $g(\theta)$.

Besides an estimation method, a test for the fit of the ICCs (Hoijtink, 1990), and an overall and item-level test for differential item functioning (DIF) are available (Hoijtink & Molenaar, 1992). The test on the fit of an

TABLE 10.2. The items of the Traffic questionnaire. The polarity in column 3 denotes whether an item is negatively (−) or positively (+) formulated towards car use.

Item		Polarity
1	People who keep driving a car are not concerned with the future of our environment.	−
2	The environmental problem justifies a tax burden on car driving so high that people quit using a car.	−
3	Putting a somewhat higher tax burden on car driving is a step in the direction of a healthier environment.	−
4	Car users should have to pay taxes per mile driven.	−
5	A cleaner environment demands sacrifices such as decreasing car usage.	−
6	It is better to deal with other forms of environmental pollution than car driving.	+
7	Considering the environmental problems, everybody should decide for themselves how often to use the car.	+
8	Technically adapted cars do not constitute an environmental threat.	+
9	Instead of environmental protection measures with respect to car use, the road system should be extended.	+
10	Car use cannot be abandoned. Some pressure on the environment has to be accepted.	+

ICC is based on the comparison of the expected and observed ICC, that is, the expected ICC given the model with the parameters inserted, and the observed ICC reconstructed from the data *and* the estimated parameters. For each node the observed and expected proportion of subjects that agree with the item is determined, after which a nodewise comparison is summarized in the SUM statistic. Although the theoretical distribution of this statistic is unknown, an extensive simulation study (Hoijtink, 1990) indicates that values smaller than 1 denote an adequate similarity between the observed and expected ICC. For values that exceed 1, the nodewise comparison indicates a substantial dissimilarity between the observed and expected ICC, implying that corresponding items should be removed from the scale.

5 The Traffic Data

The data used in this chapter concern the measurement of attitude in the car–environment issue (Doosje & Siero, 1991). These so-called Traffic Data were collected to study the effect of an information campaign (explaining the effect of car use on the environment) on the attitudes of subjects in the car–environment issue. Before and after the campaign, a group of subjects was asked to respond to the 10 statements listed in Table 10.2. The original samples consisted of 1042 and 675 subjects, respectively, from which two random samples of 300 subjects were drawn.

The items of the Traffic questionnaire (Table 10.2) are either formulated positively (Items 1 to 5) or negatively (Items 6 to 10) towards car use. The extremity of the formulations differs from a strong position against car use in favor of the environment (Items 1 and 2) to the opposite position that car use does not pose an environmental threat (Items 8 and 9).

Previous analyses of the Traffic data, as reported in the psychometric literature, observed some of the features that were hypothesized in Section 2. Although not reported as such, a gap in the measurement grid is observed for an application of the PARELLA model (Hoijtink, 1993b) and a model proposed by Verhelst and Verstralen (1993). Indications about the decreasingness and increasingness of the ICCs are reported for applications of the nonparametric MUDFOLD model (Van Schuur, 1993) and a nonparametric model proposed by Post and Snijders (1993a).

5.1 Application of the PARELLA Model

The PARELLA analysis of the Traffic Data resulted in a joint scale on which the pre- and postinformation groups can be compared. The overall likelihood ratio test for DIF gave LR $= 14.51$ with $df = 9$, which is not significant at the $\alpha = 0.05$ level and indicates that the estimated item locations and power parameter are the same in both groups. This result is confirmed by the item-level DIF tests (not presented here) and is consistent with results from a similar analysis by Hoijtink (1993b).

The results of the PARELLA analysis are given in Table 10.3. Item 10 was removed from the analysis because a preliminary analysis indicated that its ICC was too flat and its SUM statistic exceeded the critical value of 1. Considering the content of Item 10 this is not surprising because subjects against, as well as subjects in favor of car use may both agree with this item. For the remaining items the values of the SUM statistic are all below the critical value of 1, indicating that the ICCs, as predicted by the PARELLA model, adequately resemble the observed ICCs. Further information about the adequacy of the PARELLA model is obtained from the value of the power parameter. Following the guidelines of Hoijtink (1990) concerning the overall power parameter, it can be concluded that, since the estimated value is substantially larger than one, the distance between subject and item is the main determinant in the responses from the subjects to the items of the Traffic questionnaire.

Considering the order of the estimated item locations (column 2 in Table 10.3) the scale ranges from an attitude strongly against car use to protect the environment to an attitude that cars do not constitute an envi-

TABLE 10.3. Item location and power estimates, with corresponding standard error estimates, for the Traffic Data (total $n = 600$); the SUM statistic serves as a comparison of observed and expected ICCs.

Item	$\hat{\beta}_i$	SE($\hat{\beta}_i$)	SUM
1	-1.40	.05	.37
2	-1.40	.05	.77
3	-.88	.03	.22
4	-.84	.03	.30
5	-.37	.02	.20
6	1.01	.02	.34
7	1.03	.02	.62
8	1.22	.02	.17
9	1.63	.03	.52
$\hat{\gamma}$	1.80	.05	

ronmental threat. Note that the first five items on the dimension are those formulated against car use and the next four items those formulated in favor of car use.

To evaluate the measurement properties of the Traffic scale, the distributions of the pre- and postinformation groups are compared to the distribution of the item locations. This comparison is given in Figure 10.3, where the distributions of the pre- and postinformation groups are represented by histograms drawn in bold and normal typeface, respectively. The distribution of the item locations is given by the $\times$s on the horizontal axis. Comparing the histograms to the distribution of the item locations, it can be observed that the bulk of the subjects are located in the "gap" between Item 5 ($\hat{\beta}_5 = -.37$) and Item 6 ($\hat{\beta}_6 = 1.01$). For the preinformation group this involves about 83%, and for the postinformation group about 90% of the subjects. From a measurement point of view this situation is not optimal because in both groups the vast majority of subjects seem to be a homogeneous group that cannot be distinguished with respect to the latent trait (i.e., their attitude towards the car is that of an environmental threat). So, although the PARELLA model provides an adequate statistical description of the data it does not provide the necessary discriminating power that is required of a measurement instrument.

As was hypothesized in Section 2, the extremity of the item locations relative to the subject distributions indicates a bipolar latent trait for which decreasing and increasing ICCs may provide a better description. Although the tests on the adequacy of the ICCs indicate a similarity between the observed and expected ICC, they give no indication about the single-peaked shape of the ICCs. A closer look at the shape of the ICCs is obtained by

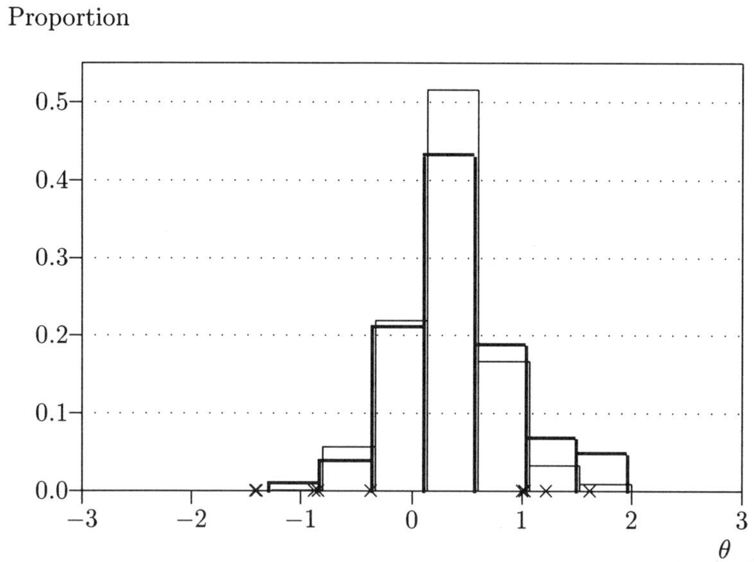

FIGURE 10.3. The PARELLA scale of the Traffic Data. The histograms represent the distributions of the subject locations and the ×s on the horizontal axis represent the estimated item locations. Bold histogram: preinformation group ($n_{pre} = 300$); normal histogram: postinformation group ($n_{post} = 300$).

considering the observed proportion of subjects that agree with the item at each node B_q. Starting with the leftmost node on the latent trait (B_1) and comparing the observed proportion to each subsequent node (B_{q+1}), an indication is obtained as to whether the observed ICC is single-peaked, increasing, or decreasing. In the case of a single-peaked ICC these observed proportions should first increase from the first node to some later node B_q, and subsequently decrease to node B_Q. An alternative way of determining the slopes of the observed ICCs was demonstrated by Post et al. (2001): they used the rows of the conditional adjacency matrix (CAM) (Post & Snijders, 1993b).

Inspection of the observed proportions indicates that most items follow either a decreasing or increasing pattern. See Table 10.4 where these proportions are given for all items. The PARELLA analysis revealed that seven nodes are sufficient to describe the subject distribution on the latent trait. Items 1 through 4 have a perfectly decreasing, and Item 9 a perfectly increasing pattern. Items 5 through 8 show a small peak, either at the beginning (Item 5) or at the end of the latent trait. The rows of the CAM (not reported here) show a similar pattern. Therefore, these data give little support for the assumption of single-peakedness of the ICCs.

From the slopes of the ICCs in Table 10.4 and the estimated item locations in Table 10.3, it can be observed that the items are perfectly sep-

TABLE 10.4. Observed proportions of subjects responding positively per item at the seven nodes B_q of the Traffic scale.

Item	B_1	B_2	B_3	B_4	B_5	B_6	B_7
1	**.97**	.67	.26	.13	.06	.03	.01
2	**.99**	.75	.33	.09	.03	.01	.00
3	**.99**	**.99**	.74	.32	.14	.09	.05
4	**.99**	**.99**	.78	.33	.13	.06	.02
5	.76	**.99**	**.99**	.76	.37	.14	.06
6	.05	.20	.43	.82	**.99**	**.99**	.84
7	.26	.22	.43	.78	**.99**	**.99**	.69
8	.06	.12	.27	.62	.95	**.99**	.88
9	.00	.03	.07	.31	.69	.98	**.99**

arated: the decreasing ICCs on the left (Items 1–5) are followed by the increasing ICCs on the right (Items 6–9). Furthermore, those parts of the subject distributions that are positioned in the gap (see Figure 10.3) are located between the last decreasing item (Item 5) and the first increasing item (Item 6). This situation explains the fit of the PARELLA model, as is also indicated by the values of the SUM statistic for each item that are all well below 1.0. The assumed single-peaked ICCs can therefore be approximated by decreasing and increasing ICCs.

The results of the application of the PARELLA model and the consistency with earlier results (Hoijtink, 1993b; Post & Snijders; 1993a; Van Schuur, 1993; Verhelst & Verstralen, 1993) indicate that an unfolding scale calibrated on these data does not possess adequate measurement properties. The signed OPLM, however, seems to be adequate for these data.

5.2 *Application of the Signed OPLM*

The information about the signs of the slopes of the ICCs, obtained in the analysis with the PARELLA model, is applied in the analysis with the signed OPLM. Prior to the analysis the item scores of the first five items in Table 10.3 are reversed in order to estimate the item locations with the available OPLM software (Verhelst, Glas, & Verstralen, 1995). The analysis resulted in a joint scale on which the pre- and postinformation groups can be compared, since the item-level DIF tests (Verhelst & Glas, 1995) indicate that DIF is absent for all items (see columns five and six of Table 10.5). The adequacy of the ICCs is tested by applying the M2 and M3 tests (Verhelst & Glas, 1995), the results of which are presented in columns 8 and 9 of Table 10.5. Extreme values for these tests indicate either an underestimate (≤ -2.5) or overestimate (≥ 2.5) of the postulated

TABLE 10.5. Item location estimates for the Traffic Data (both samples taken together, $n = 600$).

Item	α_i	$\hat{\beta}_i$	SE($\hat{\beta}_i$)	DIF	*df*	*p*	M2	M3
10	1	−1.31	.15	6.47	5	.26	.45	.78
1	−2	−.54	.08	6.50	6	.37	1.43	1.05
2	−3	−.33	.04	9.21	7	.24	−.06	−1.43
6	3	−.05	.04	10.42	9	.32	−1.41	−.28
7	3	−.01	.04	7.16	9	.62	.91	1.34
3	−3	.23	.04	12.18	10	.27	.70	.12
8	3	.24	.04	3.15	10	.98	.42	.83
4	−3	.26	.04	15.06	10	.13	−.06	−2.09
9	4	.70	.04	7.48	7	.38	−1.07	.53
5	−4	.81	.04	9.08	6	.17	−.39	−.19

slope value (Verhelst, Glas, & Verstralen, 1995). Note that the ICCs are all adequate and the absolute integer values of the slopes range from 1 (lowest discrimination) to 4 (highest discrimination). The values for the M2 and M3 tests support the observation made for the PARELLA scale, that the ICCs can be represented by either decreasing or increasing curves. This support is even more convincing when one considers that the tests for the fit of monotone increasing ICCs will detect deviations from monotonicity, such as single-peakedness.

The interpretation of the latent trait as indicated by the order of the item locations (column 3 in Table 10.5) differs from the intuitive interpretation found for the PARELLA scale. The decreasing and increasing ICCs are not perfectly separated towards both ends of the latent trait but are dispersed over the entire latent trait. This is a direct result of the change in the definition of an item location from a PARELLA ICC to an OPLM ICC. In the PARELLA model an item location is defined as the highest point of its single-peaked ICC, where $P(X_{vi} = 1 \mid \theta_v, \beta_i, \gamma) = 1$, whereas in the OPLM an item location is defined at the middle of the ICC, where $P(X_{vi} = 1 \mid \theta_v, \beta_i, \alpha_i) = 0.5$ (see Figure 10.1). By changing the definition to the OPLM ICC, the item locations for the decreasing ICCs are shifted to the right and those for the increasing ICCs are shifted to the left, as was also shown by Post et al. (2001). Application of the signed OPLM will therefore result in less extreme item locations than an application of the PARELLA model to the same ICCs. As a result, the order of the item locations has changed but the separate orders of the decreasing and increasing items remains intact. This can be confirmed by comparing Tables 10.3 and 10.5.

In Figure 10.4 the distribution of item locations and the two distributions of subject locations are given. For both the pre- and postinformation group

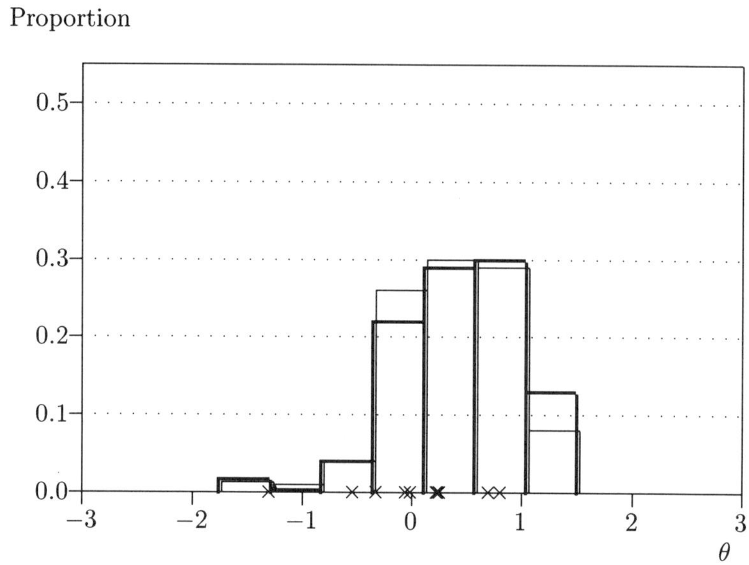

FIGURE 10.4. The signed OPLM scale of the Traffic Data. The histograms represent the distribution of subject locations and the ×s on the horizontal axis represent the estimated item locations. Bold histogram: preinformation group ($n_{pre} = 300$); normal histogram: postinformation group ($n_{post} = 300$).

the items are located within the distribution of subject locations and the gap that was found for the PARELLA scale in Figure 10.3 has disappeared. Another gap can be observed between Items 10 ($\hat{\beta}_{10} = -1.31$) and 1 ($\hat{\beta}_1 = -.54$), but that does not decrease the discriminative strength of the scale because only a small proportion of the subjects are located in it. Part of the distributions of subject locations that is not covered by any item is located to the right of Item 5 ($\hat{\beta}_5 = .81$). This involves about 25% of the subjects of the pre- and 20% of the postinformation group.

6 Summary and Discussion

In this chapter it was shown that the generally assumed distinction between unfolding and cumulative models is not exhaustive. Apart from single-peaked and monotone increasing ICCs, an empirical situation was studied in which some ICCs are decreasing and others increasing. This situation was the result of a bipolar latent attitude for which extreme negatively and positively formulated items are relatively easy to construct; for moderately formulated items this is much more difficult or even impossible. Application of single-peaked ICCs to these items resulted in a scale on which the negatively formulated items were located towards the extreme left end of the latent trait and the positively formulated items towards the extreme right end. These two clusters of items led to a gap in the measurement grid by which discrimination between different levels of the attitude was not achieved due to the large number of subjects located in that gap. Closer inspection of the observed ICCs revealed decreasingness of the negatively formulated items and the increasingness of the positively formulated items, which indicated that the extreme sides of the items are nondecreasing.

This specific situation can be modeled by a more convenient model that allows decreasing and increasing ICCs. Such a model is obtained when the cumulative OPLM is extended with the additional assumption that the slope parameters of the negatively formulated items can take on negative values. This extension is realized by reversal of the item scores of the negatively formulated items which makes the extended model equivalent to the original formulation of the OPLM. Application of the signed OPLM showed that the observed decreasing and increasing ICCs were modeled adequately and that, through the absence of a gap in the measurement grid, the majority of subjects can be distinguished with respect to their attitude towards the latent trait.

The application of the proposed extension of the OPLM is valid under the restriction that the signs of the slopes of the decreasing ICCs are known prior to the analysis. Information about these signs can be obtained in different ways. First, by considering the negative and positive formulations of the item contents with regard to the assumed unidimensional latent trait. Decreasing ICCs can be identified when the contents of the negatively formulated items can be interpreted in such a way that agreement/disagreement indicates a position to the left/right of the item. If the contents of the positively formulated items justify an opposite interpretation, then the items can be identified as either decreasing or increasing, and can be given the signs of their respective slopes. The second way involves a preanalysis of the data with the PARELLA model, or another unfolding model, to obtain the ordering of the items. Reconstruction of the observed ICCs (after inspecting the conditional adjacency matrix for this item order) provides the necessary information to determine whether an ICC is decreasing or increasing. Note that the first method is theoretically more powerful because it does not depend on empirical data, but forces researchers to consider the theoretical interpretation of the latent trait and the contents of the items.

7 Appendix: Estimating Equations of the Signed OPLM

It is shown that the estimating equations of the OPLM hold for the proposed extension of the model, given that the signs of the slopes of the ICCs are known. For the OPLM then, the likelihood of the parameters given the observed response vector is split into two parts, a part with the decreasing ICCs and a part with the increasing ICCs. Next, a data transformation is performed after which the transformed estimation equations are rewritten into the estimation equations of the OPLM.

Let X_{vi} be the random response of subject S_v, $v = 1, \ldots, n$, to item I_i, $i = 1, \ldots, k$; θ_v denotes the location of subject S_v on the latent trait, β_i the location of item I_i on the latent trait, and α_i the *absolute* slope of item I_i. The total set of k items is ordered, such that the first k_1 items have a negative slope, and the remaining $k_2 = k - k_1$ items have a positive slope. Considering (10.1), the probability of the observed response vector $\mathbf{x}_v = (x_{v1}, \ldots, x_{vk})'$ given the model parameters can be written as

$$
\begin{aligned}
P(\mathbf{X}_v = \mathbf{x}_v \mid \theta_v, \boldsymbol{\beta}, \boldsymbol{\alpha}_i) &= \\
&= \prod_{i=1}^{k_1} \frac{\exp[-x_{vi}\alpha_i(\theta_v - \beta_i)]}{1 + \exp[-\alpha_i(\theta_v - \beta_i)]} \prod_{i=k_1+1}^{k} \frac{\exp[x_{vi}\alpha_i(\theta_v - \beta_i)]}{1 + \exp[\alpha_i(\theta_v - \beta_i)]} \\
&= \frac{\exp\left(- \theta_v \sum_{i=1}^{k_1} x_{vi}\alpha_i + \sum_{i=1}^{k_1} x_{vi}\alpha_i\beta_i \right)}{\psi_1} \\
&\quad \times \frac{\exp\left(\theta_v \sum_{i=k_1+1}^{k} x_{vi}\alpha_i - \sum_{i=k_1+1}^{k} x_{vi}\alpha_i\beta_i \right)}{\psi_2} , \qquad (10.3)
\end{aligned}
$$

where

$$\psi_1 = \prod_{i=1}^{k_1} \{1 + \exp[-\alpha_i(\theta_v - \beta_i)]\} ,$$

and

$$\psi_2 = \prod_{i=k_1+1}^{k} \{1 + \exp[\alpha_i(\theta_v - \beta_i)]\} .$$

For the data transformation the random variable T_{vi} is introduced. Let

$$T_{vi} = \begin{cases} 1 - X_{vi} , & i = 1, \ldots, k_1 , \\ X_{vi} , & i = k_1 + 1, \ldots, k . \end{cases} \qquad (10.4)$$

Substitution of (10.4) in (10.3), gives the probability of the observed response vector $\mathbf{t}_v = (t_{v1}, \ldots, t_{vk})'$,

$$P(\mathbf{T}_v = \mathbf{t}_v \mid \theta_v, \boldsymbol{\beta}, \boldsymbol{\alpha}) =$$

$$= \frac{\exp\left[-\theta_v \sum_{i=1}^{k_1}(1-t_{vi})\alpha_i + \sum_{i=1}^{k_1}(1-t_{vi})\alpha_i\beta_i\right]}{\psi_1}$$

$$\times \frac{\exp\left(\theta_v \sum_{i=k_1+1}^{k} t_{vi}\alpha_i - \sum_{i=k_1+1}^{k} t_{vi}\alpha_i\beta_i\right)}{\psi_2}$$

$$= \frac{\exp\left(-\theta_v \sum_{i=1}^{k_1}\alpha_i + \sum_{i=1}^{k_1}\alpha_i\beta_i + \theta_v r - \sum_{i=1}^{k} t_{vi}\alpha_i\beta_i\right)}{\psi_1\,\psi_2}, \qquad (10.5)$$

where

$$r = \sum_{i=1}^{k} t_{vi}\alpha_i$$

is defined as the total score of subject S_v on a k-item test.

The probability of a subject obtaining total score r, given the model parameters, can be written as

$$P(r \mid \theta_v, \boldsymbol{\beta}, \boldsymbol{\alpha}) = \frac{\sum_{T \mid r} \exp\left(-\theta_v \sum_{i=1}^{k_1}\alpha_i + \sum_{i=1}^{k_1}\alpha_i\beta_i + \theta_v r - \sum_{i=1}^{k} t_{vi}\alpha_i\beta_i\right)}{\psi_1\,\psi_2}.$$

For the estimation of the model parameters it is possible to use conditional maximum likelihood (CML) because sufficient statistics for the model parameters exist. In this CML procedure the subject parameters will be conditioned out of the likelihood function. Note that the denominator in (10.5) and the first two terms in the argument of the exponent do not involve the response vector; they will therefore disappear when the conditional probability function of the transformed response pattern is considered.

$$P(\mathbf{T}_v = \mathbf{t}_v \mid r, \boldsymbol{\beta}, \boldsymbol{\alpha}) = \frac{P(\mathbf{T}_v = \mathbf{t}_v \mid \theta_v, \boldsymbol{\beta}, \boldsymbol{\alpha})}{P(r \mid \theta_v, \boldsymbol{\beta}, \boldsymbol{\alpha})}$$

$$= \frac{\exp\left[-\theta_v \sum_{i=1}^{k_1}\alpha_i + \sum_{i=1}^{k_1}\alpha_i\beta_i + \theta_v r - \sum_{i=1}^{k} t_{vi}\alpha_i\beta_i\right]}{\sum_{T \mid r} \exp\left(-\theta_v \sum_{i=1}^{k_1}\alpha_i + \sum_{i=1}^{k_1}\alpha_i\beta_i + \theta_v r - \sum_{i=1}^{k} t_{vi}\alpha_i\beta_i\right)}$$

$$= \frac{\exp \sum_{i=1}^{k} -t_{vi}\alpha_i\beta_i}{\sum_{t\,|\,r} \left(\exp \sum_{i=1}^{k} -t_{vi}\alpha_i\beta_i \right)} = \frac{\prod_{i=1}^{k} \varepsilon_i^{t_{vi}}}{\gamma_r(\varepsilon)} , \tag{10.6}$$

where

$$\exp(-\alpha_i\beta_i) = \varepsilon_i ,$$

and

$$\gamma_r(\varepsilon) = \sum_{T\,|\,r} \prod_{i=1}^{k} \varepsilon_i^{t_{vi}} .$$

Note that the conditional probability (10.6) equals the conditional probability of the OPLM (Verhelst & Glas, 1995).

The data transformation procedure (10.4) is also valid for the more general class of the OPLM for polytomously scored items resulting in equivalent estimation equations. Instead of reversing the item scores, the transformation then involves reversal of item steps.

Acknowledgments: The author wishes to express his thanks to Herbert Hoijtink and Ivo Molenaar for their valuable comments on earlier drafts of this manuscript, to Norman Verhelst for his help on the mathematical derivation of this model, to Bert Jan Doosje and Frans Siero for providing the data, and to the editors for their help and support.

References

Bock, R.D., & Aitkin, M. (1981). Marginal maximum likelihood estimation of item parameters: Application of an EM algorithm. *Psychometrika, 46*, 443–459.

Chang, L. (1995). Connotatively consistent and reversed connotatively inconsistent items are not fully equivalent: Generalizability study. *Educational and Psychological Measurement, 55*, 991–997.

Coombs C.H. (1964). *A theory of data.* Ann Arbor, MI: Mathesis Press.

Doosje, B.J., & Siero, F.W. (1991). Invloed van discrepantie tussen boodschap en eigen standpunt, kwaliteit van argumenten en motivatie tot elaboratie op de attitude ten aanzien van de auto-milieu-problematiek [Influence of the difference between message and opinion, quality of arguments and motivation to change on the attitude towards the car–environment issue]. In R.W. Meertens, A.P. Buunk, & R. van der Vlist (Eds.), *Sociale psychologie & voorlichting en maatschappelijke problemen* (pp. 1–15). 's-Gravenhage: VUGA.

Hoijtink, H. (1990). A latent trait model for dichotomous choice data. *Psychometrika, 55*, 641–656.

Hoijtink, H. (1991). The measurement of latent traits by proximity items. *Applied Psychological Measurement, 15*, 153–169.

Hoijtink, H. (1993a). PARELLA [software manual]. Groningen: ProGAMMA.

Hoijtink, H. (1993b). PARELLA: A parametric approach to parallelogram analysis. *Kwantitatieve Methoden, 42*, 93–108.

Hoijtink, H., & Molenaar, I.W. (1992). Testing for DIF in a model with single-peaked item characteristic curves: The PARELLA model. *Psychometrika, 57*, 383–397.

Jansen, P.G.W. (1983). *Rasch analysis of attitudinal data.* The Hague: RPD. (Doctoral dissertation, Catholic University Nijmegen.)

Molenaar, I.W. (1995). Some background for item response theory and the Rasch model. In G.H. Fischer & I.W. Molenaar (Eds.), *Rasch models: Foundations, recent developments, and applications* (pp. 1–14). New York: Springer-Verlag.

Post, W.J., & Snijders, T.A.B. (1993a). A nonparametric probabilistic model for parallelogram analysis. *Kwantitatieve Methoden, 42*, 55–72.

Post, W.J., & Snijders, T.A.B. (1993b). Nonparametric unfolding models for dichotomous data. *Methodika, 7*, 130–156.

Post, W.J., Van Duijn, M.A.J., & Van Baarsen, B. (2001). Cumulative or monotone tracelines? On the choice of an IRT model for scaling data. In A. Boomsma, M.A.J. van Duijn, & T.A.B. Snijders (Eds.), *Essays on item response theory* (pp. 391–414). New York: Springer-Verlag.

Schriesheim, C.A., Eisenbach, R.J., & Hill, K.D. (1991). The effect of negation and polar opposite item reversals on questionnaire reliability and validity: An experimental investigation. *Educational and Psychological Measurement, 51*, 67–78.

Schriesheim, C.A., & Hill, K.D. (1981). Controlling acquiescence response bias by item reversals: The effect on questionnaire validity. *Educational and Psychological Measurement, 41*, 1101–1114.

Thurstone, L.L. (1927). A law of comparative judgement. *Psychological Review, 34*, 278–286.

Van Schuur, W.H. (1993). MUDFOLD. *Kwantitatieve Methoden, 42*, 39–54.

Van Schuur, W.H. (1998). From Mokken to MUDFOLD and back. In C. van der Eijk, M. Fennema, & H. Schijf (Eds.), *In search of structure: Essays in social science and methodology* (pp. 45–62). Amsterdam: Het Spinhuis.

Van Schuur, W.H., & Kiers, H.A.L. (1994). Why factor analysis often is the incorrect model for analyzing bipolar concepts and what model to use instead. *Applied Psychological Measurement, 18*, 97–110.

Verhelst, N.D., & Glas, C.A.W. (1995). The one parameter logistic model: OPLM. In G.H. Fischer & I.W. Molenaar (Eds.), *Rasch models: Foundations, recent developments, and applications* (pp. 215–238). New York: Springer-Verlag.

Verhelst, N.D., Glas, C.A.W., & Verstralen, H.H.F.M. (1995). *One-parameter logistic model: OPLM.* Arnhem: CITO.

Verhelst, N.D., & Verstralen, H.H.F.M. (1993). A stochastic unfolding model derived from the partial credit model. *Kwantitatieve Methoden, 42*, 73–92.

11

Using Parameter Expansion to Improve the Performance of the EM Algorithm for Multidimensional IRT Population-Survey Models

Donald B. Rubin[1] and Neal Thomas[2]

ABSTRACT Several large-scale educational surveys use item response theory (IRT) models to summarize complex cognitive responses and relate them to educational and demographic variables. The IRT models are often multidimensional with prespecified traits, and maximum likelihood estimates (MLEs) are found using an EM algorithm, which is typically very slow to converge. We show that the slow convergence is due primarily to missing information about the correlations between latent traits relative to the information that would be present if these traits were observed ("complete data"). We show how to accelerate convergence using parameter extended EM (PX-EM), a recent modification of the EM algorithm. The PX-EM modification is simple to implement for the IRT survey model and reduces the number of required iterations by approximately one-half without adding any appreciable computation to each EM-iteration.

1 Educational Assessment Surveys Using IRT Models

The U.S. National Assessment of Educational Progress (NAEP), the U.S. Adult Literacy Survey (ALS), and the Third International Mathematics and Sciences Study (TIMSS) collect cognitive data and educational environment variables (e.g., demography, coursework) on large probability-

[1]Datametrics Research, Inc. and Harvard University, 1 Oxford Street, Cambridge, MA 02138, USA; e-mail: *rubin@stat.harvard.edu*

[2]Datametrics Research, Inc. and Bristol-Myers Squibb, 61 Dreamlake Drive, Madison, CT 06443, USA; e-mail: *snthomas99@yahoo.com*

based surveys of students. To achieve broad coverage, the surveys include small numbers of items measuring several correlated traits (e.g., geometry and algebra within mathematics). The primary goals of the surveys are to characterize academic proficiency and how it varies among different demographic groups and changes over time.

The analyses of these surveys use latent trait IRT models to summarize and report results. Multiple imputation missing data methods (Mislevy, 1991; Rubin, 1987) are used to account for the incomplete information about the latent trait due to the sparse matrix sampling. These likelihood-based multivariate methods use the observed item responses and background covariates to multiply impute the missing latent traits for each student. We focus on one step in the creation of the imputed values: the EM algorithm (Dempster, Laird, & Rubin, 1977) used to find maximum likelihood estimates (MLEs) of the unknown parameters in the population model for the latent traits.

Although the EM algorithm is a popular tool for estimation in IRT models, it can have very slow convergence. Parameter expanded EM (PX-EM; Liu, Rubin, & Wu 1998) can often make EM dramatically faster. Here, our first attempt at applying the algorithm approximately doubles the speed of the EM algorithm. PX-EM is based on a statistical principle that differs from the principles underlying most other EM acceleration methods, such as Aitken acceleration (Laird, Lange, & Stram, 1987). In particular, in addition to achieving significant improvements in the rate of convergence, PX-EM can often be applied with only simple modifications of the parent EM algorithm, while maintaining its stability and monotone convergence.

The underlying statistical principle of PX-EM is to perform a "covariance adjustment" to correct the M-step, capitalizing on extra information captured in the imputed complete data. Formally, we find an expanded complete-data model that has a larger set of parameters, but leads to the original observed-data model with the original parameters determined from the expanded parameters by a reduction function. Then PX-EM iteratively maximizes the expected loglikelihood of the expanded complete-data model, with its expanded parameters and corresponding expanded sufficient statistics. The rate of convergence of PX-EM is at least as fast as the parent EM because its M-step performs a more efficient analysis by fitting the expanded model.

2 EM and PX–EM

Here we review the EM algorithm and describe its parameter-expanded version, PX-EM. Throughout, our notation follows conventions in the EM/multiple imputation literature to make referencing that literature more straightforward.

2.1 Notation and Background for EM

Let y_{obs} be the set of observed data following the observed-data model $p(y_{\text{obs}} \mid \boldsymbol{\lambda})$, with parameter vector $\boldsymbol{\lambda}$ to be estimated by its maximum likelihood estimate $\boldsymbol{\lambda}_{\text{MLE}}$. The EM algorithm augments y_{obs} to a set of complete data y_{com}, which can be reduced to y_{obs} via a many-to-one mapping, and y_{com} follows the complete-data model $p(y_{\text{com}} \mid \boldsymbol{\lambda})$.

The EM algorithm maximizes $p(y_{\text{obs}} \mid \boldsymbol{\lambda})$ by iteratively maximizing the expected $\ln p(y_{\text{com}} \mid \boldsymbol{\lambda})$ using an expectation (E) step and a maximization (M) step at each iteration. At the $(t+1)$st iteration, the E-step computes the expected loglikelihood of the complete data model,

$$Q(\boldsymbol{\lambda} \mid \boldsymbol{\lambda}^{(t)}) = E_{y_{\text{com}}}[\ln p(y_{\text{com}} \mid \boldsymbol{\lambda}) \mid y_{\text{obs}}, \boldsymbol{\lambda}^{(t)}] ,$$

where $E_{y_{\text{com}}}(\cdot \mid y_{\text{obs}}, \boldsymbol{\lambda}^{(t)})$ denotes the expectation with respect to the conditional distribution $p(y_{\text{com}} \mid y_{\text{obs}}, \boldsymbol{\lambda} = \boldsymbol{\lambda}^{(t)})$. Then the M-step finds $\boldsymbol{\lambda}^{(t+1)}$ by maximizing $Q(\boldsymbol{\lambda} \mid \boldsymbol{\lambda}^{(t)})$ over $\boldsymbol{\lambda}$. When $p(y_{\text{com}} \mid \boldsymbol{\lambda})$ is from an exponential family, the E-step computes the conditional expectations of its sufficient statistics, and the M-step fits the complete data model using these expected statistics. Each iteration of EM increases $\ln p(y_{\text{obs}} \mid \boldsymbol{\lambda})$; see Dempster, Laird, & Rubin (1977).

2.2 Definition of PX-EM

The PX-EM algorithm expands $p(y_{\text{com}} \mid \boldsymbol{\lambda})$ to a larger model, $p(y_{\text{com}} \mid \boldsymbol{\Lambda})$, with $\boldsymbol{\Lambda} = (\boldsymbol{\lambda}^*, \boldsymbol{\kappa})$, where $\boldsymbol{\lambda}^*$ plays the same role in $p(y_{\text{com}} \mid \boldsymbol{\Lambda})$ that $\boldsymbol{\lambda}$ plays in $p(y_{\text{com}} \mid \boldsymbol{\lambda})$, and $\boldsymbol{\kappa}$ is an auxiliary parameter vector whose value is fixed at $\boldsymbol{\kappa}_0$ in the original model. Formally, certain conditions must be satisfied in the expansion. First, the observed-data model is preserved in the sense that, for all $\boldsymbol{\Lambda}$, there is a common many-to-one reduction function R, such that $y_{\text{obs}} \mid \boldsymbol{\Lambda} \sim p[y_{\text{obs}} \mid \boldsymbol{\lambda} = R(\boldsymbol{\Lambda})]$; that is, the distribution of y_{obs} given $\boldsymbol{\Lambda}$ depends on $\boldsymbol{\Lambda}$ only through the parameter $\boldsymbol{\lambda}$, which is a function of $\boldsymbol{\Lambda}$, $R(\boldsymbol{\Lambda})$. Second, the complete-data model is preserved at the null value of $\boldsymbol{\kappa}$, $\boldsymbol{\kappa}_0$, in the sense that, for all $\boldsymbol{\lambda}$,

$$p[y_{\text{com}} \mid \boldsymbol{\Lambda} = (\boldsymbol{\lambda}^*, \boldsymbol{\kappa}_0)] = p(y_{\text{com}} \mid \boldsymbol{\lambda} = \boldsymbol{\lambda}^*) .$$

These conditions imply that if $\boldsymbol{\lambda}_1 \neq \boldsymbol{\lambda}_2$ then $\boldsymbol{\Lambda}_1 \neq \boldsymbol{\Lambda}_2$, and that, for all $\boldsymbol{\lambda}$, there exists at least one $\boldsymbol{\Lambda}$ such that $y_{\text{obs}} \mid \boldsymbol{\Lambda} \sim p\,[y_{\text{obs}} \mid \boldsymbol{\lambda} = R(\boldsymbol{\Lambda})]$.

The PX-EM algorithm uses $p(y_{\text{com}} \mid \boldsymbol{\Lambda})$ to generate EM by iteratively maximizing the expected loglikelihood of $p(y_{\text{com}} \mid \boldsymbol{\Lambda})$. Specifically, let $\boldsymbol{\Lambda}^{(t)} = (\boldsymbol{\lambda}^{(t)}, \boldsymbol{\kappa}_0)$ be the estimate of $\boldsymbol{\Lambda}$ from the tth iteration with $\boldsymbol{\kappa}^{(t)} = \boldsymbol{\kappa}_0$. Then, at the $(t+1)$st iteration we have the following.

PX-E-step. Compute $Q(\mathbf{\Lambda} \mid \mathbf{\Lambda}^{(t)}) = E_{y_{\text{com}}}[\ln p(y_{\text{com}} \mid \mathbf{\Lambda}) \mid y_{\text{obs}}, \mathbf{\Lambda}^{(t)}]$.

PX-M-step. Find $\mathbf{\Lambda}^{(t+1)} = \arg\max_{\mathbf{\Lambda}} Q(\mathbf{\Lambda} \mid \mathbf{\Lambda}^{(t)})$; then apply the reduction function R to obtain $\boldsymbol{\lambda}^{(t+1)} = R(\mathbf{\Lambda}^{(t+1)})$.

Each iteration of PX-EM increases $\ln p(y_{\text{obs}} \mid \boldsymbol{\lambda})$; see Liu, Rubin, & Wu (1998). Also, since the monotone convergence of EM is maintained, conditions for the convergence of PX-EM iterates to a stationary point or a local maximum $\boldsymbol{\lambda}_{\text{MLE}}$, can be obtained following Wu (1983).

When the complete-data models are in the exponential family, the programming for implementing PX-EM beyond its parent EM algorithm only involves (1) adding to the original E-step the calculation of the expected sufficient statistics related to the expanded parameterization, which may have already been done in the original E-step, and (2) modifying the maximization over $\boldsymbol{\lambda}$ in the original M-step to include maximization over $\boldsymbol{\kappa}$, which is typically simple.

3 Statistical Model

The q background predictor variables for the ith student in a sample of size n is denoted $\mathbf{x}_i = (x_{i1}, \ldots, x_{iq})'$. Each $\mathbf{x}_i$ is composed of q scalar variables, and for modeling convenience, the first component of $\mathbf{x}_i$ is a constant intercept term. The covariates consist of demographic variables, coursework, home environment, and so on.

The variables representing the cognitive items for the ith student are denoted $\mathbf{y}_i$, and are partitioned by the test designers into p content areas $\mathbf{y}_i = (\mathbf{y}_{i1}, \ldots, \mathbf{y}_{ip})'$ corresponding to p latent traits. Each $\mathbf{y}_{ij}$ is composed of item scores $y_{ijk}, k = 1, \ldots, s_j$, that are binary or ordinal with values coded as $0, 1, \ldots, m_{jk}$, for the kth item measuring the jth trait.

A model representing the data is specified in stages. First, a latent trait vector $\boldsymbol{\theta}_i = (\theta_{i1}, \ldots, \theta_{ip})'$, $i = 1, \ldots, n$, is hypothesized for the ith student, which determines the distribution of the item scores through logistic IRT models, where conditional on the latent traits, all item responses are assumed independent of each other and the $\mathbf{x}_i$. Second, the $\boldsymbol{\theta}_i$ are assumed to follow a multivariate normal distribution conditional on the background variables.

3.1 Logistic IRT Model for Binary and Ordinal Cognitive Responses

A logistic item response model is used for the cognitive data $\mathbf{y}_i$ conditional on $\mathbf{x}_i$, $\boldsymbol{\theta}_i$, and the item parameters $\boldsymbol{\beta} = (\boldsymbol{\beta}_1, \ldots, \boldsymbol{\beta}_p)'$, where $\boldsymbol{\beta}_j = \{\boldsymbol{\beta}_{jk}, k = 1, \ldots, s_j\}$, $j = 1, \ldots, p$.

The probabilities of binary scored items are modeled by a three-parameter logistic IRT model (Birnbaum, 1968),

$$p(y_{ijk} = 1 \mid \mathbf{x}_i, \boldsymbol{\theta}_i, \boldsymbol{\beta}) = c_{jk} + \frac{1 - c_{jk}}{1 + \exp[\, a_{jk}(\theta_{ij} - b_{jk})\,]}\,, \tag{11.1}$$

where $\boldsymbol{\beta}_{jk} = (a_{jk}, b_{jk}, c_{jk})'$ are the item parameters for the kth item measuring the jth proficiency. The a_{jk}s are discrimination parameters, the b_{jk}s are difficulty parameters, and the c_{jk}s are guessing parameters. The response probabilities of an ordinal item are modeled by a partial credit model (Masters & Wright, 1984),

$$p(y_{ijk} = l \mid \mathbf{x}_i, \boldsymbol{\theta}_i, \boldsymbol{\beta}) = \frac{\exp\left[\sum\limits_{h=0}^{l} a_{jk}(\theta_{ij} - b_{jkh})\right]}{\sum\limits_{q=0}^{m_{jk}} \exp\left[\sum\limits_{h=0}^{q} a_{jk}(\theta_{ij} - b_{jkh})\right]}\,, \tag{11.2}$$

where $\boldsymbol{\beta}_{jk} = \{a_{jk}, b_{jkl}\}$, for $l = 0, \ldots, m_{jk}$.

The model invokes several independence assumptions conditional on the $\boldsymbol{\theta}_i$ and $\boldsymbol{\beta}$: (1) the item responses are independent of the $\mathbf{x}_i$, (2) the responses of a student to different items are independent (i.e., the distribution of $\mathbf{y}_i$ is the product of the probabilities in Equations 11.1 and 11.2), (3) responses from different students are independent, and (4) $\mathbf{y}_{ij}$ are independent of $\theta_{ij'}$ conditional on θ_{ij}, $j \neq j'$ (i.e., item scores depend only on the latent trait θ_{ij} to which they are assigned). With these independence assumptions, the density of $\mathbf{y}_i, i = 1, \ldots, n$ conditional on $\boldsymbol{\theta}_i$ and $\mathbf{x}_i$ can be represented as

$$\prod_{i=1}^{n} \left[\prod_{j=1}^{p} f_j(\mathbf{y}_{ij} \mid \theta_{ij}, \boldsymbol{\beta}_j)\right], \tag{11.3}$$

and each $f_j(\mathbf{y}_{ij} \mid \theta_{ij}, \boldsymbol{\beta}_j)$ is in turn the product of the response probabilities for the cognitive items given in (11.1) and (11.2). The inside product in (11.3) can be viewed as the likelihood function for $\boldsymbol{\theta}_i$ based on the $\mathbf{y}_i$ data. Items not presented to a student do not contribute to the likelihood function because they are missing at random.

Note that the parameter vectors $\boldsymbol{\beta}_j$ are included in the likelihood term $f_j(\mathbf{y}_{ij} \mid \theta_{ij}, \boldsymbol{\beta}_j)$ to denote the explicit dependence of (11.3) on each $\boldsymbol{\beta}_j$. To identify the logistic IRT model, the mean and variance of $\boldsymbol{\theta}$ in the overall population are constrained to be zero and one. Mislevy and Bock (1982) and Muraki (1992) give details of these models.

3.2 Distribution of the Latent Trait Conditional on the Background Variables

The $\boldsymbol{\theta}_i$ vectors are assumed to be normally distributed conditional on the variables $\mathbf{x}_i$. The mean of this conditional distribution is given by the multivariate multiple linear regression $\boldsymbol{\Gamma}'\mathbf{x}_i$, where $\boldsymbol{\Gamma} = \left[\gamma_1 \mid \cdots \mid \gamma_p\right]$, and

$\gamma_j, j = 1, \ldots, p$, are unknown regression parameter vectors of dimension q. The common (unknown) p-dimensional conditional variance-covariance matrix is $\mathbf{\Sigma}$ with its (j, k) element denoted by Σ_{jk}. The dimension of $\boldsymbol{\theta}_i$ has varied between one and six in applications to the educational surveys.

The distribution of $\boldsymbol{\theta}_i$ can be viewed as a normal prior distribution conditional on $\mathbf{x}_i$ and the parameters $\mathbf{\Gamma}$ and $\mathbf{\Sigma}$, before observing the cognitive data: $\boldsymbol{\theta}_i \sim \phi\left(\mathbf{\Gamma}'\mathbf{x}_i, \mathbf{\Sigma}\right)$.

3.3 Combining the Two Components

Applying the independence assumptions, the regression model and the item response model fully specify the distribution of observed data. The likelihood function for the parameters $(\boldsymbol{\beta}, \mathbf{\Gamma}, \mathbf{\Sigma})$ is the distribution of the data; that is, $(\mathbf{y}_i, \mathbf{x}_i)$, $i = 1, \ldots, n$, given $(\boldsymbol{\beta}, \mathbf{\Gamma}, \mathbf{\Sigma})$:

$$L\left(\boldsymbol{\beta}, \mathbf{\Gamma}, \mathbf{\Sigma}\right) = \prod_{i=1}^{n} \int \phi\left(\boldsymbol{\theta}_i;\ \mathbf{\Gamma}'\mathbf{x}_i, \mathbf{\Sigma}\right) \prod_{j=1}^{p} f_j\left(\mathbf{y}_{ij} \mid \theta_{ij}, \boldsymbol{\beta}_j\right) d\boldsymbol{\theta}_i\ . \tag{11.4}$$

The integrand in (11.4) is proportional to the posterior distribution of $\boldsymbol{\theta}_i$ with $\boldsymbol{\beta}$, $\mathbf{\Gamma}$, and $\mathbf{\Sigma}$ regarded as known:

$$f\left(\boldsymbol{\theta}_i; \mathbf{x}_i, \mathbf{y}_i, \boldsymbol{\beta}, \mathbf{\Gamma}, \mathbf{\Sigma}\right) = \phi\left(\boldsymbol{\theta}_i;\ \mathbf{\Gamma}'\mathbf{x}_i, \mathbf{\Sigma}\right) \prod_{j=1}^{p} f_j\left(\mathbf{y}_{ij} \mid \theta_{ij}, \boldsymbol{\beta}_j\right)\ . \tag{11.5}$$

3.4 Estimating the Model Parameters

A two-stage procedure has been used to find parameter estimates for $(\boldsymbol{\beta}, \mathbf{\Gamma}, \mathbf{\Sigma})$. During the first stage, an EM algorithm is used to find maximum likelihood estimates for the item parameters $\hat{\boldsymbol{\beta}}_j$, $j = 1, \ldots, p$, separately for each latent trait based on the likelihood function for the subset of examination data associated with the jth latent trait. In the second estimation stage, an EM algorithm is used to estimate $(\mathbf{\Gamma}, \mathbf{\Sigma})$ with $\boldsymbol{\beta}$ fixed at their first-stage estimates. For the second-stage EM algorithm, the latent $\boldsymbol{\theta}$ are regarded as missing data. The $\boldsymbol{\theta}_i$ are missing completely at random because all of them are missing. Evaluation of the expected complete-data loglikelihood function involves the calculation of the first two moments of the posterior distribution in (11.5).

The use of this two-stage procedure to estimate $\boldsymbol{\beta}$ and $(\mathbf{\Gamma}, \mathbf{\Sigma})$ has been investigated by Mislevy and Sheehan (1989), and Adams, Wilson, and Wu (1997). They found there was little difference between the two-stage procedure and the full joint maximum likelihood estimation. This two-stage procedure was also validated using simulation studies by Thomas (in press). We focus on the EM algorithm for the estimation of $\mathbf{\Gamma}$ and $\mathbf{\Sigma}$.

3.5 Normal Measurement Error Approximation

The contribution of the IRT component of the NAEP model to the likelihood function in (11.5), $\prod_{j=1}^{p} f_j(\mathbf{y}_{ij} \mid \theta_{ij}, \boldsymbol{\beta}_j)$, approaches normality as the number of items increases (Chang & Stout, 1993). A convenient representation of a normal approximation to this component of the likelihood function in terms of a familiar regression model replaces the multivariate item responses $\mathbf{y}_{ij}$ measuring the jth trait for the ith subject by a single measurement,

$$y_{ij} = \theta_{ij} + \epsilon_{ij} \ , \qquad \mathbf{y}_i = (y_{i1}, \ldots, y_{ip})' \ , \tag{11.6}$$

where $\epsilon_{ij} \sim \phi(0, \tau_{ij})$ are independent normally distributed "measurement" errors. The y_{ij} and τ_{ij} are constructed so that the likelihood function arising from (11.6) approximates the one in (11.5). Mislevy (1992) and Thomas (1993) show that useful approximations to the logistic IRT model can be obtained with a suitable choice of the y_{ij} and τ_{ij}, which are a function of the original item responses and the (assumed known) item parameters.

The variance of y_{ij} in the approximating normal model including the variance of θ_{ij} is

$$\mathrm{Var}(y_{ij}) = \Sigma_{jj} + \tau_{ij} \ ,$$

and the covariance of y_{ij} and y_{ik} is Σ_{jk} when $j \neq k$. The mean of $\mathbf{y}_i$ is $\boldsymbol{\Gamma}'\mathbf{x}_i$, the same as that of $\boldsymbol{\theta}_i$. The EM algorithm applied to this approximating model produces excellent starting distributions for more computationally intensive Markov chain Monte Carlo (MCMC) methods (Patz, 1996, 1999).

3.6 PX-EM for the Normal Measurement Error Model

For the implementation of the PX-EM algorithm, the $\boldsymbol{\theta}_i$ are regarded as missing data and the $(\mathbf{y}_i, \mathbf{x}_i)$ are the observed data just as they are with the second-stage EM algorithm in Section 3.4. We develop the accelerated algorithm for the normal approximation model from Section 3.5. The distribution of $\boldsymbol{\theta}_i$ given $\mathbf{x}_i$, $\boldsymbol{\Gamma}$, and $\boldsymbol{\Sigma}$ is $\boldsymbol{\theta}_i \sim \phi\left(\boldsymbol{\Gamma}'\mathbf{x}_i, \boldsymbol{\Sigma}\right)$ in the original model and it is unchanged in the expanded model. In the original model,

$$\mathbf{y}_i = \boldsymbol{\theta}_i + \boldsymbol{\epsilon}_i \ , \qquad \boldsymbol{\epsilon}_i \sim \phi\left[\mathbf{0}, \mathrm{diag}\left(\tau_{i1}, \ldots, \tau_{ip}\right)\right] \ ,$$

which is expanded for PX-EM to

$$\mathbf{y}_i = \boldsymbol{\kappa}'\boldsymbol{\theta}_i + \boldsymbol{\epsilon}_i \ ,$$

where $\boldsymbol{\kappa} = [\boldsymbol{\kappa}_1 \mid \cdots \mid \boldsymbol{\kappa}_p]$ and $\boldsymbol{\kappa}_j, j = 1, \ldots, p$, are p-dimensional unknown regression parameters. The parameters of the expanded model that correspond to parameters in the original model when $\boldsymbol{\kappa}$ is the identity are denoted by adding an $*$ to distinguish them from the parameters in the original model.

The resulting distribution of $\mathbf{y}_i$ given $\boldsymbol{\theta}_i$, $\boldsymbol{\Gamma}$, $\boldsymbol{\Sigma}$, $\boldsymbol{\kappa}$, and $\mathbf{x}_i$ is $\mathbf{y}_i \sim \phi[\boldsymbol{\kappa}'\boldsymbol{\theta}_i, \text{diag}(\tau_{i1}, \ldots, \tau_{ip})]$, and the distribution of $\mathbf{y}_i$ given $\boldsymbol{\Gamma}$, $\boldsymbol{\Sigma}$, $\boldsymbol{\kappa}$, $\mathbf{x}_i$ is $\mathbf{y}_i \sim \phi[(\boldsymbol{\Gamma}^*\boldsymbol{\kappa})'\mathbf{x}_i, \boldsymbol{\kappa}'\boldsymbol{\Sigma}^*\boldsymbol{\kappa} + \text{diag}(\tau_{i1}, \ldots, \tau_{ip})]$. The reduction function that maps the extended parameters into the original parameters in the observed data model is thus $R(\boldsymbol{\Gamma}^*, \boldsymbol{\Sigma}^*, \boldsymbol{\kappa}) = (\boldsymbol{\Gamma}^*\boldsymbol{\kappa}, \boldsymbol{\kappa}'\boldsymbol{\Sigma}^*\boldsymbol{\kappa})$. The null-value of $\boldsymbol{\kappa}$ is the p-dimensional identity matrix.

For calculations in the M-step, matrices of the $\mathbf{y}_i$, $\mathbf{x}_i$, and $\boldsymbol{\theta}_i$ are

$$\mathbf{y} = \begin{bmatrix} y_{11} \cdots y_{1p} \\ \vdots \ \ddots \ \vdots \\ y_{n1} \cdots y_{np} \end{bmatrix}, \qquad \mathbf{x} = \begin{bmatrix} x_{11} \cdots x_{1q} \\ \vdots \ \ddots \ \vdots \\ x_{n1} \cdots x_{nq} \end{bmatrix}, \qquad \boldsymbol{\theta} = \begin{bmatrix} \theta_{11} \cdots \theta_{1p} \\ \vdots \ \ddots \ \vdots \\ \theta_{n1} \cdots \theta_{np} \end{bmatrix},$$

and the $(n \times n)$-dimensional weight matrices are $\mathbf{W}_j = \text{diag}(\tau_{1j}, \ldots, \tau_{nj})^{-1}$, $j = 1, \ldots, p$.

E-step. The likelihood function of the expanded complete-data model can be factored as

$$p(\mathbf{y}, \mathbf{x}, \boldsymbol{\theta} \mid \boldsymbol{\Gamma}, \boldsymbol{\Sigma}, \boldsymbol{\kappa}) \propto p(\mathbf{y} \mid \boldsymbol{\theta}, \boldsymbol{\kappa})\, p(\boldsymbol{\theta} \mid \mathbf{x}, \boldsymbol{\Gamma}, \boldsymbol{\Sigma}) \,.$$

Because the last factor is the likelihood function for the original model, the sufficient statistics for $\boldsymbol{\Gamma}$ and $\boldsymbol{\Sigma}$ are unchanged. The $\mathbf{y}$ are independent given $\boldsymbol{\theta}$ and $\boldsymbol{\kappa}$, so the first factor reduces to p separate models.

The additional sufficient statistics required to estimate $\boldsymbol{\kappa}$ are

$$SS_j = \boldsymbol{\theta}'\mathbf{W}_j\boldsymbol{\theta} \,, \qquad SY_j = \boldsymbol{\theta}'\mathbf{W}_j\mathbf{y}_{.j} \,, \qquad j = 1, \ldots, p \,,$$

where $\mathbf{y}_{.j}$ is the jth column of $\mathbf{y}$. Note that there is no centering of the sums of squares because no intercept term is included.

M-step. The $\hat{\boldsymbol{\Gamma}}$ and $\hat{\boldsymbol{\Sigma}}$ are computed during each iteration using the standard normal theory estimators, as is done in the EM algorithm (Mislevy, 1985). The estimator of $\boldsymbol{\kappa}_j$ during each iteration is

$$\hat{\boldsymbol{\kappa}}_j = (SS_j)^{-1}\, SY_j \,, \qquad j = 1, \ldots, p \,.$$

4 Example

4.1 1992 U.S. NAEP Mathematics Data

The 1992 Eighth Grade National Mathematics Assessment was based on a sample of 10,291 students who were in the eighth grade and/or who were 13 years old. For the example, two of the five traits in the assessment are used, Geometry (GEOM) and Algebra (ALGE). Each student was administered forms with between 5 and 10 items measuring each trait. The item parameter estimates $\hat{\boldsymbol{\beta}}$, in Section 3.4, were copied from the actual assessment. A detailed description of the sample and its contents are in Johnson and Carlson (1994). The covariates $\mathbf{x}$ are 18 indicator variables representing six demographic questions: race, gender, region of country, parent education, private/public school, and age/grade. The model in Section 3 has $p = 2$ and $q = 19$ (which includes an intercept) for the mathematics assessment.

4.2 Rate of EM Convergence

The EM algorithm when applied to the mathematics assessment, described in Section 4.1, required 565 iterations to achieve convergence using the criteria of (1) a change of < 0.00001 in each regression parameter $\hat{\boldsymbol{\Gamma}}$ and variance parameter in $\hat{\boldsymbol{\Sigma}}$, and (2) a change of < 0.000015 for the estimated correlation between the traits. The criteria are based on experience with numerous similar data sets, and are sufficient to ensure changes resulting from additional iterations are not of practical significance. Examination of changes in the parameter estimates indicated that $\hat{\boldsymbol{\Gamma}}$ had converged within 60 iterations, and the large number of remaining iterations were necessary to ensure convergence of $\hat{\boldsymbol{\Sigma}}$. Dempster, Laird, and Rubin (1977) showed that the rate of convergence near the maximum is linear, that is, for a generic parameter denoted by λ with tth EM iteration estimate $\hat{\lambda}_t$, and a rate of convergence

$$r \approx \frac{|\hat{\lambda}_t - \hat{\lambda}_{t-1}|}{|\hat{\lambda}_{t-1} - \hat{\lambda}_{t-2}|},$$

which implies a linear relationship between t and the logarithm of the differences. Examination of the differences in the variance estimates for the GEOM trait from the final 165 iterations showed that the linear approximation is very good, but that the rate of convergence is 0.995, which explains the large number of iterations required to achieve convergence. The rate of convergence for the ALGE variance estimator and the correlation estimator are almost identical, as predicted by general theoretical results because the estimators are all correlated and thus share the same convergence rate. The $\hat{\boldsymbol{\Gamma}}$ are essentially uncorrelated with the variance estimators and thus not subject to their slow convergence rate.

The EM algorithm was also applied to estimate the parameters of the distribution of each latent trait separately ($p = 1$), for the GEOM trait using only the cognitive items associated with the GEOM trait, and similarly for the ALGE trait. The univariate procedures converged in 47 and 52 iterations with rate r for the variance estimates of 0.817 and 0.835 for the GEOM and ALGE traits, respectively. The slow convergence rate is present only when the correlation is estimated in the bivariate procedure. This finding has also been validated by examination of the eigenvalues of the missing-data information matrix multiplied by the inverse of the complete-data information matrix, which shows that the largest eigenvalue has essentially no loading on the $\mathbf{\Gamma}$ parameters, and weights of approximately $(1, 1, -2)$ for the two variance and covariance parameters.

4.3 Information when Latent Traits Are Highly Correlated

The slow convergence sometimes achieved by the EM algorithm usually occurs when a parameter is near a boundary where the distribution of the data becomes degenerate, that is, of lower dimension or a point mass. In the example in Section 4.2, the estimated correlation between the two traits is 0.91, near the boundary value of 1. The complete-data correlation estimator, which is the sample correlation of the latent traits, has (approximate) variance $(1 - \rho^2)^2/n$ that approaches 0 as ρ approaches 1, where ρ is the correlation between the traits. When the traits are measured with error, however, even if the error variances are known, the data do not effectively differentiate between ρ-values near 1, and thus the ratio of the missing-data information to the complete-data information, which determines the rate of convergence, approaches 1 as ρ approaches 1. The expanded model used in the PX-EM includes additional parameters $\boldsymbol{\kappa}$ that are poorly estimated even with complete data because high correlation between the latent traits yields near collinearity in the regression of $\mathbf{y}$ on $\boldsymbol{\theta}$, which is used to estimate the $\boldsymbol{\kappa}$. The inclusion of the $\boldsymbol{\kappa}$ parameter thus decreases the ratio of missing information relative to complete information, and leads to faster convergence.

4.4 Convergence for PX-EM

The PX-EM algorithm converged in 341 iterations, roughly one-half the 565 iterations required by the EM algorithm. There was essentially no computational overhead added to each EM-step. The rate of convergence was still relatively slow, with $r = 0.990$ for the PX-EM algorithm, compared to $r = 0.995$ for the EM algorithm. The convergence was linear, like that of the original EM algorithm. The number of iterations for each univariate EM procedure was also reduced by approximately one-half by the PX-EM modification.

5 Summary

In the application, PX-EM was faster than EM by an approximate factor of two. From one perspective, this improvement should be considered worthwhile because each iteration of PX-EM has essentially the same computational requirements as each iteration of EM. However, we were disappointed and hoped for more drastic improvements based on experience with other applications. A different, more extensive parameter expansion may lead to larger improvements over EM. Choosing parameter expansions, such as choosing the missing data to augment the observed data with EM, remains something of an art. A possibility that would permit greater flexibility for both EM and PX-EM methods would replace the IRT logistic specification with a probit specification, thereby allowing the creation of additional missing normal variables, which could be handled directly by EM or PX-EM, because of the straightforward complete-data sufficient statistics of the probit model.

Acknowledgments: This research has been supported by NCES Grant 84.902B980011 to Datametrics Research, Inc.

References

Adams, R., Wilson, M., & Wu, M. (1997). Multilevel item response models: an approach to errors in variables regression. *Journal of Educational and Behavioral Statistics, 17*, 47–76.

Birnbaum, A. (1968). Some latent trait models. In F.M. Lord & M.R. Novick, *Statistical theories of mental test scores* (pp. 397–424). Reading, MA: Addison-Wesley.

Chang, H., & Stout, W. (1993). The asymptotic posterior normality of the latent trait in an IRT model. *Psychometrika, 58*, 37–52.

Dempster, A., Laird, N., & Rubin, D. (1977). Maximum likelihood estimates from incomplete data via the EM algorithm (with discussion). *Journal of the Royal Statistical Society, Series B, 39*, 1–38.

Johnson, E., & Carlson, J. (1994). *The NAEP 1992 Technical Report.* Washington, DC: National Center for Educational Statistics.

Laird, N., Lange, N., & Stram, D. (1987). Maximizing likelihood computations with repeated measures: Application of the EM algorithm. *Journal of the American Statistical Association, 82*, 97–105.

Liu, C., Rubin, D., & Wu, Y. (1998). Parameter expansion to accelerate EM: The PX-EM algorithm. *Biometrika, 85*, 755–770.

Masters, G.N., & Wright, B.D. (1984). The essential process in a family of measurement models. *Psychometrika, 49*, 529–544.

Mislevy, R. (1985). Estimation of latent group effects. *Journal of the American Statistical Association, 80*, 358–381.

Mislevy, R. (1991). Randomization-based inferences about latent variables from complex samples. *Psychometrika, 56*, 177–196.

Mislevy, R. (1992). Scaling procedures. In E. Johnson & N. Allen (Eds.), *The NAEP 1990 Technical Report* (pp. 199–213). Washington, DC: National Center for Educational Statistics.

Mislevy, R., & Bock, D. (1982). BILOG: Item analysis and test scoring with binary logistic models [Computer software]. Mooresville, IN: Scientific Software.

Mislevy, R., Johnson, E., & Muraki, E. (1992). Estimating population characteristics from sparse matrix samples of item responses. *Journal of Educational Statistics, 17*, 131–154.

Mislevy, R., & Sheehan, K. (1989). The role of collateral information about examinees in the estimation of item parameters. *Psychometrika, 54*, 661–679.

Muraki, E. (1992). A generalized partial credit model: Application of an EM algorithm. *Applied Psychological Measurement, 16*, 159–176.

Patz, R. (1996). Markov chain Monte Carlo methods for incomplete, multiple item type educational data. *Proceedings of the Bayesian Section of the American Statistical Association*, 128–133.

Patz, R. (1999). A straightforward approach to Markov chain Monte Carlo for item response models. *Journal of Educational and Behavioral Statistics, 24*, 146–178.

Rubin, D. (1987). *Multiple imputation for nonresponse in surveys.* New York: Wiley.

Thomas, N. (1993). Asymptotic corrections for multivariate posterior moments with factored likelihood functions. *Journal of Computational and Graphical Statistics, 2*, 309–322.

Thomas, N. (in press). Assessing model sensitivity of imputation methods used in NAEP. *Journal of Educational and Behavioral Statistics.*

Wu, J. (1983). On the convergence properties of the EM algorithm. *The Annals of Statistics, 11*, 95–103.

12

Cross-Validating Item Parameter Estimation in Adaptive Testing

Wim J. van der Linden and Cees A.W. Glas[1]

ABSTRACT From a statistical point of view, adaptive testing belongs to the class of problems of selecting a best subset of random variables. As in any other instance of this problem, adaptive testing is likely to capitalize on estimation error. An empirical example of the phenomenon is discussed, and a strategy for cross-validating results from adaptive testing is proposed. Results from a simulation study on the strategy are presented.

1 Introduction

The simultaneous development of item response theory (IRT) in test theory (Birnbaum, 1968) and the arrival of the computer in the practice of educational and psychological testing have prepared the introduction of large-scale application of computerized adaptive testing (CAT) in the 1990s (Wainer, 1990). In adaptive testing the statistical properties of the items are adapted to the ability level of the individual examinees. To this end, the computer records the examinee's responses, updates his or her ability estimate continuously, and selects subsequent items to have optimal statistical properties at the last estimate. The process is stopped as soon as a fixed number of items has been administered or the ability estimator has reached a predetermined precision.

The IRT model assumed in this chapter is the two-parameter logistic model (2PLM) defined as

$$p_i(\theta) = P(X_i = 1 \mid \theta) = \frac{\exp[\alpha_i(\theta - \beta_i)]}{1 + \exp[\alpha_i(\theta - \beta_i)]} , \tag{12.1}$$

where θ is the ability of the examinee, and β_i and $\alpha_i \in [0, \infty)$ are parameters for the difficulty and discriminating power of item i, respectively. The

[1]Both authors are at the Department of Educational Measurement and Data Analysis, University of Twente, P.O. Box 217, 7500 AE Enschede, The Netherlands; e-mail: *vanderlinden@edte.utwente.nl* and *glas@edte.utwente.nl*

2PLM was introduced by Birnbaum (1968), who also laid the foundation for most of its statistical theory. When a CAT item pool is calibrated, the values of its items for the parameters in (12.1) are estimated. During the operational stage of the test, it is common to treat these estimates as true values.

To formalize the principle of adaptive testing, the following notation is used. The items in the pool are denoted $i = 1, \ldots, k$. In addition, the index $j = 1, \ldots, h$ is introduced to denote the items in the adaptive test. Thus, i_j is the index of the item in the pool administered as the jth item in the test. Furthermore, $\mathcal{I}_j = \{i_1, \ldots, i_{j-1}\}$ is defined as the set of indices of the first $j - 1$ items in the test. These items have a vector of random responses $\mathbf{X}_{j-1} = (X_{i_1}, \ldots, X_{i_{j-1}})'$ with realization $\mathbf{x}_{j-1}$. The update of the estimated value of the ability parameter after $j-1$ responses is denoted $\hat{\theta}_{\mathbf{X}_{j-1}}$. Finally, $\mathcal{L}_j = \{1, \ldots, k\} \backslash \mathcal{I}_j$ is defined as the set of items left in the pool after $j - 1$ items have been administered. Item j is selected to be optimal at $\hat{\theta}_{\mathbf{X}_{j-1}}$ among the items in this set. In the empirical example in Section 4, the adaptive test is stopped if m items have been administered.

1.1 Item Selection Criterion

Various criteria of optimality for item selection have been proposed (for a review, see Van der Linden & Pashley, 2000), but the criterion applied most frequently is the maximum-information criterion in combination with maximum likelihood (ML) estimation of the ability parameter.

The maximum-information criterion follows directly from Birnbaum's (1968) proposal to use the test information function for the assembly of fixed tests. The test information function is Fisher's information measure defined as a function of the ability parameter. For the model in (12.1), it takes the following form,

$$I_{\mathbf{X}_{j-1}}(\theta) = \sum_{d=1}^{j-1} \alpha_{i_d}^2 p_{i_d}(\theta)[1 - p_{i_d}(\theta)] \ . \tag{12.2}$$

Birnbaum's motivation for this proposal was the fact that, for increasing test length, the variance of the maximum likelihood estimator of θ is known to converge to the reciprocal of (12.2). In addition, as follows from (12.2), information functions are additive in the items and easy to calculate.

The maximum-information criterion selects the jth item in the adaptive test to maximize (12.2) at $\theta = \hat{\theta}_{\mathbf{X}_{j-1}}$. Formally, it can be presented as

$$i_j = \underset{i \in \mathcal{L}_j}{\arg\max} \ I_{\mathbf{X}_{j-1}, X_i}(\hat{\theta}_{\mathbf{X}_{j-1}}) \ .$$

Because of the additivity of the information function, the criterion boils down to

$$i_j = \underset{i \in \mathcal{L}_j}{\arg\max} \ I_{X_i}(\hat{\theta}_{\mathbf{X}_{j-1}}) \ . \tag{12.3}$$

In the empirical study reported below, Warm's (1989) weighted likelihood estimator (WLE) was used. After $j-1$ items are administered, the WLE is defined as

$$\hat{\theta}_{\mathbf{x}_{j-1}} = \arg\max_{\theta} \left[w_{j-1}(\theta) L(\theta \mid x_{i_1} \ldots, x_{i_{j-1}}) \right] , \tag{12.4}$$

where $L(\theta \mid x_{i_1} \ldots, x_{i_{j-1}})$ is the likelihood function after $j-1$ items are administered and $w_{j-1}(\theta)$ is a weight function defined such that

$$\frac{\partial \ln w_{j-1}(\theta)}{\partial \theta} = \frac{J_{j-1}(\theta)}{2 I_{\mathbf{x}_{j-1}}(\theta)} , \tag{12.5}$$

with

$$J_{j-1}(\theta) = \sum_{d=1}^{j-1} \alpha_{i_d}^2 \frac{\partial^2 p_{i_d}(\theta)}{\partial \theta^2} , \tag{12.6}$$

The WLE has the advantage of being unbiased to order m^{-1} (Warm, 1989).

1.2 Item Exposure Control

Early experiences with CAT in high-stakes testing have shown that, unless special precaution is taken, examinees are able to memorize items and tend to share them with each other. Sympson and Hetter (1985; see also Hetter & Sympson, 1997) introduced a probabilistic method of item exposure control that has been effective in protecting the item pool from being compromised. The method uses a probability experiment to determine whether an item selected by the CAT algorithm should be actually administered or removed from the pool for the examinee.

Let B_i be the event of the algorithm selecting item i for an arbitrary examinee from the population, which has probability $P(B_i)$. In addition, the event of administering item i to an arbitrary examinee is denoted A_i. This event has probability $P(A_i)$, which is also known as the exposure rate of item i. Because A_i implies B_i, it follows from probability calculus that

$$P(A_i) = P(A_i, B_i) = P(A_i \mid B_i) P(B_i) . \tag{12.7}$$

The conditional probability $P(A_i \mid B_i)$ is known as the exposure control parameter for item i. For a given CAT algorithm, exposure rate $P(A_i)$ can be controlled by setting the value of the exposure control parameter at an appropriate level.

In practice, the values of the control parameter for the items are set prior to the operational stage of the adaptive tests through an iterative adjustment process consisting of the following cycle: (1) simulating a large number of test administrations for random examinees from the population, (2) estimating $P(B_i)$ and $P(A_i)$, and (3) adjusting the values of $P(A_i \mid B_i)$.

Let t denote the steps in the iterative process and r the target value below which the exposure rates of the items should be. The adjustments can be summarized by the following iterative relation,

$$P(A_i^{(t)} \mid B_i^{(t)}) = \begin{cases} 1\,, & \text{if } P(B_i^{(t-1)}) \geq r\,, \\ r/P(B_i^{(t-1)})\,, & \text{if } P(B_i^{(t-1)}) < r\,. \end{cases} \tag{12.8}$$

Because $B_i^{(t)}$ is dependent on $A_d^{(t-1)}$ for some items d in the pool, (12.7) only holds within each iterative step. Nevertheless, the adjustments in (12.8) are based on the assumption of (12.7) holding between steps. In practice, however, for a reasonable choice of r and a well-designed item pool, most adjustment processes do converge to a stable set of values for the exposure control parameters.

Several extensions of the Sympson–Hetter method have been proposed and used in practice. A sensible extension is to adjust the item exposure rates conditionally on θ. This extension may discourage examinees to try assessing what items in a CAT program have high exposure rates at their own ability level. A review of all current extensions of the Sympson–Hetter method is given in Stocking and Lewis (2000). In the empirical study discussed in Section 4, the unconditional version of the Sympson–Hetter method defined by (12.8) was used.

2 Capitalization on Item Calibration Error

If the application of a statistical model involves optimizing a function of its parameters, it may show a tendency to favor combinations of parameter values with large estimation errors. This capitalization on chance is known to exist, for example, when selecting a best subset of random variables. Typical cases include the selection of a set of predictors to have maximum multiple correlation in linear regression (Neter & Wasserman, 1974, Sect. 11.6), classical item analysis with the goal to select a set of items with maximum reliability or validity (Lord & Novick, 1968, Sect. 15.4), and the assembly of tests to have optimum information functions in IRT (Hambleton & Jones, 1994; Hambleton, Jones, & Rogers, 1993).

The danger of capitalization on chance is large when (Lord & Novick, 1968, Sect. 15.4)

1. the errors in the parameter values are large,

2. the selection ratio is small, and

3. the true parameter values have small variability.

The selection ratio is defined as the ratio of the number of variables selected to the number available. The effects of the first two factors were

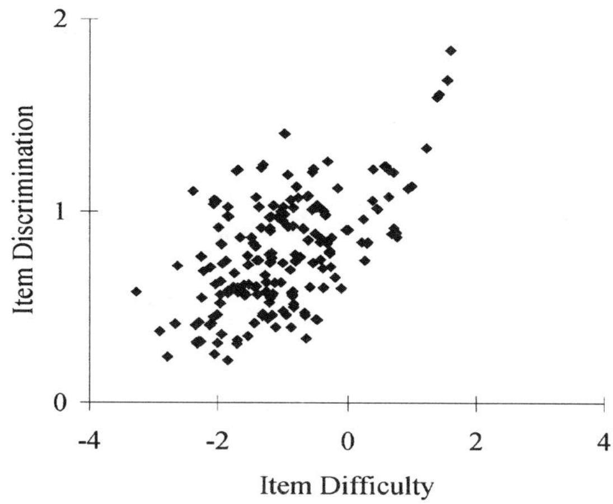

FIGURE 12.1. Distribution of difficulty and discrimination parameters in the item pool.

confirmed in test assembly studies by Hambleton et al. (1993; 1994). They showed that optimal test assembly may lead to overestimation of the height of the test information function, and that this overestimation tends to increase when the size of the errors in the item parameter values is increased or the selection ratio decreased.

Another example of capitalization on chance relates to adaptive testing. Item selection criterion (12.3) is a direct function of the squared value of α_i. Item selection is thus sensitive to large positive errors in the estimates of this value, especially if the value of the estimated difficulty parameter β_i is close to $\hat{\theta}_{\mathbf{x}_{j-1}}$. On the other hand, it is less sensitive to the factor $p_{i_j}(\theta)[1 - p_{i_j}(\theta)]$ in (12.2), which is known to be robust to variation in the item parameter values over a large neighborhood of $\theta = \beta_i$.

The phenomenon of capitalization on item calibration error in CAT (i.e., error in the item parameter estimates) was shown to exist in a previous study (Van der Linden & Glas, 2000). They simulated adaptive tests from item pools calibrated on subject samples of four different sizes ($n = 250$, 500, 1500, 2500). Each pool was manipulated to have the same (joint) distribution of true item parameter values; only the size of the pools varied. The true values were derived from an empirical item pool with a distribution of item parameter values shown in Figure 12.1.

The items in this pool were relatively easy; the values for the difficulty parameter ranged from -3.59 to 1.62, with a mean equal to -0.97. Also, the pool had large values for the discrimination parameter only at the upper end of the θ scale.

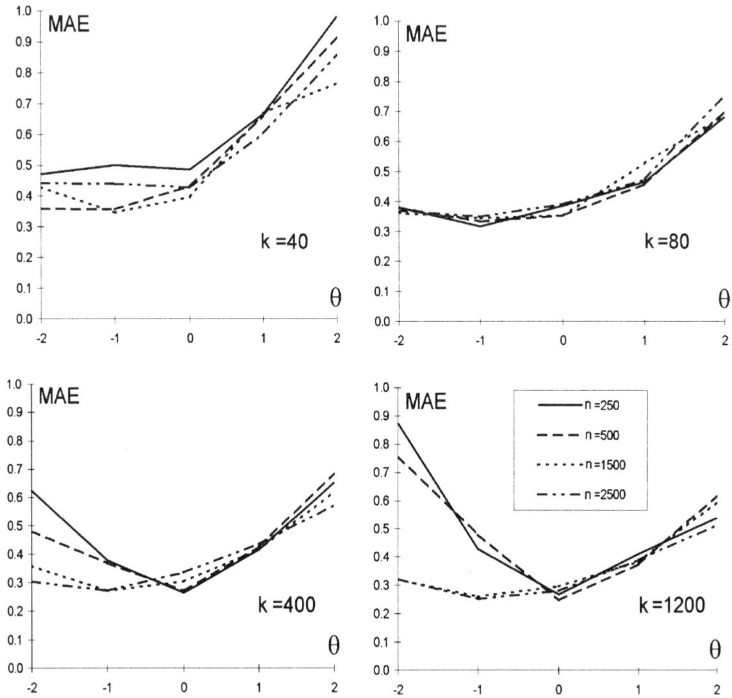

FIGURE 12.2. Mean absolute error in the ability estimates for item pools calibrated after $m = 20$ items.

One of the results from this study is reproduced in Figure 12.2, which shows the mean absolute error (MAE) in the final ability estimator as a function of θ for tests of $m = 20$ items. The first panel in this figure shows the MAE for a pool of $k = 40$ items. As expected, calibration of the pool on the smaller samples yielded a larger MAE. In addition, for all sample sizes, the MAE at the upper end of the θ scale was larger than at the lower end. This phenomenon follows from the distribution of the true item parameter values in Figure 12.1. The second panel shows that increasing the item pool to a size of $k = 80$ items yielded better MAE curves across the scale. Also, only minor differences between the MAE curves for the four sample sizes n were observed. The third panel shows surprising results for the item pool of $k = 400$ items. Although one would expect the trend in the first two panels to continue and see lower MAE curves, at the lower end of the scale the curves for the smaller calibration samples actually went up. However, the most shocking results were obtained for the pool of $k = 1200$ items. At the lower end of the scale, the curves of the two smallest samples were approximately twice as high as those for the pool of $k = 40$ items. That is, increasing the size of the item pool by a factor equal to 30—intuitively one of the best things that could happen to a CAT program—did not lead to any improvement in ability estimation, but to a redoubling of estimation error across half the range of θ values!

Observe that the above findings show an interaction between size of the item pool and size of the calibration sample. The combinations with larger pools and smaller calibration samples yielded the largest errors in the estimated values of the ability parameter. Therefore, capitalization on calibration error by the CAT algorithm explains the results: the smaller calibration samples yielded item pools with larger errors for the item parameter values. Because the larger item pools had smaller selection ratios, those among them with large estimation errors showed a strong tendency to capitalize on them.

The role of the selection ratio also explains the different behavior of the MAE curves at the lower and upper ends of the scale. The portions of the curves at the upper end were affected only by the size of the calibration sample, but at the lower end both the size of the sample and the size of the item pool had an impact. The reason is that the strongly skewed distribution of the difficulty of the items in the pool resulted in huge differences in the selection ratios at the two ends of the scale. Finally, Figure 12.1 also shows that the true values of the item discrimination parameters showed less variability at the lower end of the scale, a condition listed above as the third factor known to induce capitalization on chance. For other data showing the existence of capitalization on item calibration error in adaptive testing, see Van der Linden and Glas (2000).

3 Cross-Validating Item Parameter Estimation

The danger of capitalization on chance seems much larger in CAT than in the other selection problems referred to above. The reason is the extremely low selection ratio involved in CAT. In prediction problems, variables are usually selected from a set at most twice as large. In classical item selection, typically a few more items than needed are available, and the goal is only to weed out the worst items. However, in IRT-based test assembly, a fixed test is assembled from an item pool that is often at least 10 times larger than the length of the test. Adaptive tests, however, are much shorter than fixed tests, implying an even more unfavorable selection ratio.

A well-known recommendation to neutralize the effects of capitalization on chance is to use the technique of cross-validation. To cross-validate results from selection problems, two samples are drawn. One sample is used to select the variables and the other to reestimate the values of their parameters. Lord and Novick (1968, Sect. 13.2) call these two samples the screening and calibration sample, respectively.

Although straightforward in other selection problems, the technique of cross-validation seems difficult to implement in CAT. Because ability estimation is in real time, it is impossible to recalibrate the item selected for administration by the CAT algorithm during the test. However, it does appear possible to apply the following technique: when the item pool is calibrated, the sample of examinees in the pretest is split into two parts and the items are calibrated on both parts separately. In the operational stage, one set of estimates is then used to select the items from the pool and the other set to update the ability estimate $\hat{\theta}_{\mathbf{X}_{j-1}}$. If the split is random, both sets of estimates are independent and capitalization on chance cannot occur.

Any attempt to cross-validate results from selection problems is bound to face a trade-off between error due to capitalization on chance and sample size. If the sample is split into two parts and different parts are used to calibrate the items for item selection and ability estimation, the error in the ability estimator due to capitalization on chance is removed. However, because ability is now estimated using items calibrated on a smaller number of examinees, the estimators of the item parameters, and hence the ability estimator, are expected to have greater mean squared errors. Both effects can be manipulated by changing the size of the two subsamples. Thus, from the point of view of sample design, the question is which split would result in an optimal net effect. Although this question has received a good deal of attention in predictive validity studies (e.g., Raju, Bilgic, Edwards, & Fleer, 1997), guidelines for splitting samples in cross-validating item parameter estimation in IRT cannot be given.

It is the purpose of this chapter to study empirically the effects of the proposed cross-validation technique for two different sample designs. In one design, the sample is split into two halves and item selection and ability estimation are performed on different halves. In the other design, the sample is also split into two halves but, reminiscent of the idea of double cross-validation (Mosier, 1951), item selection is performed on one half and ability estimation on the full sample. In the former, reduction of error due to capitalization on chance is maximal but the efficiency of the ability estimator is based on items calibrated using half the data; in the latter, the effects of capitalization on chance are only partially reduced but maximum efficiency of ability estimation is maintained.

To make the study realistic, both designs were applied to CAT with and without the item exposure control defined in (12.8). Several large-scale testing programs use exposure control, and its presence can be expected to have a mitigating effect on the effects of capitalization on item calibration error as well. If the Sympson–Hetter method of item exposure control is applied, items are no longer ordered solely by the maximum-information criterion but also by a random component independent of the estimation errors in their parameter values.

4 Empirical Study

An empirical study was conducted to assess the extent to which cross-validation would be able to remove the effects of capitalization on item calibration error on ability estimation in CAT. To make the results comparable with those in Figures 12.1 and 12.2, the design of the previous study was replicated for the two pool sizes that showed capitalization on chance in the previous study ($k = 400$ and $k = 1200$). These factors were thus manipulated:

1. the size of the calibration sample ($n = 250,\ 500,\ 1500,\ 2500$);
2. the size of the item pool ($k = 400,\ 1200$);
3. cross-validation with ability estimation on half or full sample sizes;
4. item selection with and without item exposure control.

4.1 Method

The item pools and calibration samples in the study were generated from a data matrix collected in a Dutch national examination of English as a foreign language. The matrix contained the responses of 1000 examinees to 100 items. To manipulate the size of the calibration sample and item pool without confounding these factors with possible differences between the ability or item parameter distribution, respectively, the following setup was used. First, the values of the item parameters were estimated from the data matrix assuming the 2PLM (12.1). The method of estimation was marginal maximum likelihood with a $\mathcal{N}(0,1)$ distribution for the ability parameter. Second, item pools of required sizes were obtained by sampling items from the data matrix. Sampling was at random and with replacement. Each item was assumed to have true parameter values equal to the values estimated from the data matrix. Third, errors in the estimates of the item parameter values were simulated using the assumption of normally distributed error with mean equal to the true parameter values and covariance matrix equal to the inverse of the Fisher information matrix. The information matrix is additive in the contributions by the examinees. Therefore, the information matrices for the various sizes of the calibration sample could be obtained by sampling examinees until the required sample sizes were reached and then adding their contributions. Again, sampling was at random and with replacement. Fourth, the errors in the estimated values of the item parameters were sampled randomly from the appropriate normal distribution and added to the true item parameter values.

For each condition, administrations of adaptive tests of $m = 20$ items were simulated. Ability was estimated using the WLE (12.4). Items were

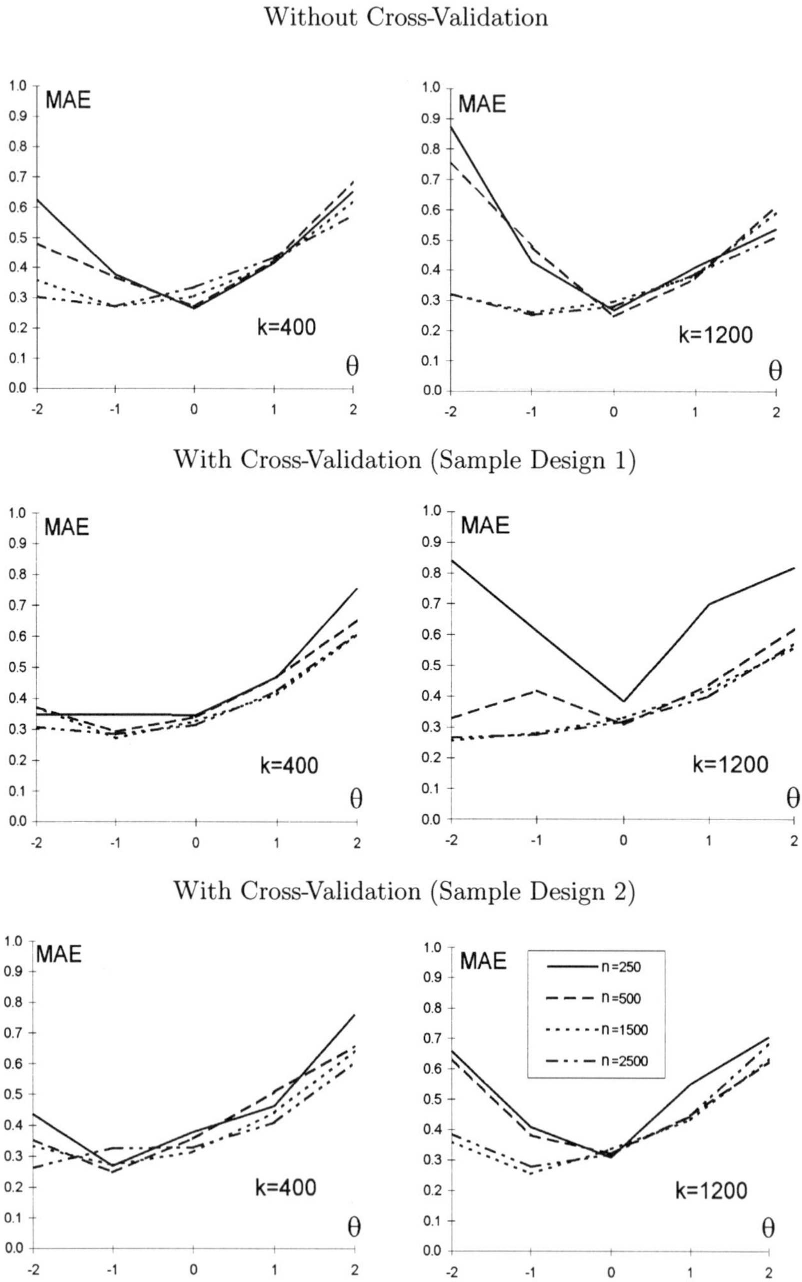

FIGURE 12.3. Mean average error for item pools without item exposure control ($m = 20$).

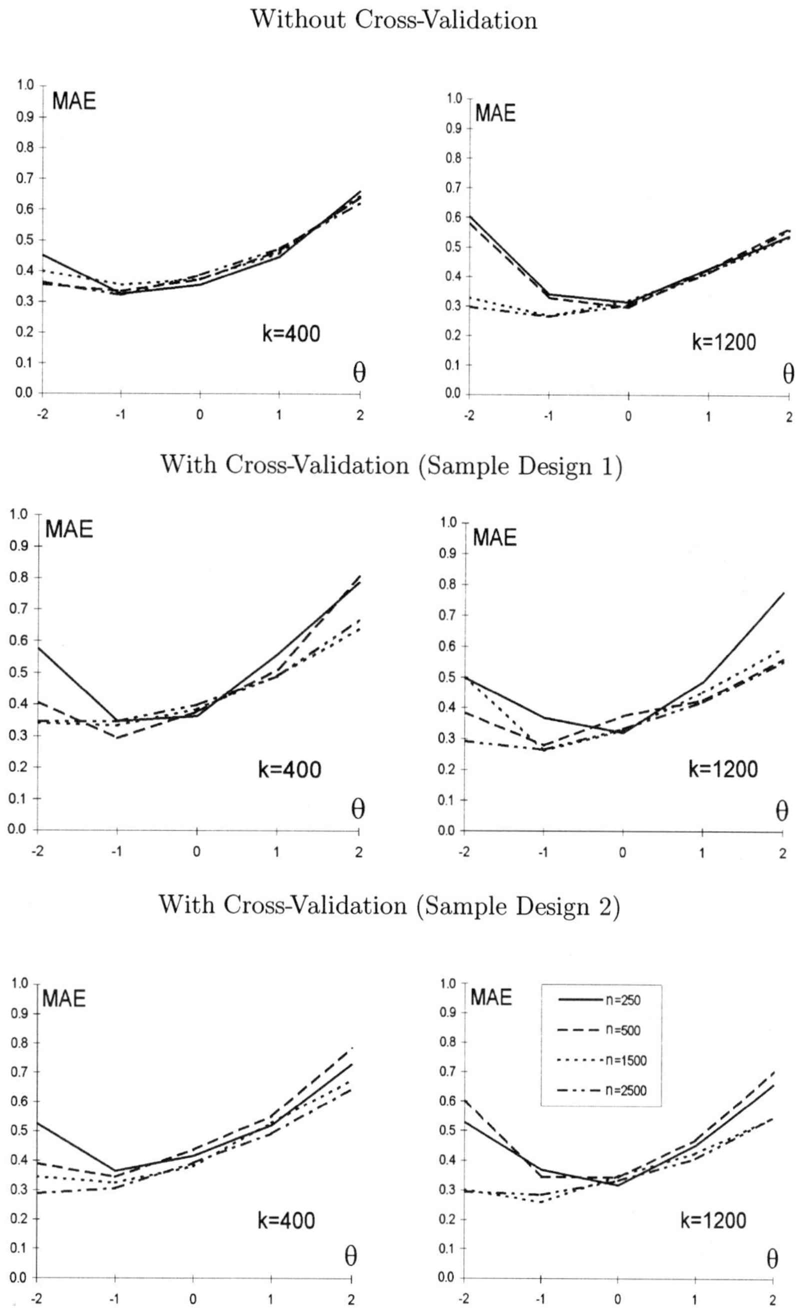

FIGURE 12.4. Mean average error for item pools with item exposure control ($m = 20$).

selected using the maximum-information criterion in (12.3). Each administration was replicated 1000 times for $\theta = -2.0,\ -1.0,\ 0.0,\ 1.0$, and 2.0. For each examinee, the absolute error in $\hat{\theta}$ for $k = 20$ was recorded. Mean absolute errors were calculated as a function of the θ values in the study. The number of replications was assumed to be large enough to obtain stable estimates of the mean absolute error.

Cross-validation was implemented using the two sample designs discussed above. Adaptive tests were simulated both with and without the Sympson–Hetter method of item exposure control. In the conditions with item exposure control, the adjustment process based on iterative relation (12.8) yielded exposure rates smaller than 0.23 for all items.

4.2 Results

The results for the item pools of $k = 400$ items without item exposure control are given in the three panels on the left-hand side of Figure 12.3. The panels show the MAE curves for the ability estimation without cross-validation, with cross-validation and ability estimation updated on half the sample (Sample Design 1), and with cross-validation but ability estimation updated on the full sample (Sample Design 2). Cross-validation of item parameter estimation appeared to have a positive effect. The mean absolute estimation errors at the lower end of the scale were reduced, and now of the same size as those near the middle of the scale. However, the differences between the two sampling designs were small and did not seem to be systematic.

The results for the item pool with $k = 1200$ items, displayed in the three panels on the right-hand side of Figure 12.3, show that cross-validation of item parameter estimation appeared to have a favorable impact, in particular for the calibration sample of size $n = 500$ and Sample Design 1. Generally, for this pool size, Sample Design 1 seems to be preferred over Sample Design 2. Hardly any effects were obtained for the calibration samples with $n = 250$ examinees. Apparently, splitting samples of this size results in a reduction of estimation efficiency too large to be compensated for by cross-validation. The same effect was observed in all other cases.

The most remarkable result for $k = 400$ in Figure 12.4 is in the first panel, which shows that nearly all effects of capitalization have already been removed by the use of the Sympson–Hetter method of item exposure control. As already noted, this method adds a random component to the item selection mechanism. As a consequence, the items are no longer selected only according to the criterion of maximum-information in (12.3), and the tendency to capitalize on calibration error is reduced. The second and third panels show that adding cross-validation to item exposure control did not produce any extra reduction. In fact, for the case of sample size $n = 250$ and Sample Design 1, the results at the lower end of the scale for $k = 400$ in Figure 12.4 are even worse than for $k = 400$ in Figure 12.3.

The first panel on the right-hand side of Figure 12.4 shows that for the larger item pool of $k = 1200$ items, use of the Sympson–Hetter method of item exposure control is not enough to remove all effects of capitalization on chance. The reason for this phenomenon is the much more unfavorable selection ratio for this pool size compared to a pool size of 400. For the calibration sample of size $n = 500$, cross-validation seems to have an additional positive effect on the presence of capitalization on chance, but for Sample Design 1 and $n = 250$ no systematic effects were observed.

5 Concluding Remarks

Capitalization on calibration error is a potential danger in adaptive testing. As shown by the simulation results in this chapter, its effects can be removed by cross-validating item parameter estimation. The same appears to hold for the method of item exposure control. Either method was approximately equally effective. The combination of both methods did not offer any additional benefits.

Also, as a general recommendation, to avoid capitalization on calibration error, it seems necessary to choose the calibration sample larger than 250. For this sample size, the effects of cross-validation were not strong enough to compensate for the loss of efficiency due to sample split.

Two different sample designs for cross-validating item parameter estimation were studied. Although definitive conclusions based on this study are impossible, it seems that for the combination of cross-validation with the Sympson–Hetter method of item exposure control Sample Design 1 is to be preferred over Sample Design 2.

A different solution to the problem of calibration estimation error in adaptive testing would be through a Bayesian approach. Generally, a Bayesian approach does not use point estimates for the item parameters but treats them as random. It would estimate the examinee's ability by its marginal posterior distribution, averaging out the item parameters. The approach has been described for ability estimation from fixed tests in Tsutakawa and Johnson (1990). A modification for application in adaptive testing is given in Van der Linden and Pashley (2000). Unfortunately, due to the increasing number of item parameters, the approach is numerically not yet feasible for adaptive testing. Therefore, for the time being, it is recommended to choose the size of the calibration sample in adaptive testing not too small, or to use one of the two methods to reduce the effects of capitalization on calibration error studied in this chapter.

Acknowledgments: This study received funding from the Law School Admission Council (LSAC). The opinions and conclusions contained in this chapter are those of the authors and do not necessarily reflect the position or policy of LSAC.

References

Birnbaum, A. (1968). Some latent trait models and their use in inferring an examinee's ability. In F.M. Lord & M.R. Novick, *Statistical theories of mental test scores* (pp. 395–479). Reading, MA: Addison-Wesley.

Hambleton, R.K., & Jones, R.W. (1994). Item parameter estimation errors and their influence on test information functions. *Applied Measurement in Education, 7,* 171–186.

Hambleton, R.K., Jones, R.W., & Rogers, H.J. (1993). Influence of item parameter estimation errors in test development. *Journal of Educational Measurement, 30,* 143–155.

Hetter, R.D., & Sympson, J.B. (1997). Item exposure control in CAT-ASVAB. In W.A. Sands, B.K. Waters, & J.R. McBride (Eds.), *Computerized adaptive testing: From inquiry to operation* (pp. 141–144). Washington, DC: American Psychological Association.

Lord, F.M., & Novick, M.R. (1968). *Statistical theories of mental test scores.* Reading, MA: Addison-Wesley.

Mosier, C.I. (1951). Symposium: The need and means of cross validation. *Educational and Psychological Measurement, 11,* 5–11.

Neter, J., & Wasserman, W. (1974). *Applied linear statistical models: Regression, analysis of variance, and experimental design.* Homewood, IL: Irwin.

Raju, N.S., Bilgic, R., Edwards, J.E., & Fleer, P.F. (1997). Methodology review: Estimation of population validity and cross validity, and the use of equal weights in prediction. *Applied Psychological Measurement, 21,* 291–306.

Stocking, M.L., & Lewis, C. (2000). Methods of controlling the exposure of items in computerized adaptive testing. In W.J. van der Linden & C.A.W. Glas (Eds.), *Computerized adaptive testing: Theory and practice* (pp. 163–182). Boston, MA: Kluwer.

Sympson, J.B., & Hetter, R.D. (1985, October). *Controlling exposure rates in computerized adaptive testing* [Transparencies]. Paper presented at the 27th Annual Conference of the Military Testing Association. San Diego, CA: Military Testing Association.

Tsutakawa, R.K., & Johnson, J.C. (1990). The effect of uncertainty on item parameter estimation on ability estimates. *Psychometrika, 55*, 371–390.

Van der Linden, W.J., & Glas, C.A.W. (2000). Capitalization on item calibration error in adaptive testing. *Applied Measurement in Education, 13*, 35–53.

Van der Linden, W.J., & Pashley, P.J. (2000). Item selection and ability estimation in adaptive testing. In W.J. van der Linden & C.A.W. Glas

(Eds.), *Computerized adaptive testing: Theory and practice* (pp. 1–25). Boston, MA: Kluwer.

Wainer, H. (Ed.). (1990). *Computerized adaptive testing: A primer.* Hillsdale, NJ: Erlbaum.

Warm, T.A. (1989). Weighted likelihood estimation of ability in item response theory. *Psychometrika, 54,* 427–450.

13

Imputation of Missing Scale Data with Item Response Models

Mark Huisman and Ivo W. Molenaar[1]

ABSTRACT Confronted with incomplete data due to nonresponse, a researcher may want to impute missing values to estimate latent properties of respondents. In this chapter the results of a simulation study are presented, investigating the performance of several imputation techniques. Some imputation procedures are based on item response theory (IRT) models, which can also be used to estimate latent abilities directly from the incomplete data by using incomplete testing designs when data are missing by design. This strategy has some serious disadvantages in the case of item nonresponse, because nonresponse is assumed to be ignorable and computational problems arise in scales with many items. In a second simulation study, the performance of some imputation techniques is compared to the incomplete design strategy, in the case of item nonresponse. The latter strategy results in slightly better ability estimates, but imputation is almost as good, especially when it is based on IRT models.

1 Introduction

Inferences about latent properties of subjects are usually made by analyzing the responses of these subjects to a set of items. Each single item does not cover all aspects of the latent trait, and a subject's position on the latent trait can only be inferred indirectly by investigating the responses to all items in the scale. In empirical research, measurement of latent abilities is frequently hampered by *item nonresponse*. This results in missing item responses, which make estimation of the abilities a difficult task.

Missing values can be dealt with in several ways. One popular technique is *imputation*. Imputation procedures estimate the missing responses and the estimates are substituted for the missing entries. This results in a completed data set that can be analyzed with standard measurement models. In this chapter, procedures to impute missing item responses are compared.

[1]Both authors are at the Department of Statistics and Measurement Theory, University of Groningen, Grote Kruisstraat 2/1, 9712 TS Groningen, The Netherlands; e-mail: *m.huisman@ppsw.rug.nl* and *w.molenaar@ppsw.rug.nl*

Measurement models to estimate the latent properties of subjects can be found in item response theory (IRT). A frequently used missing data strategy within the framework of IRT is to treat the missing values as if they resulted from an incomplete testing design. Such designs are created by the researcher who administers subsets of items to different respondents. The resulting incomplete data can be analyzed with IRT models and estimates of latent abilities obtained.

In Section 2 the handling of missing data in incomplete designs is discussed. Although it seems a profitable strategy to handle missing item responses, it cannot be used in every situation. Frequently other procedures are needed, and imputation is a good alternative. There are, however, many different ways to impute the missing data, and choosing the best technique is not always easy. In Section 3 several imputation techniques are described and their performance is investigated by a simulation study. The design and results of the study are presented in Section 4. Some of the investigated techniques are based on IRT models.

In Section 5 the two strategies, treating missing item responses as resulting from incomplete designs and imputation, are compared. The results of a second, small simulation study, in which latent abilities were estimated following both strategies, are presented. In the last section all results are summarized and the usefulness of missing data strategies is discussed.

2 Incomplete Testing Designs

2.1 Missing by Design

The use of incomplete testing designs stems from a major advantage of IRT: latent abilities can be estimated independent of the item set that is used. This means that different subsets of items can be administered to individuals to estimate their position on a common latent trait (Eggen, 1993). The mechanism generating the resulting incompleteness is specified by the researcher, who determines which subset of items is presented to which subject, and the data are called *missing by design.* In Figure 13.1 an example of missing data patterns in incomplete designs is given.

The reasons for administering only subsets of the available item pool can be of a practical nature, for example, when testing time is limited. However, incomplete designs are often constructed for reasons of efficiency with respect to estimation. Measurement of latent properties is more efficient when, for example, knowledge about the difficulty of tests and the abilities of respondents is used to determine which respondent is administered which subset of items (Eggen, 1993).

Incomplete testing designs can be analyzed with standard IRT software. In the program OPLM (Verhelst, Glas, & Verstralen, 1995) incomplete designs are handled by defining "booklets," that is, subsets of items adminis-

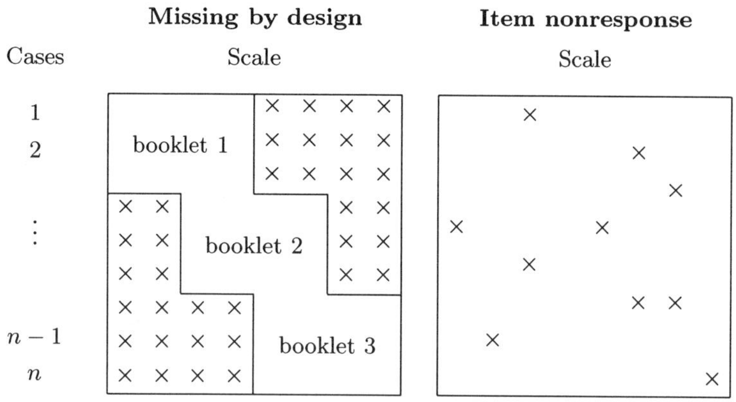

FIGURE 13.1. Responses missing-by-design and item nonresponse in test data ($\times$ denotes missing).

tered to different respondents (see Figure 13.1). OPLM can simultaneously analyze up to 70 booklets and estimate latent abilities on a common scale. One restriction is that the design must be linked. This means that either booklets have some items in common (as in Figure 13.1), or the ability parameters of the respondents of the booklets come from the same ability distribution.

2.2 *Ignorable Missing Data Mechanisms*

Procedures to handle missing data are usually based on the assumption that the data are *missing at random* (MAR; Rubin, 1976). This means that the missingness is only related to variables that are observed for both respondents and nonrespondents, and given these variables the mechanism that generates the missing responses is a random process. If the data are MAR and the parameters of the measurement model and the missingness mechanism are functionally unrelated, ignoring the process that causes the missing data gives correct likelihood inferences about the latent abilities (cf. Eggen & Verhelst, 1992).

In incomplete testing designs, the mechanism underlying the incompleteness is controlled by the researcher, and the ignorability of the missing data mechanisms can be assessed. Mislevy and Wu (1996) proved that for some routinely used incomplete designs, the process generating the missing data is ignorable under direct likelihood inference for the latent ability θ. Eggen and Verhelst (1992) investigated the estimation of item parameters in incomplete testing designs. They showed that Rubin's ignorability principle can directly be applied to justify the incomplete designs for marginal maximum likelihood estimation. For conditional likelihood estimation, however, the ignorability principle can never safely be applied, which means that the design mechanism explicitly has to enter the analysis.

2.3 Item Nonresponse

In the case of item nonresponse, the respondents "control" the mechanisms that generate the missingness, and ignorability cannot easily be verified. Mislevy and Wu (1996) even argued that the missing data mechanisms are generally not ignorable in the case of intentionally omitted items, and should be included in the analysis to prevent results from becoming biased.

Missing responses caused by item nonresponse will appear to be randomly distributed over the data set, as depicted in Figure 13.1. Although the data are not missing by design, the patterns of missingness may be treated as if they resulted from an incomplete design. For example, latent abilities can be estimated with the program OPLM, by defining booklets with subsets of observed items. The advantage of this strategy is that all information in the data set is used in the estimation process. Moreover, no artificial information is included, which means that no extra uncertainty is created as is the case with imputation.

This procedure has, however, some major disadvantages. First, it heavily depends on the assumption of ignorability, which is probably invalid in the case of item nonresponse. The second disadvantage is of a practical nature. In situations where the respondent is administered a moderate to large number of items and/or the percentage of missing values is large, defining booklets for every possible nonresponse pattern may cause problems (e.g., 10 items result in $2^{10} - 1 = 1023$ possible patterns). With realistic numbers of items, the number of patterns will probably exceed the limit of OPLM (70 booklets), which implies that ability estimates cannot be obtained for every respondent. Even with 30 booklets, the iterative estimation process may encounter computational problems. Therefore, it is problematic to raise the limit currently imposed by the software.

When the proportion of missing responses is small, the number of booklets will be smaller, and missing data due to item nonresponse can be handled as if resulting from an incomplete design. However, in this case of small nonresponse percentages, other missing data treatments may work just as well and may be easier to implement. Because of the disadvantages of the incomplete design strategies, imputation procedures can be good alternatives to handle missing item responses. In the next sections the performance of some imputation procedures is investigated.

3 Imputation of Missing Item Responses

Imputation, used to adjust for item nonresponse in surveys, can be viewed from different approaches. The first one is the *randomization approach*, deriving inferences from the distribution used to select units into the sample. The second approach is *model-based*, deriving inferences from the model specified for the random population values (cf. Rubin, 1983).

3.1 Randomization Approach: Adjustment Cells

In the randomization approach, population values are treated as fixed and no model is specified. In the presence of missing data, however, some modeling assumptions have to be made about the response mechanism to obtain unbiased estimates of parameters of interest (Little & Rubin, 1987). This extension of the randomization approach is named *quasirandomization* inference. A well-known class of missing data adjustments from this approach are weighting procedures.

Sande (1982) and Little and Rubin (1987, Chap. 4) discuss the general problem of nonresponse and imputation of the resulting missing data from the randomization point of view. They present techniques like *mean imputation* and *hot deck imputation* as randomization-based adjustments. From this notion of quasirandomization stems the idea of the construction of *adjustment cells* to correct for bias caused by nonresponse. It is closely related to poststratification, a frequently used weighting technique to handle nonresponse (cf. Bethlehem, 1988).

Adjustment cells are subsamples of respondents with equal values for some completely observed covariates, and contain both complete and incomplete cases. Correction for nonresponse is most effective when the adjustment cells are homogeneous with respect to the variables of interest and/or to the probability of response (Bethlehem, 1988). The reduction of the bias will, however, only be complete when the nonresponse is ignorable given the completely observed covariates (MAR). This makes the choice of the covariates an important, but difficult task. Also in the case of nonignorable nonresponse, the nonresponse bias may be reduced substantially by using adjustment cells. Rubin, Stern, and Vehovar (1995) showed that when the covariates are strongly related to the variable of interest (and therefore to the response probability), analysis under the MAR assumption may yield satisfying results.

In this chapter, two imputation techniques from the quasirandomization approach are investigated: *hot deck nearest neighbor* and *corrected item mean imputation* within adjustment cells. Both techniques were investigated earlier without adjustment cells by Huisman (1999). The data sets to impute consist of responses of $v = 1, \ldots, n$ respondents to $i = 1, \ldots, k$ items ($\mathbf{X} = [x_{vi}]$) and values of $h = 1, \ldots, q$ covariates $\mathbf{Z} = [z_{vh}]$, which are assumed to be completely observed.

Hot deck nearest neighbor (HD). With hot deck imputation techniques a completely observed donor case from the same sample as the incomplete case is used to impute the missing values of the incomplete case. Within each adjustment cell a donor case is sought for which the distance function

$$d_{v,v'} = \sum_{i \in obs(v)} (x_{vi} - x_{v'i})^2 \tag{13.1}$$

is minimized for the collection of observed items $obs(v)$, of a given incomplete case v, over all complete cases v' in the same adjustment cell as v. The response pattern of case v' for which $d_{v,v'}$ is minimized resembles the observed responses of the incomplete case best, and therefore case v' is used as the donor case. If ties occur, the complete case that is closest to the incomplete case with respect to position in the data set is used as donor.

Corrected item mean (CM). The method of corrected item mean imputes an item mean that is corrected for the ability of the respondent. The item mean is multiplied by a weight reflecting the ratio between the total score of the respondent on the observed items and the total of average scores on these items. The corrected item means are computed in each adjustment cell separately:

$$\mathrm{CM}_{vi} = w_v \bar{x}_{.i}^{(i)} = \left(\frac{\sum_{i \in obs(v)} x_{vi}}{\sum_{i \in obs(v)} \bar{x}_{.i}^{(i)}} \right) \bar{x}_{.i}^{(i)} , \tag{13.2}$$

where $\bar{x}_{.i}^{(i)}$ is the mean score on item I_i for the nonmissing cases in an adjustment cell, and $obs(v)$ is the collection of observed items for subject S_v in the same adjustment cell. The values of CM_{vi} are rounded to the nearest integer to create an imputed data set which contains adequate scores that could have been observed. Note that this procedure assumes that the items are measured on a ratio scale. In practice this means that item scores should range from 0 to some maximum value.

3.2 Model-Based Approach: IRT Models

In the model-based approach, population values are treated as random variables and inferences are based on the probability model that is specified for these random variables. Little (1988) advocated the use of the model-based approach to missing data adjustments. Because imputation is a form of prediction, imputations should be based on the predictive distribution of the missing values, given the observed data. He argued that "*a model underlies this distribution, so a systematic approach to imputation requires modeling the data*" (p. 288). Conditions for good model-based imputation techniques are: (1) the models should take into account contextual knowledge about the items and avoid excessive extrapolation, (2) imputations should be drawn from the predictive distribution, and (3) appropriate methods for computing sampling errors should be included.

The obvious class of models to use for estimation of missing item responses is the class of IRT models. In this chapter, several IRT imputation procedures are investigated, based on two IRT models: the nonparametric *Mokken scaling model* (Mokken, 1971) and the *one-parameter logistic model* (Verhelst & Glas, 1995).

Mokken scaling (MOK). In Mokken scale analysis the cumulative response mechanism of the scale is used to order the subjects and the items on the latent trait θ. Because this is a nonparametric scaling technique, imputation based on the principles of the Mokken analysis can also be discussed from the randomization perspective. The imputation procedure is presented here as a model-based approach, however, to stress its relation with Mokken scaling. The procedure is summarized below. The item scores are denoted $x_1, x_2, \ldots, x_k$, and have the value 1 if correct, 0 if incorrect, and 9 if missing. Apply the following rules to impute the missing values.

1. Order the items decreasingly according to the proportion of correct responses based on the available cases. The items are thus ordered from "easy" to "difficult." In case of polytomous items, the responses should be transformed into dichotomous item steps first (Molenaar, 1997). These item steps represent imaginary thresholds between two adjacent response options. The value 1 indicates that the respondent crossed the threshold, and 0 if not.
2. x_i is imputed by the first of these rules for which the corresponding condition is satisfied:
   - 2.1. if $x_{i+1} = 1$, then $x_i = 1$;
   - 2.2. if $x_{i-1} = 0$, then $x_i = 0$;
   - 2.3. if the number of scores 0 is larger than or equal to the number of 1s among $x_1, \ldots, x_{i-1}$, then $x_i = 0$;
   - 2.4. if the number of scores 1 is larger than or equal to the number of 0s among $x_{i+1}, \ldots, x_k$, then $x_i = 1$;
   - 2.5. draw x_i at random according to its marginal distribution.
3. In the case of polytomous items, transform the item steps back into the original response options. Take into account the deterministic dependence between the item steps belonging to the same item (Molenaar, 1997): all item steps preceding a score 1 also have value 1, and all item steps following a score 0 also have the value 0.

Note that the rules, especially 2.3 and 2.4, are somewhat arbitrary; other rules could be formulated. However, they perform reasonably well because they keep the number of (Guttman) errors as small as possible.

OPMISS–Imputing with the one-parameter logistic model. Based on the observed responses, estimates of the subject and item parameters in the one-parameter logistic model can be computed with the program OPLM (Verhelst, Glas, & Verstralen, 1995). With these parameter estimates the distribution of the responses in every cell of the data matrix can be determined; $\hat{\pi}_{ij}(\hat{\theta}_v)$ is the estimated probability that subject S_v gives response j on item I_i. An ad hoc module called OPMISS was added to OPLM to impute missing values with the estimated probabilities; the module was described by Nap (1994).

With the estimated distributions, OPMISS can impute the missing responses in three different ways: imputing the modal category, imputing the expected value, or imputing a random draw from the distribution. In several simulation studies Nap (1994) showed that the best results were obtained with the random draw option, and that imputation of the expected value is as good as, and in many cases even better than, imputing the modal category. Therefore, the techniques investigated in this chapter are *imputing the expected value* and *imputing random draws* from the estimated distributions. These techniques are now described.

OPMISS–Expected value (OEV). The expected response of subject S_v for item I_i is substituted for the missing entry:

$$\mathrm{OEV}_{vi} = \sum_{j=0}^{c} j\hat{\pi}_{ij}(\hat{\theta}_v) , \tag{13.3}$$

where $j = 0, \ldots, c$ are the $c+1$ response options. All items in one scale are assumed to have the same number of response options. The values OEV_{vi} are rounded to the nearest integer.

OPMISS–Single draw (OSD). The values used for the imputation of the missing cells are randomly drawn from the estimated distribution of each cell, with probabilities $\hat{\pi}_{i0}(\hat{\theta}_v), \ldots, \hat{\pi}_{ic}(\hat{\theta}_v)$. Every missing data entry is imputed only once.

OPMISS–Multiple draws (OMD). Instead of imputing the missing item responses once, the OSD procedure is repeated five times for every incomplete data matrix; that is, every missing value is imputed five times (multiple imputation; Rubin, 1987). This results in five completed data sets. These multiply imputed data sets are investigated separately, and the five results are combined to yield estimates and standard errors that formally incorporate missing data uncertainty. The variation among the five imputations reflects the uncertainty due to both missingness and estimation of missing values from observed data.

4 Effects of Imputation: A Simulation Study

In this section the design of the simulation study to investigate the performance of the imputation techniques is described and the results are presented. Four independent factors were used: data (d), sample size (n), missing data mechanism (m), and proportion of missing values (p). In every cell of the design, incomplete data matrices were generated and imputed.

4.1 *Simulation Design*

Data (d). Four data sets were used. Three sets consisted of empirical data from behavioral science research: a physical functioning scale, RAND(PF), a mental health scale, RAND(MH), and a social interactions scale, NHP(SI) (see Huisman, 1999, pp. 97–98). The fourth data set, RM(DI) was an artificial one generated using the OPLM, following a Rasch scale. The data sets are presented in Table 13.1. Although actual field data will provide a better picture of the accuracy of imputation methods in practice, the RM(DI) data are included because some of the imputation techniques are based on the OPLM. This allows comparison of the methods under ideal circumstances, without the additional problem of misfit of the measurement model. Another reason to include RM(DI) is that not all data sets can be modeled equally well with the OPLM, which makes the results less comparable. Two aspects of the quality of measurement presented in Table 13.1, namely, reliability (Cronbach's alpha) and scalability of the items (Loevinger's H), are discussed later.

Sample size (n). Samples of $n = 200$, 400, and 800 cases were randomly drawn without replacement from the original data sets.

TABLE 13.1. Data sets serving as basis for data used in simulation. Total number of cases (n), number of items (k), response options (Category), item weight (w_i), Cronbach's alpha, and Loevinger's H (Huisman, 1999, pp. 97–98).

	Scale	n	k	Category	w_i	Alpha	H
$d1$	RAND(PF)	2456	10	$0-2$	5.0	0.94	0.76
$d2$	RAND(MH)	2517	5	$0-3$	6.7	0.85	0.65
$d3$	NHP(SI)	955	5	$0-1$	20.0[1]	0.74	0.60
$d4$	RM(DI)[2]	2000	10	$0-1$	10.0	0.63	0.25

[1] Average of unique weights: 22.01, 19.36, 20.13, 22.53, and 15.97.

[2] Item parameters: $-2.0, -1.7, -0.9, -0.5, -0.1, 0.2, 0.6, 1.1, 1.5, 1.8$.

Missing data mechanism (m). Missing values were created in the samples, using three different mechanisms.

$m1$: MCAR–Missing completely at random. The probability of response of subject S_v on item I_i is a random number between 0 and 1.

$m2$: NRX–Nonrandomness depending on **X**. The probability of response of subject S_v on item I_i is a logistic function of the scale score r_v and the mean item score $\bar{x}_{.i}$:

$$P(M_{vi} = 0 \mid r_v, \bar{x}_{.i}) = \frac{\exp(\alpha_0 + \alpha_1 r_v + \delta \bar{x}_{.i})}{1 + \exp(\alpha_0 + \alpha_1 r_v + \delta \bar{x}_{.i})}, \tag{13.4}$$

where $M_{vi} = 0$ indicates an observed response and α_0, α_1, and δ are suitably chosen scalar parameters (Huisman, 1999, p. 100).

$m3$: NRXZ–Nonrandomness depending on **X** and **Z**. The probability of response depends on the scale score r_v, the mean item score $\bar{x}_{.i}$, and the covariates sex and age, in the same way as for mechanism NRX.

The parameter values are fixed at specific levels such that the probability of response is small for subjects with low sum scores, women, older respondents, and for items with low mean value ("bad performance"). An observation is classified missing ($M_{vi} = 1$) when the probability of response is small, that is, smaller than a random number from a uniform distribution.

Proportion of missing values (p). Three proportions of missing cells in the data matrices were created, $p = 0.05$, 0.12, and 0.20, by setting the parameters of the missing data mechanisms at specific levels.

Adjustment cells. For the construction of adjustment cells two covariates were chosen that were present in all four data sets. Moreover, to make the nonresponse adjustment most effective, the covariates should be related to the latent abilities and/or the probability to respond. Because the covariates sex and age were used to create the missing responses, these two were also used in the construction of the adjustment cells. The continuous variable age was categorized, such that two different sets of adjustment cells were obtained: a set of four adjustment cells (two sex and two age groups), and a finer categorization of age, resulting in eight adjustment cells (two sex and four age groups). The advantage of the finer categorization is that more homogeneous groups are formed, which may result in better imputations. Its disadvantage is that the groups may be too small and do not contain complete cases. When that is the case, two adjacent groups are taken together and missing values are imputed within this new adjustment cell. The hot deck (HD) and corrected item mean (CM) techniques are investigated for both categorizations, and labeled HD4, HD8, CM4, and CM8, respectively.

4.2 Performance of Imputation Techniques

The effectiveness of the imputation techniques is evaluated using criteria from the practice of psychometrics. They are discussed on a subject level and a scale level, subsequently.

Subject level: Latent ability estimates. On the subject level, we are interested in the position of the respondents on the latent dimension. The latent ability θ_v is often estimated as the simple sum of item responses, $r_v = \Sigma_i x_{vi}$. Actually, in the OPLM r_v is a sufficient statistic for θ_v (if item discriminations are equal), and in Mokken scaling r_v itself is used as an estimate of θ_v. Therefore, the behavior of imputation techniques is judged by comparing the respondent's sum score before deletion of datapoints r_v, and after imputation of the missing values with procedure t, $r_v(t)$.

For the comparison of the imputation techniques, a weighted sum score $r_v = \Sigma_i w_i x_{vi}$ is used instead of the simple sum score. By multiplying every item with an item weight w_i, the sum scores of the four scales have the same range $(0 - 100)$. In three scales, the item weights equal $100/(k \times c)$, with k the number of items in the scale and c the difference between the largest and smallest response option. For these scales the weights are independent of the items (see also Table 13.1). For the NHP(SI) scale weights are used that differ per item, but the average weight equals $100/(k \times c)$. The weights of the RAND and NHP are actually the ones recommended in the literature (see Huisman, 1999, pp. 97–98).

It should be noted that the estimation of the latent abilities θ_v with the OPLM software is based on the simple, unweighted sum scores. Because the weights are constant in three of the four scales, the unweighted sum score is a sufficient statistic for θ_v. However, for the NHP data in which the weights differ for every item, the simple sum score is not sufficient for θ_v. Because of the high correlation between the weighted and the unweighted sum scores, this is considered not to distort the analyses.

On the subject level two evaluation measures were used: the mean $\bar{d}(t)$ of the deviation $d_v(t) = r_v(t) - r_v$ for all cases v with incomplete data, and the *root mean-squared deviation*

$$\text{RMSD}(t) = \left[\frac{1}{n_{mis}} \sum_{v=1}^{n_{mis}} \left(\sum_{i \in mis(v)} w_i \left[x_{vi}(t) - x_{vi} \right] \right)^2 \right]^{1/2}, \tag{13.5}$$

where $x_{vi}(t)$ is the imputed value for subject S_v and item I_i with imputation technique t, $mis(v)$ is the collection of missing item responses for subject S_v, and n_{mis} is the number of subjects with missing data. With $\bar{d}(t)$ the occurrence of systematic under- or overestimation of the sum scores is investigated, while RMSD(t) gives an indication of the general accuracy of the procedures.

Because of the item weights $w_i = 100/(k \times c)$ the values of RMSD(t) reflect certain things. First, incorrectly imputing an item in a short scale (small k) causes more bias in $r_v(t)$ than an item in a long scale. Second, incorrectly imputing an item with few response options (small c) causes more bias in $r_v(t)$ than an item with many response options. For example, incorrectly imputing a RAND(PF) item causes a change in $r_v(t)$ of 5 when the difference between the imputed and original value equals 1, whereas the same error in the NHP(SI) causes on average a change of 20.

Note that the differences in the sum score are on the subject level, which means that possible differences on items-within-subject level can be canceled out (e.g., imputing a 2 for item I_i while the original score was 3, and imputing a 4 for item I_j where the original score also was 3, does not affect the sum score of that particular incomplete case). This is considered acceptable because r_v is a sufficient statistic for the latent ability. When, however, the OPLM is used with the imputed data to estimate subject abilities, some complications arise. These are discussed in Section 5.

Scale level: Scale quality. On a scale level, we are interested in the ability of the techniques to preserve the relationship between the items. For this purpose, two measures of internal consistency of a scale are investigated: Cronbach's alpha, a lower bound for the classical concept of reliability, and Loevinger's H-coefficient, a measure of the scalability of items used in Mokken scaling (Mokken, 1971; Molenaar, 1997). The original values are presented in Table 13.1. The difference between the original values and those after imputation is investigated: $d_\alpha(t) = alpha(t) - alpha$ and $d_H(t) = H(t) - H$.

4.3 Results: Recovering Latent Ability Estimates

The design of the simulation is a complete factorial one, with four levels of d, three levels of n, three levels of m, and three levels of p, resulting in $4 \times 3 \times 3 \times 3 = 108$ cells. Within each cell 100 data sets with missing data were generated and imputed with the eight imputation techniques.

Main effects. In Table 13.2 the values of RMSD(t) for all imputation techniques averaged over all independent factors can be found. It can be seen that across all factors the OPMISS multiple draws method (OMD) performs best when estimating the scale score of respondents, followed by imputing an expected value (OEV). For each separate factor, OMD also has the smallest values for the main effect of RMSD(t).

Inspection of the mean deviations in sum scores, $\bar{d}(t)$—not reported here; see Huisman (1999, pp. 142–143)— shows that on average the r_vs are slightly overestimated. This means that the imputed values are larger than the deleted ones. This is due to the patterns of missingness for the NRX and NRXZ mechanisms, which delete small values with high probability. For the case of MCAR data such overestimation is not found.

TABLE 13.2. Main effects of the independent factors for the imputation techniques: RMSD(t) values averaged over all other factors.

	HD4	HD8	CM4	CM8	MOK	OEV	OSD	OMD
All	11.6	12.1	8.9	9.0	9.2	8.6	9.3	8.3
RAND(PF)	8.6	9.6	5.6	5.6	5.6	4.9	5.5	4.8
RAND(MH)	10.5	10.9	8.0	8.1	8.2	7.8	8.8	7.7
RM(DI)	10.4	10.7	7.7	7.7	8.4	7.9	8.6	7.3
NHP(SI)	16.9	17.2	14.5	14.5	14.4	13.8	14.2	13.4
$n = 200$	12.0	12.6	8.9	9.0	9.1	8.5	9.3	8.2
$n = 400$	11.5	12.0	8.9	9.0	9.2	8.6	9.3	8.3
$n = 800$	11.2	11.6	9.0	9.0	9.2	8.7	9.3	8.3
MCAR	7.4	7.6	6.5	6.5	6.7	6.3	7.0	5.9
NRX	12.9	13.3	9.1	9.1	9.3	8.5	9.2	8.2
NRXZ	14.4	15.3	11.2	11.2	11.4	11.0	11.7	10.7
$p = 0.05$	9.6	10.2	7.5	7.6	7.7	7.0	7.8	6.8
$p = 0.12$	11.7	12.1	9.0	9.0	8.6	8.6	9.3	8.3
$p = 0.20$	13.5	14.0	10.3	10.3	10.6	10.2	10.8	9.8

The effects of the four independent factors are as expected. Incorrect imputation in a short scale is more severe than in a long scale, as is the case with a scale consisting of dichotomous items compared with polytomous items. Second, the more complicated the missingness mechanism, the harder it is to impute correctly. And third, the higher the proportion of missing responses, the more mistakes are made by all techniques. The effect of sample size is negligible.

Interaction effects. Interaction effects show the same general picture: when the situation becomes uglier, the performance of the techniques worsens. Uglier is defined here as a scale that is more difficult to impute, which may be due to: higher item weights (i.e., few items and/or a small number of response options), a more complicated missing data mechanism, and/or more missing values.

In Figures 13.2 and 13.3, the RMSD(t) values of three-way interaction effects $d \times m \times p$ are presented. The values are averaged over sample size n, because the effect of n is small and not considered to be relevant. More specifically, the results for the scales RAND(PF) and RAND(MH) are presented in Figure 13.2, and the results for RM(DI) and NHP(SI) in Figure 13.3. The rows of the complete figures show the effect of the

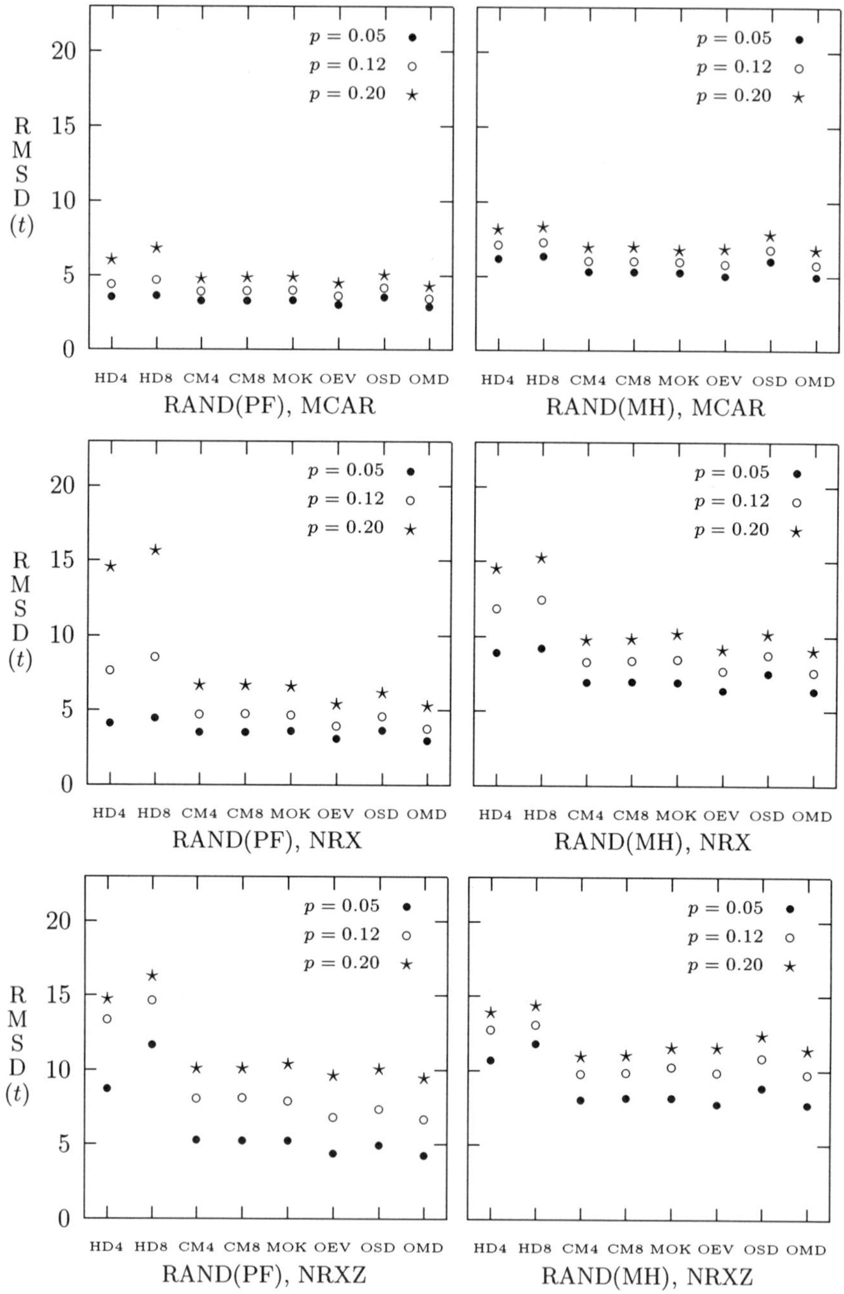

FIGURE 13.2. Comparison of imputation techniques based on RMSD(t). Data sets RAND(PF) and RAND(MH).

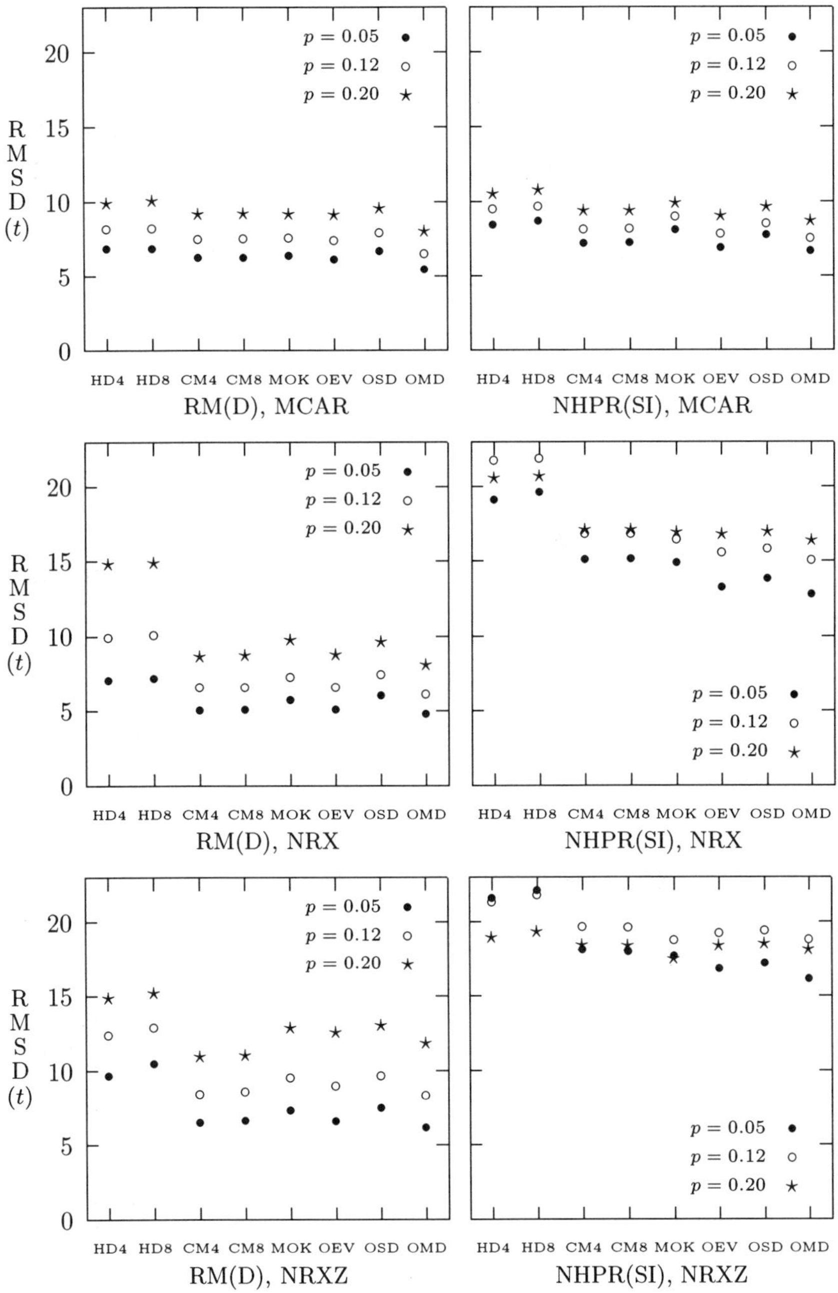

FIGURE 13.3. Comparison of imputation techniques based on RMSD(t). Data sets RM(DI) and NHP(SI).

different scales, and the columns—from top to bottom—the effect of the missing data mechanisms.

From Figures 13.2 and 13.3 it follows that OMD is also generally the best technique here, as it was in Table 13.2. It is closely followed (in varying order) by OEV and CM; the MOK procedure also performs well. The hot deck methods have the highest RMSD(t) values, that increase rapidly when missingness mechanisms become more complicated. Hot deck was also the only procedure on which sample size had a noticeable, albeit small effect. Using a finer categorization in constructing adjustment cells in hot deck methods does not lead to better results. Three-way interactions $n \times m \times p$ —not reported here, see Huisman (1999, p. 145)—show that in some cases the construction of adjustment cells slightly improves the performance.

The artificially created data RM(DI) shows some deviations from the general picture. In this data set, the number of errors made by imputation with an IRT model decreases when the missingness mechanism becomes more complicated. The imputation techniques perform equally well, and most of the time even better, when the mechanism is NRX rather than MCAR. This result is caused by the combination of logistic models for both observed and missing data in which response depends on item difficulty and subject ability. If the missingness mechanism depends on r_v, it is implicitly included in the analysis when IRT models are used for imputation. Imputation with an IRT model is in this case an appropriate solution. This result is only found in the RM(DI) data, because in the other data sets the fit of the OPLM is not particularly good.

4.4 Results: Assessment of Scale Quality

To assess the ability of the imputation techniques to preserve the relationship between the items, the change in Cronbach's alpha and Loevinger's H-coefficient is investigated. In this section the main results are summarized.

The most noticeable result is the overestimation of both alpha and H of the scales after imputation with the corrected item mean (CM), MOK, and OPMISS techniques. The overestimation is largest for OEV, where even in the case of MCAR data the coefficients are severely overestimated, and smallest for the MOK procedure. The overestimation of scale quality is not surprising. When IRT models are used for both imputation and analysis of the imputed data, the model fit will increase. This means that values are imputed that "fit the scale." The two quality criteria can be seen as fit measures for scales, and will therefore be overestimated. The hot deck techniques, on the other hand, underestimate alpha and H, probably due to the underestimation of covariances (Little & Rubin, 1987, pp. 60–71).

There is no generally best method among the imputation techniques, although OEV results in the largest overestimation. Also, the underestimation by hot deck is in some cases smaller than the overestimation by the other procedures. The effects of mechanism and proportion of missing values is as found for r_v: when the situation becomes uglier, the performance of the techniques worsens.

There is, however, an important difference between the two scale criteria. The absolute values of the deviations in alpha are all smaller than the absolute deviations in H, indicating that the values of alpha are more robust against imputation than H. Moreover, scale length and number of response options have a large influence on the differences in alpha, and not on the differences in H: the larger the scale and/or the larger the number of response options, the smaller the deviations in alpha, but the deviations in H are not affected (cf. Molenaar & Sijtsma, 1984).

5 Incomplete Designs versus Imputation

As discussed in Section 2, two strategies can be used to estimate latent abilities θ of respondents in the case of item nonresponse: the strategy using incomplete testing designs and imputation. However, in the presence of scales with many items and large percentages of missing values, the former procedure may be unsuitable to estimate θs for all respondents. Still, it has the advantage over imputation that all information in the data is used without creating extra uncertainty by estimating missing item responses.

The question rises as to which is the best strategy to use in cases where both strategies will result in estimates of the latent abilities of all respondents. In this section, the incomplete design strategy for item nonresponse is compared with three imputation techniques from Section 3: hot deck and corrected item mean imputation within adjustment cells (HD4, CM4), and OPMISS multiple draws (OMD). The results of a small simulation study are presented, in which the four strategies are used to handle missing item responses and latent abilities θ are estimated with the OPLM.

5.1 Simulation Design

The design of this simulation study consists of four cells of the simulation study described in Section 4. Two data sets (d) are used: RAND(MH) and NHP(SI). Both consist of only 5 items, which means that all possible nonresponse patterns ($2^5 - 1 = 31$) can be assigned to different booklets. From the original data sets, samples of $n = 400$ cases are drawn in which 12% of the datapoints were deleted ($p = 0.12$). The mechanisms (m) used to create the missing response were MCAR and NRXZ. The former randomly deletes values; in the latter missingness is related to the item responses and the covariates age and sex.

In all four cells, 100 data sets were drawn from the original data and estimates of the latent ability $\hat{\theta}_v$ of each subject S_v were obtained with the program OPLM. Next, missing data were generated and the 100 incomplete data sets were analyzed with the OPLM by defining booklets for every pattern of missingness, assuming the missingness resulted from an incomplete testing design. This resulted in estimates $\hat{\zeta}_v$ of the latent abilities. Finally, the incomplete data were imputed with the three imputation techniques, and from the imputed data estimates of latent abilities were obtained: $\hat{\eta}_{1v}$, $\hat{\eta}_{2v}$, and $\hat{\eta}_{3v}$, for the data sets imputed by HD4, CM4, and OMD, respectively.

The performance of the four strategies was evaluated by investigating plots of the estimated abilities from the original data against the estimated abilities from the incomplete or imputed data, for example, plots of $\hat{\theta}_v$ against $\hat{\zeta}_v$. Also, the differences between the estimated abilities, for example, $(\hat{\eta}_{3v} - \hat{\theta}_v)$, were investigated by computing the root mean-squared deviation (13.5).

5.2 *Results*

In Figure 13.4 plots of estimated abilities based on the original data against estimates based on the four missing data strategies are presented; only the four plots concerning the RAND(MH) data and the MCAR mechanism are depicted. Specifically, going from top left to bottom right, plots of $(\hat{\theta}_v, \hat{\zeta}_v)$, $(\hat{\theta}_v, \hat{\eta}_{v1})$, $(\hat{\theta}_v, \hat{\eta}_{v2})$, $(\hat{\theta}_v, \hat{\eta}_{v3})$ are shown for the incomplete design strategy, hot deck within adjustment cells imputation, corrected item mean imputation, and OPMISS multiple draws imputation, respectively.

From the plots it follows that even if the data are missing completely at random and the percentage of missingness is moderate, the differences between the estimated abilities can be considerably large. When there are no differences between estimated abilities from the original data and those from the incomplete data, the pairs of estimates would be on the main diagonal. From Figure 13.4 it follows that the points scatter around this line, but the variation is fairly large.

To investigate the association between the estimates and the usefulness of a missing data strategy for the estimation of θ, values of the squared multiple correlation coefficient R^2 were computed for the regression of the estimates resulting from one of the four missing data strategies on the estimated abilities from the original data; for example, $\hat{\theta} = \alpha + \beta\hat{\zeta} + \epsilon$. Note that the estimated regression lines are not necessarily equal to the main diagonal. The values of R^2 of all four cells of the simulation design are presented in Table 13.3. In this table the RMSD(t) values of the estimates are given also.

Comparing the techniques using the RMSD(t) is not completely fair, because analyzing the incomplete or imputed data with the OPLM not only gives new estimates of the latent abilities, but also new estimates of the

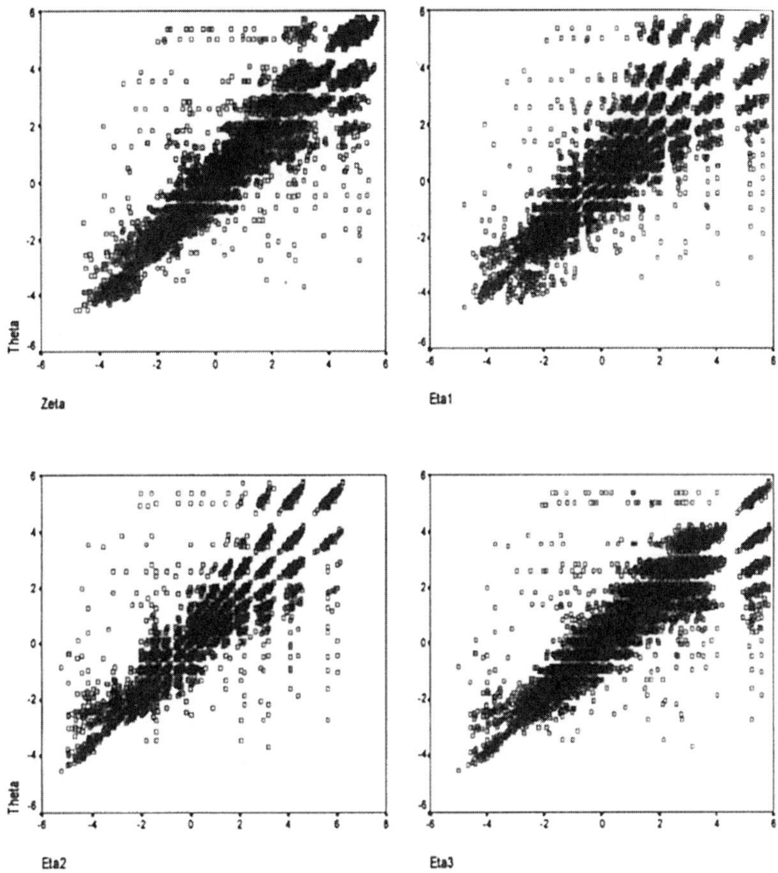

FIGURE 13.4. Comparison of estimated latent abilities from the complete data ($\hat{\theta}$) with estimates based on the incomplete design strategy ($\hat{\zeta}$) and imputed data (HD4: $\hat{\eta}_1$, CM4: $\hat{\eta}_2$, OMD: $\hat{\eta}_3$); RAND(MH) scale, $n = 400$, $p = 0.12$, and MCAR mechanism.

item parameters of the model. The estimates of the latent abilities are based on the estimated item difficulties, and therefore, systematic changes in the item parameter estimates change the ability estimates. This means that the RMSD(t) reflects the extra deviation due to changed item parameters on top of the differences between latent abilities. The same is true for the parameters α and β in the regression model. Therefore, the R^2 values are used, which indicate the usefulness of the techniques to estimate θ.

The values of R^2 and RMSD(t) in Table 13.3 show that in three of the four cells the incomplete design strategy results in the best estimates of the latent abilities. Only in the case of MCAR dichotomous data, some

TABLE 13.3. R^2 values for regressions of $\hat{\zeta}_v$ on $\hat{\theta}_v$ and $\hat{\eta}_v$s on $\hat{\theta}_v$, and RMSD(t) of the estimated abilities for the missing data adjustments in the four cells of the simulation study.

	R^2				RMSD			
	Design	HD4	CM4	OMD	Design	HD4	CM4	OMD
	$\hat{\zeta}_v$	$\hat{\eta}_{v1}$	$\hat{\eta}_{v2}$	$\hat{\eta}_{v3}$	$\hat{\zeta}_v$	$\hat{\eta}_{v1}$	$\hat{\eta}_{v2}$	$\hat{\eta}_{v3}$
d2m1	0.91	0.87	0.89	0.90	0.59	0.77	0.72	0.70
d2m3	0.89	0.78	0.86	0.83	0.71	1.01	0.86	1.04
d3m1	0.82	0.81	0.88	0.87	0.66	0.63	0.51	0.50
d3m3	0.69	0.52	0.64	0.65	1.06	1.21	1.20	1.34

imputation procedures perform better. The difference, however, between the R^2 and RMSD(t) values, from the incomplete design strategy and from CM4 and OMD, on the other hand, are never large. It should be noted that all ability estimates (also those from the original data) have large standard errors due to the small number of items in both scales.

5.3 Comparison of Imputation Techniques

Our comparison of imputation techniques showed that the ordering of the techniques based on the estimated abilities is not always the same as the ordering based on the sum scores. From the results of Section 4.3 it follows that imputation with the OPLM gives the best estimates of the sum score r_v (OMD performs best). When the R^2 values are used to order the imputation techniques, the ordering is generally the same: the differences between OMD and CM4 are small, and HD4 is always worst. Inspection of RMSD(t) values of estimated abilities, however, showed a change in the ordering, because in the case of nonignorable missing data, OMD performs worst.

This different ordering of the imputation techniques is caused by the change in item parameter estimates. For implicit imputation models (HD4, CM4) the changes are small and, more important, probably random. With the OMD procedure, however, the changes in the estimated item parameters are systematic and result in considerable changes in the estimated abilities. This is caused by the combination of a nonignorable mechanism that deletes points for difficult items and subjects with low sum scores (NRXZ), and using the OPLM for both imputation and analysis of the data. The results of Section 4.4 already showed this disadvantage: IRT imputation models overestimated the quality of the scale. The data set completed with OMD thus results in systematic changes of the item parameter estimates, which leads to systematic changes in the estimated abilities.

The nonignorable missingness mechanism also leads to decreasing performance of the other three procedures. Only OMD suffers extra from the nonignorability, due to a systematic change in the ability estimates by using the OPLM twice, for imputation and estimation consecutively. However, it should be noted that the differences in the estimated abilities mainly occur in the tails of the ability distribution where ability estimation already is a difficult task, especially in short scales. Still, when the $\hat{\eta}_3$s are used to predict the $\hat{\theta}$s, OMD performs almost as well as the incomplete design strategy and better than the other imputation procedures, as is shown by the R^2 values in Table 13.3.

6 Summary and Conclusions

In this chapter, techniques to impute missing responses to scale items were investigated. Moreover, the performance of some imputation techniques was compared with a strategy frequently used within the framework of IRT to analyze missing data caused by incomplete testing designs.

The likelihood-based incomplete design procedure has the advantage that all information in the data set is used for the estimation of the latent abilities of respondents, without generating extra uncertainty caused by estimating missing values. Therefore, it seems preferable over imputation. The results of Section 5 show that this is indeed the case in short scales ($k = 5$) with a medium percentage of nonignorable missing data (12%). When the missingness mechanism is ignorable, the performance of the two strategies is equally good. However, in the case of item nonresponse, the use of the incomplete design strategy is limited, because many different patterns of missingness may emerge that cannot be modeled separately. Moreover, imputation has the advantage that it can be used in any situation and for any scale. Also, information from data collection can be included to correct for the nonignorability of the missingness mechanisms (e.g., with adjustment cells).

The results of the simulations presented in this chapter give an indication of the performance of several imputation techniques. When the sum scores r_v of the subjects are investigated, imputing with the parametric OPLM proved to be best. If only the mean sum scores are considered, imputing the expected value from the estimated response distributions (OEV) is best; if the variation between the sum scores is taken into account, multiply imputing random draws from the distributions (OMD) is best. The performance of the nonparametric methods CM and MOK is only slightly worse, but they have the advantage that they are easy to use.

The capability of the imputation techniques to recover the relationships that exist between the items was investigated by looking at differences between the reliability of the scale (Cronbach's alpha) and the scalability of the items (Loevinger's H) before and after creating and imputation of missing values. The results showed a disadvantage of using IRT models for imputation: they overestimate the quality of the scale. This overestimation is due to the usage of IRT models for both imputation and analysis of the completed data. This results in an increased model fit and therefore in a too optimistic presentation of the internal structure of the test. This also holds for imputing a corrected item mean, which can be considered as a very simple IRT imputation model, but the overestimation is generally smaller.

Inspection of estimated latent abilities $\hat{\theta}_v$ from the OPLM before and after imputation, reveals a second disadvantage of IRT imputation models. Successive estimation of the model parameters, first to generate imputations, next to measure latent abilities, leads to systematic changes in both item and subject parameter estimates. This results in an overestimation of abilities after OMD imputation, especially in the right tail of the ability distribution. Note that this occurs when the mechanism is nonignorable (dependent on sum and item scores), and that comparison of the estimated abilities is influenced by differences in item parameter estimates.

As expected, the proportion of missing data, the missingness mechanism, and characteristics of the scale largely influence the success of the imputation techniques. Sample size, on the other hand, does not seem to affect the performance of the imputation techniques to a large extent. Exceptions are the hot deck procedures, the performance of which slightly improves when the sample size increases, and imputation within adjustment cells.

The construction of adjustment cells does not seem to improve the performance of the HD and CM procedures. These unsatisfactory results can be explained by the sometimes weak association of the covariates used to construct the adjustment cells with the latent abilities. However, Huisman (1999, pp. 163–164) gives examples of similar incomplete data sets (same scale and/or latent trait) for which adjustment cells based on sex and age clearly improve the hot deck procedure. These examples are based on a follow-up study of nonrespondents, in which the mechanisms underlying the data are largely unknown, and also indicate that in the simulation study described in this chapter, the adjustment cells are too weakly related to the probability of response to improve the performance of the techniques.

Unfortunately, no generally best solution exists for handling item nonresponse. Still, good results can be obtained with imputation when good imputation models are available and the nonignorability of missing data mechanisms is taken into account. In such cases the performance of imputation techniques is as good as that of the incomplete design strategy. Among all investigated methods, multiple draws generated with the OPLM is generally the best strategy. It is, however, not always easy to use. Therefore, when the researcher is not in the position to multiply impute missing responses with an IRT imputation model, the computationally much easier corrected item mean and Mokken procedures are to be recommended for use in practical research.

Acknowledgments: The authors are grateful to Anne Boomsma, Marijtje van Duijn, and two reviewers for their comments on an earlier version of this chapter.

References

Bethlehem, J.G. (1988). Reduction of nonresponse bias through regression estimation. *Journal of Official Statistics, 4*, 251–260.

Eggen, T.J.H.M. (1993). Itemresponstheorie en onvolledige gegevens [Item response theory and incomplete data]. In T.J.H.M. Eggen & P.F. Sanders (Eds.), *Psychometrie in de praktijk* (pp. 239–284). Arnhem: CITO.

Eggen, T.J.H.M., & Verhelst, N.D. (1992). *Item calibration in incomplete testing designs* (Measurement and Research Department Reports, 92–3). Arnhem: CITO.

Huisman, M. (1999). *Item nonresponse: Occurrence, causes, and imputation of missing answers to test items* (M&T Series 32). Leiden: DSWO Press.

Little, R.J.A. (1988). Missing-data adjustments in large surveys. *Journal of Business & Economic Statistics, 6*, 287–296.

Little, R.J.A., & Rubin, D.B. (1987). *Statistical analysis with missing data.* New York: Wiley.

Mislevy, R.J., & Wu, P.-K. (1996). *Missing responses and IRT ability estimation: Omits, choice, time limits, and adaptive testing* (Research Report RR–96–30–ONR). Princeton, NJ: Educational Testing Service.

Mokken, R.J. (1971). *A theory and procedure of scale analysis: With applications in political research.* The Hague: Mouton.

Molenaar, I.W. (1997). Nonparametric models for polytomous responses. In W.J. van der Linden & R.K. Hambleton (Eds.), *Handbook of modern item response theory* (pp. 369–380). New York: Springer-Verlag.

Molenaar, I.W., & Sijtsma, K. (1984). Internal consistency and reliability in Mokken's nonparametric item response model. *Tijdschrift voor Onderwijsresearch, 9*, 257–268.

Nap, R.E. (1994). *OPMISS: Handling missing data in OPLM* (Heymans Bulletin HB–94–1173–IN). University of Groningen, Department of Statistics & Measurement Theory.

Rubin, D.B. (1976). Inference and missing data. *Biometrika, 63*, 581–592.

Rubin, D.B. (1983). Conceptual issues in the presence of nonresponse. In W.G. Madow, I. Olkin, & D.B. Rubin (Eds.), *Incomplete data in sample surveys, Vol. II: Theory and bibliographies* (pp. 123–142). New York: Academic Press.

Rubin, D.B. (1987). *Multiple imputation for nonresponse in surveys.* New York: Wiley.

Rubin, D.B., Stern, H.S., & Vehovar, V. (1995). Handling "don't know" survey responses: The case of the Slovenian plebiscite. *Journal of the American Statistical Association, 90*, 822–828.

Sande, I.G. (1982). Imputation in surveys: Coping with reality. *The American Statistician, 36*, 145–152.

Verhelst, N.D., & Glas, C.A.W. (1995). The one parameter logistic model. In G.H. Fischer & I.W. Molenaar (Eds.), *Rasch models. Foundations, recent developments, and applications* (pp. 215–237). New York: Springer-Verlag.

Verhelst, N.D., Glas, C.A.W., & Verstralen, H.H.F.M. (1995). One-parameter logistic model: OPLM [Software manual]. Arnhem: CITO.

Part II

Nonparametric Item Response Theory

14

On the Interplay Between Nonparametric and Parametric IRT, with Some Thoughts About the Future

Brian Junker [1]

ABSTRACT This chapter reviews some of the important research in nonparametric and parametric item response theory (IRT) today, and considers some current measurement challenges in education and cognitive psychology. This leads to assessment models that do not look very much like today's IRT models, but for which the tools and conceptual framework of nonparametric and parametric IRT are still quite well suited.

1 Introduction

Item response theory (IRT; e.g., Fischer & Molenaar, 1995; Van der Linden & Hambleton, 1997) is a psychometric approach to modeling data from social surveys and educational and psychological tests, dating back at least to Lord (1952) and Rasch (1960), and to the work of Loevinger and Guttman before them. IRT enables us to study the characteristics of test or survey items across multiple respondent populations, and to study respondents' propensities to answer positively across various items. IRT has arguably been one of the most successful and widely used techniques in psychometrics, with applications in developmental, social, educational, and cognitive psychology, for example, as well as in medical research, demography, and other social science settings.

This chapter briefly summarizes some of the important research in nonparametric and parametric IRT today, emphasizing the interplay between these different but complementary approaches to modeling sequences of individual responses to test, survey, and scale items. The chapter shows that a broad understanding of IRT, together with an appreciation of some of the current measurement challenges in education and cognitive psychology,

[1]Department of Statistics, Carnegie Mellon University, 232 Baker Hall, Pittsburgh, PA 15213, USA; e-mail *brian@stat.cmu.edu*

lead us to assessment models that do not look very much like today's IRT models, but for which the tools and conceptual framework of nonparametric and parametric IRT are particularly well suited.

2 Nonparametric IRT: Scale Construction

To save space and preserve focus, this summary of nonparametric IRT concentrates on the scaling theory techniques introduced by Mokken and pursued by Molenaar, Sijtsma, and their colleagues. Important related work on nonparametric essential unidimensionality (e.g., Stout 1987, 1990; Stout, Habing, Douglas, Kim, Roussos, & Zhang, 1996; and Zhang & Stout, 1999), nonparametric regression estimates of item response functions and test response surfaces (especially Ramsay, 1991, 1995, 1996), and related parametric and nonparametric work (e.g., Meijer, 1996; Cliff & Donoghue, 1992; Drasgow, Levine, Tsien, Williams, & Mead, 1995; Samejima, 1997) are not considered in detail.

2.1 Monotone Homogeneity / Strict Unidimensionality

The Mokken (1971) model of *monotone homogeneity (MH)* starts with very few assumptions. A collection of item response variables $\mathbf{X} = (X_1, \ldots, X_k)'$ satisfies this model if θ, the latent trait, is a real-valued random variable (unidimensionality); if each item step response function (ISRF) $P(X_i > c \mid \theta)$ is nondecreasing in θ for each item response variable X_i and each real threshold c (monotonicity); and if

$$P(X_1 > c_1, \ldots, X_k > c_k \mid \theta) = \prod_{i=1}^{k} P(X_i > c_i \mid \theta) , \qquad (14.1)$$

for all possible cutoffs c_i (local independence). When the response is dichotomous ($X_i = 0$ or 1) let $P_i(\theta) = P(X_i = 1 \mid \theta)$ denote the item response function (IRF). In that case, (14.1) reduces to the familiar form

$$P(X_1 = x_1, \ldots, X_k = x_k \mid \theta) = \prod_{i=1}^{k} P_i(\theta)^{x_i} [1 - P_i(\theta)]^{1-x_i} . \qquad (14.2)$$

The data we observe when examinees or subjects respond to test or survey items can be thought of as i.i.d. samples from the marginal discrete multivariate distribution,

$$P(X_1 > c_1, \ldots, X_k > c_k) = \int P(X_1 > c_1, \ldots, X_k > c_k \mid \theta) \, dF(\theta) , \qquad (14.3)$$

where the integrand comes from either (14.1) or (14.2), and $dF(\theta)$ represents the distribution of θ in the population of interest. The assumptions of unidimensionality, monotonicity, and local independence can be relaxed in various ways (e.g., Sections 2.2, 3.2, and 4.1), but this basic model has been the foundation of much nonparametric scale construction.

Nonparametric scale construction. For dichotomous items, a nonparametric theory of scale construction has existed at least since Mokken (1971, 1997; see also Molenaar, 1991, 1997). The principal tools of that theory are adaptations of Loevinger's (1948) H-coefficients, comparing the marginal covariance $\mathrm{Cov}\,(X_i, X_j)$ of each item pair with the maximum covariance $\mathrm{Cov}_{\max}(X_i, X_j)$ possible, preserving the margins of the observed $X_i \times X_j$ table. The bound $\mathrm{Cov}_{\max}(X_i, X_j)$ is obtained by adjusting the table to remove Guttman errors (e.g., Molenaar, 1991); and indeed the original formulae for the H-coefficients were expressed as ratios of Guttman errors (Mokken, 1997).

A related modeling condition for dichotomous items is *invariant item ordering* which says that IRFs do not cross: for every pair i and j, either $P_i(\theta) \leq P_j(\theta)$ or $P_j(\theta) \leq P_i(\theta)$ holds uniformly for all θ. Rosenbaum (1987a, 1987b) and Sijtsma and Junker (1996, 1997) explore and extend this idea and provide scale construction examples. Mokken's *double monotonicity* model incorporates both the monotone homogeneity and invariant item ordering assumptions.

The H-coefficients are directly sensitive only to high or low correlations between items, rather than to local independence given θ as in (14.1) and (14.2). If the correlations are near zero, we may be unsatisfied to assume that such a θ exists; see, for example, the discussion of comonotonicity in Junker and Ellis (1997). While a perfect Guttman scale would produce H-coefficients equal to one, large H-coefficients provide only indirect evidence of a θ "explaining" covariation in the item responses in the sense of local independence.

Stout, Ramsay, and their students and colleagues have attacked the nonparametric scaling problem by conditioning on a proxy for θ such as the total test score. Stout's essential unidimensionality approach (1990; and e.g., Stout et al., 1996) combines conditional inter-item covariances and related quantities into nonparametric tests and indices of latent trait dimensionality. Ramsay (1991) constructs nonparametric regression estimates of item (step) response functions, and Ramsay (1995) constructs local smoothing estimates of the joint response surface of all items on the test. These methods are generally more computationally complex, and seem to require larger examinee and item sample sizes than the Mokken methods outlined above. Thus the Mokken techniques have been more widely used in smaller social survey and experimental psychology settings.

Molenaar (1991) provided a direct generalization of the H-coefficients to the case of polytomous responses, and developed an effective computational method for obtaining $\mathrm{Cov}_{\max}(X_i, X_j)$ in the polytomous case. Hemker, Sijtsma, and Molenaar (1995) apply these ideas to scaling polytomous items. Generalizing invariant item ordering to the polytomous case turns out to be somewhat delicate. Sijtsma and Hemker (1998) propose preserving the order of expected item scores, for example, $E(X_i \,|\, \theta) \le E(X_j \,|\, \theta)$ uniformly in θ for fixed i and j, and compare this approach with other approaches (e.g., Molenaar, 1997; Scheiblechner, 1995) that impose order restrictions directly on the ISRFs $P(X_i > c \mid \theta)$, uniformly in θ.

Another approach to understanding scaling by the Mokken model—in dichotomous, polytomous, and more general settings—was initiated by Holland (1981), Rosenbaum (1984), and Holland and Rosenbaum (1986; see also Meredith, 1965). Junker (1993), Ellis and Junker (1997), and Junker and Ellis (1997) continued this approach; they combine Holland and Rosenbaum's (1996) *conditional association* (CA) condition

$$\mathrm{Cov}\,[f(\mathbf{Y}), g(\mathbf{Y})] \mid h[(\mathbf{Z})] \ge 0\;, \tag{14.4}$$

for all partitions $\mathbf{X} = (\mathbf{Y},\, \mathbf{Z})$, for all f, g nondecreasing, and for all $h(\mathbf{Z})$, with a *vanishing conditional dependence* condition

$$\begin{array}{l}\text{as } m \to \infty,\ (X_1, \ldots, X_k)' \text{ become independent,} \\ \text{given } (X_{k+1}, \ldots, X_{k+m})',\ \text{for all } k,\end{array} \tag{14.5}$$

to obtain a complete characterization of an infinite-item-pool formulation of the basic monotone homogeneity model, in which θ is both genuinely latent and consistently estimable, in terms of the joint distribution of observable item responses.

Stochastic ordering. A side effect of attempts to characterize the monotone homogeneity model has been a collection of interesting necessary conditions, such as Junker's (1993) "manifest monotonicity" property for dichotomous items,

$$P(X_i = 1 \mid X_+^{(-i)} = s) \text{ is nondecreasing in } s. \tag{14.6}$$

This property, and examples showing that it does not hold when the restscore $X_+^{(-i)} = \Sigma_{j \ne i} X_j$ is replaced by the total score $X_+ = \Sigma_{i=1}^k X_i$, are included in unpublished work of Ivo Molenaar and Tom Snijders (Junker, 1993; Junker & Sijtsma, 2000).

Existing proofs of (14.6) hinge on establishing a stochastic ordering property for θ, given the total score X_+ (or equivalently the restscore $X_+^{(-i)}$):

$$P(\theta > c \mid X_+ = s) \text{ is nondecreasing in } s, \text{ for all } c. \tag{14.7}$$

Hemker, Sijtsma, Molenaar, and Junker (1996, 1997) call this property SOL (stochastic ordering of the latent trait by the sum score; see also Hemker, 2001, and Sijtsma & Van der Ark, 2001), and show that, surprisingly, this property does not generalize to "most" nonparametric ordered-polytomous response IRT models. Thus, for example, rules based on cutoffs for X_+

need not be most powerful for "mastery decisions" in the sense of $\theta > c$; on the other hand, such cutoff rules for X_+ *are* most powerful for mastery decisions in the nonparametric dichotomous response case (Grayson, 1988; Huynh, 1994).

In the process of developing these stochastic ordering ideas, Hemker et al. (1997; see also Van der Ark, 1999) have developed a taxonomy of nonparametric and parametric item response models, that usefully complements the taxonomy of Thissen and Steinberg (1986). The Hemker taxonomy is based on the cumulative, continuation-ratio, and adjacent-category logits that are commonly used to define parametric families of polytomous IRT models. Common forms of graded response models (GRM; Samejima, 1997, e.g.), sequential models (SM; Tutz, 1990; Mellenbergh, 1995; Samejima, 1969, 1995), and partial credit models (PCM; Masters, 1982) assume, respectively, that the logit functions $\text{logit}\, P(X_i > c \mid \theta)$, $\text{logit}\, P(X_i > c \mid X > c-1, \theta)$, and $\text{logit}\, P(X_i = c+1 \mid X_i \in \{c, c+1\}, \theta)$ are *linear* in θ. Hemker's analogous nonparametric model classes (np-GRM, np-SM, and np-PCM), assume only that these logits are *nondecreasing* in θ.

This taxonomy is a powerful way to organize ideas about model definition and model development in applications of both parametric and nonparametric IRT. It turns out that the np-PCM class is nested within the np-SM class, which is nested within the np-GRM class; moreover all three linear-logit families above (GRM, SM, and PCM) are in fact subclasses of the np-PCM class; see also Hemker (2001). As Van der Ark (1999) shows, this approach also highlights links between polytomous IRT and the machinery of generalized linear models (McCullagh & Nelder, 1989), just as it has long been realized that parametric dichotomous IRT is basically multivariate mixed effects logistic regression (e.g., Douglas & Qui, 1997; see also Lee & Nelder, 1996). It is important to realize that of all the models studied by Hemker and his colleagues, only the parametric PCM and its special cases have the stochastic ordering property SOL; see (14.7).

2.2 Some Interesting Questions

An ongoing question in this area is developing adequate data analysis methodology. Sijtsma (1998) provides a survey of nonparametric IRT approaches to the analysis of dichotomous item scores; Molenaar (1997) and Sijtsma and Van der Ark (2001) survey extensions to the polytomous case. Most of this Mokken methodology is encoded in the computer program MSP (Molenaar & Sijtsma, 2000). Snijders (2001) introduces Mokken scaling tools for multilevel data. Ellis (1994) has reexamined Mokken's hypothesis testing framework for the H-coefficients, and proposed new tests based on the theory of order-restricted inference of Robertson, Wright, and Dykstra (1988). The same methods may be useful to develop tests of manifest monotonicity. Yuan and Clarke (1999) have developed asymptotic theory for testing the conditions of Junker (1993), that should be adaptable to the conditions of Junker and Ellis (1997). A more direct application of the theory of order-restricted inference to testing conditional association (14.4) and related conditions is given by Bartolucci and Forcina (in press).

Sijtsma and Van der Ark (2001) discuss progress on several problems related to the lack of SOL (14.7) in ordered polytomous IRT models and to the sensitivity and specificity of the manifest monotonicity condition (14.6) for detecting (violations of) the monotone homogeneity model. In dichotomous IRT, the nonparametric theory assures us, under very general conditions, that simple summaries of the data are informative about inferences we wish to make. But no such assurances have emerged for polytomous IRT. Understanding the impact this has on nonparametric polytomous IRT modeling will surely entail facing and solving the problems that Sijtsma and Van der Ark discuss.

Finally, some of the monotone homogeneity machinery outlined above seems ready to be applied to characterize probabilistic unfolding models, and multidimensional nonparametric IRT models, on the basis of observable data. For example, conditional association greatly sharpens the well-known fact that interitem correlations are nonnegative under monotone homogeneity. Post (1992; Post & Snijders, 1993) has similarly shown that the interitem correlation matrix of a class of nonparametric probabilistic unfolding models has a band of positive correlations near the main diagonal, bordered by bands of negative correlations. Is there a sharpening of the Post result analogous to conditional association? Could this be combined with the vanishing conditional dependence condition (14.5) to characterize Post's models? Could such a result follow from a folding of the monotone homogeneity model to produce nonparametric unfolding models (Verhelst & Verstralen, 1993; Andrich, 1996)? In another direction, much of the work in Ellis and Junker (1997) and Junker and Ellis (1997) does *not* depend on unidimensionality of the latent variable θ. For example, Junker and Ellis (1997) conjecture that their item-step "true scores" $P[X_i > c|\tau(X)]$ should form a manifold of the same dimension as the underlying latent variable θ. A characterization theorem may again result, if a weakening of conditional association to accommodate multidimensional θ could be developed.

3 Parametric IRT: Modeling Dependence

Parametric IRT, as surveyed, for example, in the edited volumes of Fischer and Molenaar (1995) and Van der Linden and Hambleton (1997), is a well-established, wildly successful statistical modeling enterprise. A basic and familiar model in this area is the two-parameter logistic model (2PLM) for dichotomous item response variables (e.g., Van der Linden & Hambleton, 1997, Chap. 1). The 2PLM is given by the monotone homogeneity assumptions in Section 2 and the assumption of a logistic form for the item response functions, $P(X_{vi} = 1 \mid \theta_v, \alpha_i, \beta_i) = 1/\{1 + \exp[-\alpha_i(\theta_v - \beta_i)]\}$, describing the dichotomous response of examinee S_v to item I_i. The discrimination parameter α_i controls the rate of increase of this logistic curve, and is directly related to the Fisher information for estimating θ_v, and the difficulty parameter β_i is the location on the θ_v scale at which this information is maximal; note also that at $\theta_v = \beta_i$, $P(X_{vi} = 1) = 1/2$. The three-parameter logistic model (3PLM) extends the 2PLM by adding a nonzero lower asymptote to each item response function; the Rasch or one-parameter logistic model (1PLM) is a restriction of the 2PLM obtained by setting α_i identically equal to some constant, usually 1.

Such parametric IRT models, extended by hierarchical mixture/Bayesian modeling and estimation strategies, make it possible to incorporate covariates and other structures. Many violations of the basic local independence assumption of IRT models are in fact due to unmodeled heterogeneity of subjects and items, which can now be explicitly modeled using these methods. These models have greatly extended the data analytic reach of psychometricians, social scientists, and educational measurement specialists.

The main purpose of this section is to introduce a general modeling framework and highlight a few developments in parametric IRT, some old and some new, that are relevant to the discussion of applying IRT and related models to cognitive assessment problems in Section 4. The summary of parametric IRT is even less complete, relative to the vast parametric IRT literature, than the summary of nonparametric IRT.

3.1 Two-Way Hierarchical Structure

The estimation of group effects, and the use of examinee and item covariates in estimating item parameters, play an important role in the analysis of large multisite educational assessments such as the National Assessment of Educational Progress (NAEP; e.g., Algina, 1992; Johnson, Mislevy, and Thomas, 1994; Zwick 1992). These efforts, which go back at least to Mislevy (1985; see also Mislevy & Sheehan, 1989), can be recognized as the wedding of hierarchical linear or multilevel modeling methodology with standard dichotomous and polytomous IRT models. The general model is a two-way hierarchical structure for n individuals and k response variables, as in Table 14.1, where $P(\theta_v; \gamma_i)$ is the IRF, depending on person parameters θ_v and item parameters γ_i (e.g., $\gamma_i = (\alpha_i, \beta_i)'$ in the 2PLM above), and where independence is assumed between items I_i conditional on θ_v at the first level and between subjects S_v at the second level. Terms in the first level are multiplied together to produce the usual joint likelihood for the $(n \times k)$ item response matrix $[X_{vi}]$; the second and third levels are used to impose constraints on the latent variable and first-level parameters, to introduce covariates, and so on. λ_f and λ_g represent sets of hyperparameters needed to specify these person distributions f_v and item distributions g_i, with hyperprior distributions ϕ_f and ϕ_g, respectively. The model in

TABLE 14.1. Two-way hierarchical structure for n individuals and k dichotomous response variables.

First Level:	$X_{vi} \sim P(\theta_v; \gamma_i)^{X_{vi}}[1 - P(\theta_v; \gamma_i)]^{1-X_{vi}}$, $v = 1, \ldots, n;\ i = 1, \ldots, k.$
Second Level:	$\theta_v \sim f_v(\theta \mid \lambda_f)$, for each v, $\gamma_i \sim g_i(\gamma \mid \lambda_g)$, for each i.
Third Level:	$\lambda_f \sim \phi_f(\lambda_f)$, $\lambda_g \sim \phi_g(\lambda_g)$.

Table 14.1 is expressed for dichotomous items, but can easily be generalized to polytomous items, or combinations of item types (see, e.g., Patz & Junker, 1999a, 1999b). It is also usual to assume for $f_v(\theta)$ a single latent trait distribution not depending on v, and similarly for g_i.

Alternatively, we may allow the distributions $f_v(\cdot)$ of θ to depend hierarchically on examinee covariates, to model multigroup population heterogeneity, as in Mislevy (1985) and Bock and Zimowski (1997), or to reflect hierarchical linear structure as in Fox and Glas (1998). We may also elaborate $g_i(\cdot)$, for instance, by building linear structure into the item parameters. For example, in the 2PLM, where $\gamma_i = (\alpha_i, \beta_i)'$, we might take

$$(\beta_1,\ \beta_2,\ \ldots,\ \beta_{k-1},\ \beta_k)' = Q(\psi_1,\ \psi_2,\ \ldots,\ \psi_m)' , \tag{14.8}$$

where Q is an appropriate design matrix of full column rank, to reflect common sources (ψ_ℓs) of item difficulty (β_is) across items. In the case of Rasch (1PLM) IRFs, this is the linear logistic test model (LLTM; Scheiblechner, 1972; Fischer, 1973). This model and its various generalizations (e.g., Glas & Verhelst, 1989; Patz & Junker, 1999b) continues to be used for psychological experiments with multiple outcomes per subject (e.g., Fischer & Molenaar, 1995, and the references therein) and for research in cognitively motivated test design (Embretson, 1995, 1999). There is no reason to restrict attention to the βs, and, for example, Embretson (1999) has explored a similar decomposition of the αs in a 2PLM.

These generalizations of the basic IRT model both simplify and unify parametric approaches to many thorny test analysis questions, including differential item functioning and item parameter drift, nonequivalent groups and vertical equating, two-stage testing, and matrix-sampled educational assessment survey work. The computer programs ConQuest (Wu, Adams, & Wilson, 1997) and BILOG-MG (Zimowski, Muraki, Mislevy, & Bock, 1997) provide fairly general EM-based solutions when the underlying

IRT model is the 1PLM (ConQuest), or the 2PLM/3PLM (BILOG-MG). Fox and Glas (1998) and Patz and Junker (1999a, 1999b) give two different Markov chain Monte Carlo (MCMC) approaches to the problem. Generalization to polytomous items, facets-style rated response models, and mixtures of item types is conceptually, and often computationally, straightforward; see, for example, Glas and Verhelst (1989) and Patz and Junker (1999a, 1999b).

3.2 Some Multidimensional Models

Research in multidimensional IRT models has concentrated on additive and conjunctive combinations of multiple traits to produce probabilities of response. Additive models, known as *compensatory* models in much of the literature, replace the unidimensional latent trait θ in a parametric model such as the 2PLM, with an item-specific, known (e.g., Stegelmann, 1983; Embretson, 1991; Kelderman & Rijkes, 1994; and Adams, Wilson, & Wang, 1997) or unknown (e.g., Reckase, 1985; Wilson, Wood, & Gibbons, 1983; Fraser & MacDonald, 1988; Muraki & Carlson, 1995) linear combination of components $a_{i1}\theta_1 + \cdots + a_{id}\theta_d$ of a d-dimensional latent trait vector. Béguin and Glas (1998) survey the area well and give an MCMC algorithm for estimating these models; Gibbons and Hedeker (1997) pursue related developments in biostatistical and psychiatric applications.

Conjunctive models are often referred to as *noncompensatory* or *componential* models in the literature. These models (e.g., Embretson, 1997) combine unidimensional models for components of response multiplicatively, so that $P(X_{vi} = 1 \mid \theta_{v1}, \ldots, \theta_{vd}) = \prod_{\ell=1}^{d} P_{i\ell}(\theta_{v\ell})$ where $P_{i\ell}(\theta_{v\ell})$ are parametric unidimensional dichotomous response functions. Usually the $P_{i\ell}(\theta_{v\ell})$ represent skills or subtasks all of which must be performed correctly in order to generate a correct response to the item itself. Janssen and De Boeck (1997) give a recent application.

Compensatory structures are attractive because of their conceptual similarity to factor analysis models. They have been very successful in aiding the understanding of how student responses can be sensitive to major content and skill components of items, and in aiding parallel test construction when the underlying response behavior is multidimensional (e.g., Ackerman, 1994). Noncompensatory models are largely motivated from a desire to model cognitive aspects of item response, a topic to which we return in Section 4. Embretson (1997) blends these two approaches in her general component latent trait models (GLTM).

3.3 Models that Accommodate Extra Behavioral Features of Assessment

In addition to providing a way to model dependence of item responses on specific examinee and item covariates, the hierarchical or multilevel approach to IRT also allows us to model extra behavioral features of assessment. This largely unexplored area is worth further study, since these features can affect both the certainty with which we make inferences from assessment data and the kinds of inferences we make; current models and methods largely relegate them to the "error distribution" of the model.

One example of this sort of work involves recent efforts to more elaborately model the behavior of raters in rated item response data. When only one rater rates each item, it may be sufficient to treat each rating as a different, locally independent pseudo item—so that the first level in Table 14.1 contains one factor for each rater × item × examinee combination—and to model the rater effect as a linear influence on the item's difficulty parameter β_i. Mathematically this is equivalent to the LLTM sketched above, but it has come to be known in this setting as the facets model (e.g., Linacre, 1989; Engelhard, 1994).

For both formative and summative evaluation of raters, a number of multiple-read rating designs are now commonplace (Wilson & Hoskens, 1999), including designs with as many as six raters per item. These ratings are mutually dependent in the marginal model of the observed rating data, but the usual facets model likelihood is the product of LLTM-style factors for each rating of each item, ignoring a part of this dependence. Patz, Junker, and Johnson (2000) suggest that for this model, even with only one item response, in the limit as the number of raters grows, the standard error for estimating θ apparently goes to zero, contradicting the notion (e.g., Junker, 1993) that the number of items should tend to infinity in order to make the error of estimation of θ vanishingly small.

Wilson and Hoskens' (1999) rater bundle model (RBM) attacks this problem by replacing the facets product across raters for each item in level one of Table 14.1 with a log-linear model that models the dependence between ratings of the item, conditional on θ. Their approach seems very useful for, for example, modeling "table effects" and other rater dependence phenomena that follow when raters are allowed or encouraged to discuss ratings amongst themselves to increase rating quality and interrater reliability.

Patz, Junker, and Johnson's (2000) hierarchical rater model (HRM) directly reflects the idea that multiple ratings can be used to learn about the single item response being rated, and then multiple item responses are used to estimate θ. The HRM replaces X_{vi} in Table 14.1 with an unobserved, latent rating ξ_{vi}. A new zeroth layer is added to the model to connect raters' observed ratings X_{vir} to the corresponding latent ratings ξ_{vi} through a simple signal-detection model. This produces a model that behaves, for multiple discrete ratings, much as a standard generalizability theory model would behave for multiple continuous ratings (see Verhelst & Verstralen, 2001, for a similar development). The HRM also makes possible calibration and monitoring of individual rater behavioral effects such as, in the case of polytomous responses, rater severity and rater precision.

3.4 Some Current Questions

Almost any assessment phenomenon—from between-examinee dependence due to institutional or sociological factors, to behavioral aspects of raters, to the analysis of item responses into requisite examinee skills or item features—can be expressed in the hierarchical mixture/Bayes modeling framework, because of its conceptual simplicity. Recent advances in computation, and MCMC methods in particular, have made it possible to estimate a vastly wider variety of these models than would have been imaginable even 10 years ago. Questions motivating this work inevitably involve identifying phenomena that are worth detailed parametric modeling, and seeing if the computational machinery can be pushed to estimate models of these phenomena. Recent examples include multidimensional (Gibbons & Hedeker, 1992) and hierarchical (Bradlow, Wainer, & Wang, 1999) modeling of testlets, blending IRT and hierarchical linear models, and behavioral models such as the rater models described above.

Speeding up the computations with approximations (including formal and informal applications of Laplace's method such as Rigdon & Tsutakawa, 1983, and Kass, Tierney, & Kadane, 1990; blends of Monte Carlo and EM approaches as surveyed in Tanner, 1996; and variational methods, e.g., Jaakkola & Jordan, 2000) continues to be an essential and fruitful avenue of research. An avenue that has not been explored as much in IRT work is choosing models for which sufficient statistics are simple and interpretable. The most familiar basic model of this sort is of course the Rasch model, but as we show in the next section some current cognitively motivated measurement problems may require a new basic model.

4 Measurement Challenges Posed by Cognitive and Embedded Assessments

In recent years, as cognitive theories of learning and instruction have become richer, and computational methods and machinery to support assessment have become more powerful, there has been increasing pressure to make assessments truly criterion referenced, that is, to "report" on student achievement relative to theory-driven lists of examinee skills, beliefs, and other cognitive features needed to perform tasks in a particular assessment domain. For example, Baxter and Glaser (1998) and Nichols and Sugrue (1999) present compelling cases that assessing examinees' cognitive characteristics can and should be the focus of assessment design. In a similar vein, Resnick and Resnick (1992) advocate standards-referenced or criterion-referenced assessment closely tied to curriculum, as a way to inform instruction and enhance student learning.

Appropriate criterion-referenced testing can also be an effective teaching tool when embedded directly in teaching practice. Indeed there is substantial argument and evidence (e.g., Bloom, 1984), that part of what distinguishes higher student achievement in mastery learning and individualized tutoring settings as opposed to the conventional classroom, is the use of frequent and relatively unobtrusive formative tests coupled with student feedback and corrective interventions by the instructor, and follow-up tests to evaluate the interventions. This approach continues to be advocated as part of a natural and effective apprenticeship style of human instruction (e.g., Gardner, 1992), and it is the basis of many computer-based intelligent tutoring systems (ITSs; e.g., Anderson, 1993; and more broadly Shute & Psotka, 1996). Here too, a decomposition of assessment items into appropriate cognitive attributes is important: feedback and/or corrective action in a mastery class or from an ITS depends on knowing which cognitive attributes the examinee has and has not mastered.

Cognitive assessment models generally address a more complex goal than linearly ordering examinees, or partially ordering them in a low-dimensional Euclidean space, which is what IRT has been designed and optimized to do. The goal of cognitive assessment can be thought of producing, for each examinee, a list of skills or other cognitive attributes that the examinee may or may not possess, based on the evidence of tasks performed by the examinee. This list of cognitive features generally comes from an analysis of the cognitive attributes needed to successfully perform each task in a domain of interest. For a particular set of tasks, this analysis can be encoded in an incidence matrix, the *Q-matrix* (e.g., Tatsuoka, 1990), which we write as a $(k \times d)$ matrix $Q = [Q_{i\ell}]$ of 0s and 1s with entries

$$Q_{i\ell} = \begin{cases} 1\,, & \text{if attribute } \ell \text{ is required by task } i, \\ 0\,, & \text{if not.} \end{cases} \tag{14.9}$$

While the Q-matrix does not capture all of the structure we may be interested in (Q treats the skills in a flat, non-time-ordered manner, and there may be both hierarchical and time-order structure in the skills as they are applied to a task), it is a useful bookkeeping device.

Many attempts (e.g., as surveyed by Roussos, 1994) to blend IRT and cognitive measurement are based on a linear decomposition of item parameters, as in the LLTM, or on a linear decomposition of the latent trait θ, as in the multidimensional compensatory IRT models. Compensatory IRT models, like factor analysis models, can be sensitive to relatively large components of variation in examinees' propensity to answer items correctly. LLTM-style models can be sensitive to finer components of variation *among items* but are not at all sensitive to components of variation *among examinees.* Noncompensatory approaches (e.g., Embretson, 1997) are intended to be sensitive to finer variations among examinees, in situations in which several cognitive components are required simultaneously for successful task performance. Of course, whether cognitive assessment data actually support models that track this finer level of variation is an empirical matter.

4.1 Three Approaches to Cognitive Assessment

To illustrate some differences between traditional IRT and cognitively motivated approaches to assessment, we consider two published models intended to deal with the same data: task performance by students learning the LISP programming language using a computer-based intelligent tutoring system developed by John R. Anderson and his colleagues at Carnegie Mellon University (e.g., Anderson, Corbett, Koedinger, & Pelletier, 1995). The first model is the assessment model actually embedded in the tutoring software, as described by Corbett, Anderson, and O'Brien (1995); the second is an IRT model for essentially the same data, as presented by Draney, Pirolli, and Wilson (1995). Then we consider a third model to illustrate that, although the modeling traditions in IRT are not much like those in cognitive assessment modeling, the fundamental techniques and concepts from IRT modeling are quite useful in cognitive assessment.

The Corbett/Anderson/O'Brien model (CAOM). The knowledge tracing model embedded in the LISP tutor software (Corbett et al., 1995) treats successful performance of a task at time t, $t = 1, 2, 3, \ldots$, as an uncertain indicator of possession of the necessary underlying skills at time t, according to the model

$$P[X_{vi}(t) = 1] = P[\xi_{vi}(t) = 1](1 - s) + \{1 - P[\xi_{vi}(t) = 1]\}g \ , \qquad (14.10)$$

where $X_{vi}(t) = 1$ or 0 indicates whether examinee S_v performed task I_i correctly at time t, $\xi_{vi}(t) = 1$ or 0 indicates whether examinee S_v possesses the requisite skills for task I_i at time t, and parameters s and g in (14.10) are universal "slipping" and "guessing" probabilities that accommodate uncertainty in predicting task performance $X_{vi}(t)$ from $\xi_{vi}(t)$. $P[\xi_{vi}(t) = 1]$ follows a simple conjunctive model (although more complex expressions along the lines of Mislevy, 1996, could be imagined):

$$P[\xi_{vi}(t) = 1] = \prod_{\ell=1}^{d} P[\alpha_{v\ell}(t) \geq Q_{i\ell}] \ , \qquad (14.11)$$

with latent variables $\alpha_{v\ell}(t) = 1$ or 0, indicating whether student S_v possesses skill ℓ at time t, $t = 1, 2, 3, \ldots$. The $Q_{i\ell} = 0$ or 1 to indicate whether skill ℓ is needed for task i, and skills combine conjunctively to predict task performance. The $\xi_{vi}(t)$s (or the $\alpha_{v\ell}(t)$s) can be interpreted as playing the role of Maris's (1995) latent responses (they can also be interpreted in terms of the method of data augmentation in Bayesian computation, e.g., Tanner, 1996). Given current estimates of $P(\alpha_{v\ell}(t) \geq Q_{i\ell})$, the tutor can both identify which skills need additional practice, and select items of suitable difficulty that exercise those skills.

Corbett et al. (1995) were particularly interested in how to gather evidence about $P[\alpha_{v\ell}(t)]$ as the number of opportunities t to apply rule ℓ

increases; that is, they were interested in *learning*. To account for the order in which correct and incorrect actions are observed when skill ℓ is called for, they treat each $\alpha_{v\ell}(t)$ as a two-state hidden Markov chain in t, with an absorbing state at $\alpha_{v\ell}(t) = 1$, and corresponding observable evidence $a_{v\ell}(t) = 1$ or 0 that the correct action was performed when calling for skill ℓ. For brevity, let $\omega = P[\alpha_{v\ell}(t) = 1 \mid \alpha_{v\ell}(t-1) = 0]$, and let $\alpha_{v\ell}(t)$ denote $\alpha_{v\ell}(t) = 1$ and $\bar{\alpha}_{v\ell}(t)$ denote $\alpha_{v\ell}(t) = 0$. The following recursions, derived from Bayes' Rule, relate the observable data to the hidden Markov states.

$$P[\alpha_{v\ell}(t) \mid a_{v\ell}(t)] = P[\alpha_{v\ell}(t-1) \mid a_{v\ell}(t)] + \{1 - P[\alpha_{v\ell}(t-1) \mid a_{v\ell}(t)]\}\, \omega\ ,$$

$$P[\alpha_{v\ell}(t) \mid \bar{a}_{v\ell}(t)] = P[\alpha_{v\ell}(t-1) \mid \bar{a}_{v\ell}(t)] + \{1 - P[\alpha_{v\ell}(t-1) \mid \bar{a}_{v\ell}(t)]\}\, \omega\ ,$$

$$P[\alpha_{v\ell}(t-1) \mid a_{v\ell}(t)] = \frac{(1-s)P[\alpha_{v\ell}(t-1)]}{(1-s)P[\alpha_{v\ell}(t-1)] + gP[\bar{\alpha}_{v\ell}(t-1)]}\ ,$$

$$P[\alpha_{v\ell}(t-1) \mid \bar{a}_{v\ell}(t)] = \frac{sP[\alpha_{v\ell}(t-1)]}{sP[\alpha_{v\ell}(t-1)] + (1-g)P[\bar{\alpha}_{v\ell}(t-1)]}\ . \qquad (14.12)$$

Given a priori fixed values of ω, s, g, and a probability p_0 that each skill is in the learned state before tutoring begins, we may substitute p_0 for $P[\alpha_{v\ell}(t-1)]$ when $t = 1$, and the above formulae (14.12) give an algorithm for recursively updating $P[\alpha_{v\ell}(t)]$ on the basis of the observed sequence of correct and incorrect actions in the first t opportunities to apply rule ℓ. This is a particularly simple and fast computation, capable of updating the tutor's model of the student's skills in real time as the student works with the tutor.

Interestingly, in order to obtain good model fit, Corbett et al. (1995) allowed the probabilities ω, s, g, and p_0 to be perturbed differently from overall population values for each student S_v; effectively, they allowed these parameters to become random effects. Thus, in addition to the individual differences in skills acquisition that the model had been designed to detect, there were substantial individual differences in starting knowledge of the students, in tendency to slip or guess, and in the rate of learning, under this model.

The Draney/Pirolli/Wilson model (DPWM). Draney, Pirolli, and Wilson (1995) developed an LLTM-style model to analyze essentially the same data. They begin with an indicator $a_{vi\ell}(t) = 1$ if student S_v performs correctly when the tth opportunity to use skill ℓ occurs under condition i; and $a_{vi\ell}(t) = 0$ otherwise. These $a_{vi\ell}(t)$ differ from the $a_{v\ell}(t)$ in the CAOM only in that the task in which the skill is performed is allowed to affect the difficulty of skill performance. In the DPWM, the skill response functions are given by

$$P[a_{vi\ell}(t) = 1 \mid \theta_v, \tau_i, \delta_\ell, \gamma] = \frac{1}{1 + \exp[-\theta_v + \tau_i + \delta_\ell - \gamma \ln(t)]} .$$

This model decomposes the difficulty parameter β in the Rasch model according to a Q matrix, as in (14.8) and (14.9), in terms of parameters for task τ_i, skill δ_ℓ, and slope of the learning curve γ. The logarithmic dependence on t is intended to follow the development of Anderson (1993, Appendix to Chap. 3). If the decomposition of tasks into skills is complete, and the skills are of a suitable granularity, Anderson's ACT-R theory predicts that skill performances will be approximately independent of one another, given the relevant difficulty and student parameters. This is essentially a statement of local independence, so that the "skill response functions" above may be multiplied together as usual to form an IRT likelihood.

It is interesting to compare the two modeling approaches. For example, for a data set similar to that analyzed by Draney et al. (1995), the assessment model of Corbett et al. (1995, p. 32; see also Draney et al., 1995, p. 115) would employ essentially four continuous latent variables, and 33 dichotomous latent attribute indicators, *per student tested*, in addition to 132 parameters to characterize features of the skills being assessed. Draney et al. (1995) provided equivalent or better fit to learning curves, employing *one* continuous latent variable per student tested (op. cit., p. 109), and 36 parameters for the skills being tested (op. cit., p. 115). Thus, if the goal is to model learning curves, clearly the more complex CAOM is not needed.

However, the uses to which the two models can be put, and the substantive interpretations of estimated parameters in the two models, are very different. The CAOM is immediately useful for diagnosing individual differences in student task performance behavior by relating it directly to specific skills in the task decomposition of their task domain, but can only indirectly assess the validity and reliability of the tasks in the assessment, through fit to learning curves or other summaries of student task performance.

The DPWM is not immediately useful for student diagnosis, since its student parameter is one-dimensional. After fitting the model, Draney et al. (1995) go back and rank students based on (empirical Bayes) estimated θs (in the context of learning curve analysis, the θs essentially code students' initial facility in the task domain prior to tutoring, much like p_0 in the CAOM), and compare estimated skill performance difficulties to the students' aggregate θ distribution; see, for example, their Figure 5.1, p. 112. Such displays help predict which skills a "typical" student with some fixed value of θ might be expected to perform correctly, and are very useful communication devices. However, detailed cognitive diagnosis of individual students on the basis of the θs is not possible without a post hoc analysis of some sort. Indeed, for individual cognitive diagnosis, Draney et al. (1995) replace the IRT model with a Bayesian inference network that is focused on inferring the probability that an individual has learned each skill, using priors constructed from the fitted IRT model, and updated with data from further attempts to perform the skill(s).

The utility of the DPWM for analyzing important task performance features, should not be minimized however. For example, Junker, Koedinger, and Trottini (2000) are using essentially the same model to develop a semi-automatic stepwise variable construction/model selection procedure, with the goal of identifying skills that are either too narrowly or too broadly defined in a cognitive tutor (these result in stylized deviations from the theoretically predicted learning curves for the skills). Similarly, Huguenard, Lerch, Junker, Patz, and Kass (1997; see also Patz, Junker, Lerch, & Huguenard, 1996) applied a polytomous version of the LLTM to study the relationship between task features and working memory load, using the IRT θ parameter to soak up residual between-subjects variation not modeled by experimental and working memory factors.

An IRT-like cognitive assessment model. As we have shown, traditional IRT may not be well suited to individual assessment tasks in computer-based intelligent tutoring systems. The same issues arise in assessments focused on presence or absence of specific skills (rather than sorting examinees along a linear scale) based on performance on a fixed set of tasks given as a standalone test.

To illustrate this we consider a modified version of the original CAOM. We omit the hidden-Markov learning model and assume that examinee behavior is only observed at the task level, not the skill level. The new model, whose components are summarized in Table 14.2, is interpretable as a simplified version of Embretson's (1997) multicomponent latent trait (MLTM) model. A related, more general class of models is the class of probability matrix decomposition models (Maris, De Boeck, & Van Mechelen, 1996).

The goal is to try to estimate the $\alpha_{v\ell}$s, or more precisely $P(\alpha_{v\ell} = 1)$, from the task performance data. Our basic response model (level one in the hierarchy of Table 14.1) is

$$P(X_{vi} = 1 \mid \boldsymbol{\xi}, \mathbf{s}, \mathbf{g}) = \xi_{vi}(1 - s_i) + (1 - \xi_{vi})g_i = (1 - s_i)^{\xi_{vi}} g_i^{1-\xi_{vi}} ,$$

and so for the entire examinees by tasks matrix $[X_{vi}]$ of task responses,

$$\begin{aligned} P\left([X_{vi}] = [x_{vi}] \mid \boldsymbol{\xi}, \mathbf{s}, \mathbf{g}\right) \\ &= \prod_v \prod_i \left[(1 - s_i)^{\xi_{vi}} g_i^{1-\xi_{vi}}\right]^{x_{vi}} \left[1 - (1 - s_i)^{\xi_{vi}} g_i^{1-\xi_{vi}}\right]^{1-x_{vi}} \\ &= \prod_v \prod_i \left[(1 - s_i)^{x_{vi}} s_i^{1-x_{vi}}\right]^{\xi_{vi}} \left[g_i^{x_{vi}}(1 - g_i)^{1-x_{vi}}\right]^{1-\xi_{vi}} . \qquad (14.13) \end{aligned}$$

To see how estimation of the parameters depends on the data, it is instructive to set up the first stages of an estimation algorithm for the model.

TABLE 14.2. Components of a simple cognitive assessment model for a stand-alone test, with data collected at the task level X_{vi} rather than at the skill level $\alpha_{v\ell}$.

$X_{vi} = 1$ or 0 ,	indicating whether student S_v performed task I_i correctly,
$Q_{i\ell} = 1$ or 0 ,	indicating whether task I_i requires skill ℓ,
$\alpha_{v\ell} = 1$ or 0 ,	indicating whether student S_v possesses skill ℓ,
$\xi_{vi} = \prod_{\ell:\ Q_{i\ell}=1} \alpha_{v\ell}$,	indicating whether student S_v has the skills needed for task I_i,
$s_i = P(X_{vi} = 0 \mid \xi_{vi} = 1)$,	a per-problem slip parameter,
$g_i = P(X_{vi} = 1 \mid \xi_{vi} = 0)$,	a per-problem guessing parameter.

In particular, assume that prior distributions have been chosen (for levels two and three in the hierarchy in Table 14.1), so that $s_i \sim \pi_s(s_i)$, $g_i \sim \pi_g(g_i)$, $\alpha_{v\ell} \sim \pi_\ell^{\alpha_{v\ell}}(1-\pi_\ell)^{1-\alpha_{v\ell}}$, and perhaps $\pi_\ell \sim \pi(\pi_\ell)$. We calculate the so-called complete conditional distributions (e.g., Gelman, Carlin, Stern, & Rubin, 1995) of each parameter, given the data and the rest of the parameters. Such a calculation is directly useful in setting up an MCMC algorithm for Bayesian estimation of the model parameters (e.g., Patz & Junker, 1999a, 1999b), and is also useful in some versions of the EM algorithm, such as ECME (e.g., Liu & Rubin, 1998). However, even when EM is "possible" for a model such as (14.13), it need not be practical, due to the need to sum over all configurations of the latent $\boldsymbol{\alpha}$ vector (see Equation 14.14 below). For cases in which relatively few αs carry most of the latent trait distribution, MCMC can be a more efficient way to estimate the model. This phenomenon has been found in other models with complex discrete latent structure as well (e.g., Seltman, 1999; Snijders & Nowicki, 1997; Ter Hofstede, Steenkamp, & Wedel, 1999).

As usual in setting up an MCMC calculation, we note that the complete conditional distribution for each parameter or latent variable is proportional to the product of only those likelihood and prior factors in Table 14.1 depending on that parameter. Thus we obtain, for example,

$$p(s_i \mid \text{rest}) \;\propto\; (1-s_i)^{\sum_v x_{vi}\xi_{vi}} s_i^{\sum_v (1-x_{vi})\xi_{vi}} \pi_s(s_i) \;,$$

$$p(g_i \mid \text{rest}) \;\propto\; g_i^{\sum_v x_{vi}(1-\xi_{vi})} (1-g_i)^{\sum_v (1-x_{vi})(1-\xi_{vi})} \pi_g(g_i) \;,$$

for $i = 1, \ldots, k$, where "rest" stands for the data and the rest of the parameters in the model. From these it is easy to see the intuitively plausible facts that the slip parameter s_i is sensitive only to successes and failures of those examinees S_v who we hypothesize (through the values of ξ_{vi} upon which we have conditioned) to have the requisite skills to perform task I_i, and similarly the guessing parameter g_i is sensitive only to successes and failures of examinees who we hypothesize not to have the requisite skills.

More central to the goal of assessing examinees, we see that the complete conditional distributions for $\alpha_{v\ell}$ are of the form

$$p(\alpha_{v\ell} \mid \text{rest}) \propto \prod_{i:\ Q_{i\ell}=1} \left[(1-s_i)^{x_{vi}} s_i^{1-x_{vi}}\right]^{\alpha_{v\ell}\xi_{vi}^{(-\ell)}} \left[g_i^{x_{vi}}(1-g_i)^{1-x_{vi}}\right]^{1-\alpha_{v\ell}\xi_{vi}^{(-\ell)}} \times \pi_\ell^{\alpha_{v\ell}}(1-\pi_\ell)^{1-\alpha_{v\ell}},$$

where $\xi_{vi}^{(-\ell)} = \prod_{m\neq\ell:Q_{im}=1} \alpha_{vm}$, which indicates presence of all skills needed for task i, *except* for skill ℓ. From these complete conditionals we can see that when $\xi_{vi}^{(-\ell)} = 1$, the suggested conditional model for α_{vi} is some sort of Bernoulli, which makes sense. Also, when there are no tasks such that both $Q_{j\ell} = 1$ and $\xi_{vi}^{(-\ell)} = 1$, then $\alpha_{v\ell}$ is sensitive only to its prior distribution $\pi_\ell^{\alpha_{v\ell}}(1-\pi_\ell)^{1-\alpha_{v\ell}}$; that is, no learning from data occurs. This observation is really a version of the credit/blame problem (e.g., VanLehn & Niu, in press): we cannot assign blame for incorrect task performance to $\alpha_{v\ell}$, if we are hypothesizing that another needed skill is still unlearned; $\alpha_{v\ell}$'s status is masked by the unlearned status of the other skill.

There are essentially two ways to avoid the credit/blame problem. In some situations, skills can be scored directly; this is possible, for example, within Anderson's ITSs for LISP, algebra, and geometry, because students are required to successfully perform each subgoal/skill, with hints and repeated attempts if necessary, before moving on the next subgoal in the task. If one cannot score the tasks subgoal by subgoal, one can try to design the assessment (by hand or using methods from traditional statistical experimental design) so that the tasks, taken together, do inform us about each skill. VanLehn, Niu, Siler, and Gertner (1998) illustrate the inferential difficulties that can result when item sets are *not* constructed with the goal of designing around the credit/blame problem.

Finally, if we want to estimate the skill base rates π_ℓ in the population (a measure of skill difficulty) we may include a fourth set of complete conditionals $p(\pi_\ell \mid \text{rest}) \propto \pi_\ell^{c_\ell}(1-\pi_\ell)^{N-c_\ell}\pi(\pi_\ell)$, where $c_\ell = \sum_v \alpha_{v\ell}$ is the number of students who are presently estimated to have skill ℓ.

4.2 A Role for Nonparametric IRT Methods?

Such models as (14.13) may seem very far removed from the setting in which nonparametric IRT methods are familiar. Here several ways are suggested, however, in which they are not so far removed.

First, suppose that the skill variables $\alpha_{v\ell}$ vary independently of one another in a population of students or examinees, as would be consistent with, for example, Anderson's (1993) ACT-R theory, and suppose that the slip and guessing parameters s_i and g_i are fixed and satisfy the plausible inequality $(1 - s_i) \geq g_i$. A model for the task performance of a randomly sampled examinee from the population would then be

$$P(\mathbf{X} = \mathbf{x}) = \sum_{\boldsymbol{\alpha}} \prod_i P_i[\xi_i(\boldsymbol{\alpha})]^{x_i} \left\{1 - P_i[\xi_i(\boldsymbol{\alpha})]\right\}^{1-x_i} p(\boldsymbol{\alpha}) \;, \qquad (14.14)$$

where $p(\boldsymbol{\alpha})$ is a product measure over the space of binary skills $\boldsymbol{\alpha} = (\alpha_1, \ldots, \alpha_d)'$, and

$$P_i[\xi_i(\boldsymbol{\alpha})] = (1 - s_i)^{\xi_i(\boldsymbol{\alpha})} g_i^{1-\xi_i(\boldsymbol{\alpha})}$$

is a monotone function of $\xi_i(\boldsymbol{\alpha})$. Hence $P_i[\xi_i(\boldsymbol{\alpha})]$ is also monotone in the coordinates α_ℓ of $\boldsymbol{\alpha}$ (the index v has been dropped from $\alpha_{v\ell}$ to indicate that we are only working with the marginal model in what follows). Note the similarity of (14.14) to (14.2) and (14.3).

It follows from Lemma 2 of Holland and Rosenbaum (1986) that for any nondecreasing summary $g(\mathbf{X})$ of $\mathbf{X} = (X_1, \ldots, X_k)'$, $E[g(\mathbf{X}) \mid \boldsymbol{\alpha}]$ is nondecreasing in each coordinate α_ℓ of $\boldsymbol{\alpha}$. This implies the SOM (stochastic ordering of the manifest score X_+ by the latent trait) property of Hemker et al. (1997), namely, that $P(X_+ > c \mid \boldsymbol{\alpha})$ is nondecreasing in each coordinate α_ℓ of $\boldsymbol{\alpha}$. Not much is known about the inverse and more useful property SOL (14.7) when the latent trait is multidimensional. We might conjecture, for example, that $P(\alpha_\ell = 1 \mid \sum_{i:Q_{i\ell}=1} X_i = s)$ would be nondecreasing in s. As in conventional dichotomous IRT models we hope that such a property is true, because it means that a most powerful test of mastery of skill ℓ can be based on a simple total of correctly performed tasks involving skill ℓ.

It also follows from Theorem 8 of Holland and Rosenbaum (1986) that since $\boldsymbol{\alpha}$ is a collection of independent, or more generally, associated, random variables, the collection of response variables $\mathbf{X}$ is also associated, that is, for any two nondecreasing summaries $f(\mathbf{X})$ and $g(\mathbf{X})$, that $\mathrm{Cov}\,[f(\mathbf{X}), g(\mathbf{X})] \geq 0$. Obtaining conditional association (14.4), or a suitable generalization in the case of multidimensional $\boldsymbol{\alpha}$, for $\mathbf{X}$ is a more difficult challenge.

An invariant item ordering property, such as $P_i[\xi_i(\boldsymbol{\alpha})] \leq P_j[\xi_j(\boldsymbol{\alpha})]$ uniformly in $\boldsymbol{\alpha}$, will follow if we assume, for example, that $1 - s_i$, g_i, and $\xi_i(\boldsymbol{\alpha})$, are comonotone as i varies, for all $\boldsymbol{\alpha}$. This is a kind of Guttman scaling condition on the latent ξ_i that says, for example, that easier guessing is associated with lower skill requirements, and vice versa. Thus, our cognitive assessment model (14.13) provides fertile ground for broadening our thinking about invariant item ordering beyond continuous unidimensional IRT.

More broadly, Hoijtink and Molenaar (1997) show how features such as conditional association (14.4) and manifest monotonicity (14.6) can be directly relevant to assessing model fit in a parametric Bayesian setting. Given a complete and interesting set of nonparametric model features for models like (14.13), a similar approach to model fit may be taken here.

Finally, among some who work in the nonparametric IRT traditions of Mokken, Molenaar, Sijtsma, Holland, Rosenbaum, and their colleagues, it is the source of some bemusement that we work so hard to establish that scores such as X_+ (which are perfectly good in the Rasch model) also suffice to order examinees, assess monotonicity properties of the underlying model from observable data, make mastery decisions, and the like, in general nonparametric settings. Certainly comparisons between Rasch and Mokken scaling are not new (e.g., Meijer, Sijtsma, & Smid, 1990), and aspects of both Holland's (1990) "Dutch identity" work and Scheiblechner's (1995; see also Junker, 1998) work on the isotonic ordinal probabilistic (ISOP) model can be seen partly as attempts to formalize the connection between Rasch and nonparametric IRT methods.

I believe that the connection is actually rather simple, and is nearly obvious from Holland's (1990) work: the Rasch model is a very well-behaved exponential family model with immediately understandable sufficient statistics for items, given person parameters, and immediately understandable sufficient statistics for persons, given item parameters. Much of the work on monotone homogeneity models and their cousins is directed at understanding just how generally these understandable, but no longer formally sufficient, statistics yield sensible inferences about examinees and items. The model (14.13) provides us with a different basic model for cognitive assessment, in which parameter estimates depend on immediately understandable summaries of the data, as illustrated by the complete conditional calculations above. We may ask how complex the relationship between the examinee skill parameters $\boldsymbol{\alpha}$ and the task performance data $\mathbf{X}$ can get, while retaining fairly simple summaries of the data that are informative about guessing, slips, presence or absence of skills, and so on. We may also ask whether this is the "right" parametric model on which to base a nonparametric theory of cognitive assessment. For example, other possible models we might consider instead of (14.13) as a starting point for such a nonparametric theory include the constrained latent class model of Haertel (1989) and the hybrid model of Yamamoto and Gitomer (1993).

In addition to the intrinsic interest of this enterprise, the resulting nonparametric theory of cognitive assessment may have some practical utility. The relative ease with which the original CAOM can be estimated is a consequence of *both* its careful tailoring to the psychological theory, *and* the fact that data could be collected at the skill level (same granularity as the psychological theory), rather than at the task level. But other assessment settings may not permit such tight binding of data collection design and psychological theory. Our discussion of the model (14.13) shows that even relatively minor modifications along these lines can make the inferential task more difficult. Estimation probably requires EM, MCMC, or some other computationally intensive method, and great care must be taken so that every parameter is identifiable from the data. Models being proposed in the psychometric literature today to help unify the traditional IRT and discrete-cognitive-attributes approaches, such as the Unified Model of DiBello, Jiang, and Stout (2000), also appear to require such treatment. In many cases, the estimation method, while feasible, may well be too slow for use in real-time feedback, as with computer-based intelligent tutoring systems, and too complicated for teacher scoring of assessments embedded in instruction. A clear theory of which faster data summaries are relevant to the cognitive inferences we wish to make, over a wide variety of cognitive assessment models, would be an important contribution from the interface between nonparametric and parametric IRT models.

Acknowledgments: Some of the work reported here was performed in the course of preparing a commissioned paper for the National Research Council Committee on Educational and Psychological Foundations of Assessment, United States National Academy of Sciences. Preparation of this chapter was supported in part by National Science Foundation grants #SES-99.07447 and #DMS-97.05032, and the hospitality of the Learning Research and Development Center at the University of Pittsburgh. Thanks to Klaas Sijtsma, Lou Mariano, Matt Johnson, L. Andries van der Ark, and the editors of this volume for their comments and suggestions.

References

Ackerman, T.A. (1994). Using multidimensional item response theory to understand what items and tests are measuring. *Applied Measurement in Education, 7,* 255–278.

Adams, R.J., Wilson, M., & Wang, W.-C. (1997). The multidimensional random coefficients multinomial logit model. *Applied Psychological Measurement, 21,* 1–23.

Algina, J. (1992). The National Assessment of Educational Progress (Editor's Note) [Special issue]. *Journal of Educational Measurement, 29,* 93–94.

Anderson, J.R. (Ed.). (1993). *Rules of the mind.* Hillsdale, NJ: Erlbaum.

Anderson, J.R., Corbett, A.T., Koedinger, K.R., & Pelletier, R. (1995). Cognitive tutors: Lessons learned. *The Journal of the Learning Sciences, 4,* 167-207.

Andrich, D. (1996). A hyperbolic cosine latent trait model for unfolding polytomous responses: Reconciling Thurstone and Likert methodologies. *British Journal of Mathematical and Statistical Psychology, 49,* 347–365.

Bartolucci, F., & Forcina, A. (in press). A likelihood ratio test for MTP2 within binary variables. *The Annals of Statistics.*

Baxter, G.P., & Glaser, R. (1998). Investigating the cognitive complexity of science assessments. *Educational Measurement: Issues and Practice, 17,* 37–45.

Béguin, A.A., & Glas, C.A.W. (1998). *MCMC estimation of multidimensional IRT models.* (Research Report 98-14). University of Twente, Department of Educational Measurement and Data Analysis, The Netherlands.

Bloom, B.S. (1984). The 2-sigma problem: The search for methods of group instruction as effective as one-to-one tutoring. *Educational Researcher, 13,* 4–16.

Bock, R.D., & Zimowski, M.F. (1997). Multi-group IRT. In W.J. van der Linden & R.K. Hambleton (Eds.), *Handbook of modern item response theory* (pp. 433–448). New York: Springer-Verlag.

Bradlow, E.T., Wainer, H., & Wang, X. (1999). A Bayesian random effects model for testlets. *Psychometrika, 64,* 153–168.

Cliff, N., & Donoghue, J.R. (1992). Ordinal test fidelity estimated by an item sampling model. *Psychometrika, 57,* 217–236.

Corbett, A.T., Anderson, J.R., & O'Brien, A.T. (1995). Student modeling in the ACT programming tutor. In P.D. Nichols, S.F. Chipman, & R.L. Brennan (Eds.), *Cognitively diagnostic assessment* (pp. 19–41). Hillsdale, NJ: Erlbaum.

DiBello, L., Jiang, H., & Stout, W.F. (2000). *A multidimensional IRT model for practical cognitive diagnosis.* Manuscript submitted for publication.

Douglas, J., & Qui, P. (1997). *Generalized linear factor analysis with Markov chain Monte Carlo.* Unpublished manuscript, University of Wisconsin, Department of Biostatistics and Medical Informatics, Madison, WI.

Draney, K. L., Pirolli, P., & Wilson, M. (1995). A measurement model for a complex cognitive skill. In P.D. Nichols, S.F. Chipman, & R.L. Brennan (Eds.), *Cognitively diagnostic assessment* (pp. 103–125). Hillsdale, NJ: Erlbaum.

Drasgow, F., Levine, M.V., Tsien, S., Williams, B., & Mead, A.D. (1995). Fitting polytomous item response theory models to multiple-choice tests. *Applied Psychological Measurement, 19,* 143–165.

Ellis, J.L. (1994). *Foundations of monotone latent variable models.* Nijmegen: Nijmegen Institute for Cognition and Information.

Ellis, J.L., & Junker, B.W. (1997). Tail-measurability in monotone latent variable models. *Psychometrika, 62,* 495–523.

Embretson, S.E. (1991). A multidimensional latent trait model for measuring learning and change. *Psychometrika, 56,* 495–515.

Embretson, S.E. (1995). Developments toward a cognitive design system for psychological tests. In D. Lubinski & R.V. Dawis (Eds.), *Assessing individual differences in human behavior: New concepts, methods and findings* (pp. 17–48). Palo Alto, CA: Davies-Black Publishing.

Embretson, S.E. (1997). Multicomponent response models. In W.J. van der Linden & R.K. Hambleton (Eds.), *Handbook of modern item response theory* (pp. 305–321). New York: Springer-Verlag.

Embretson, S.E. (1999). Generating items during testing: Psychometric issues and models. *Psychometrika, 64,* 407–433.

Engelhard, G., Jr. (1994). Examining rater errors in the assessment of written composition with many-faceted Rasch models. *Journal of Educational Measurement, 31,* 93–112.

Fischer, G.H. (1973). Linear logistic test model as an instrument in educational research. *Acta Psychologica, 37,* 359–374.

Fischer, G.H., & Molenaar, I.W. (Eds.). (1995). *Rasch models: Foundations, recent developments, and applications.* New York: Springer-Verlag.

Fox, G.J.A., & Glas, C.A.W. (1998). *A multi-level IRT model with measurement error in the predictor variables* (Research Report 98-16). University of Twente, Department of Educational Measurement and Data Analysis, The Netherlands.

Fraser, C., & McDonald, R.P. (1988). NOHARM: Least squares item factor analysis. *Multivariate Behavioral Research 23,* 267–269.

Gardner, H. (1992). Assessment in context: The alternative to educational testing. In B.R. Gifford & M.C. O'Connor (Eds.), *Changing assessments: Alternative views of aptitude, achievement, and instruction* (pp. 77–119). Norwell, MA: Kluwer.

Gelman, A., Carlin, J.B., Stern, H.S., & Rubin, D.B. (1995). *Bayesian data analysis.* New York: Chapman and Hall.

Gibbons, R.D., & Hedeker, D.R. (1992). Full-information item bi-factor analysis. *Psychometrika, 57,* 423–436.

Gibbons, R.D., & Hedeker, D.R. (1997). Random effects probit and logistic regression models for three-level data. *Biometrics, 53,* 1527–1537.

Glas, C.A.W., & Verhelst, N.D. (1989). Extensions of the partial credit model. *Psychometrika, 54,* 635–659.

Grayson, D.A. (1988). Two-group classification in latent trait theory: Scores with monotone likelihood ratio. *Psychometrika, 53,* 383–392.

Haertel, E.H. (1989). Using restricted latent class models to map the skill structure of achievement items. *Journal of Educational Measurement, 26,* 301–321.

Hemker, B.T. (2001). Reversibility revisited and other comparisons of three types of polytomous IRT models. In A. Boomsma, M.A.J. van Duijn, & T.A.B. Snijders (Eds.), *Essays on item response theory* (pp. 277–296). New York: Springer-Verlag.

Hemker, B.T., Sijtsma, K., & Molenaar, I.W. (1995). Selection of unidimensional scales from a multidimensional item bank in the polytomous Mokken IRT model. *Applied Psychological Measurement, 19,* 337–352.

Hemker, B.T., Sijtsma K., Molenaar, I.W., & Junker, B.W. (1996). Polytomous IRT models and monotone likelihood ratio of the total score. *Psychometrika, 61,* 679–693.

Hemker, B.T., Sijtsma K., Molenaar, I.W., & Junker, B.W. (1997). Stochastic ordering using the latent trait and the sum score in polytomous IRT models. *Psychometrika, 62,* 331–347.

Hoijtink, H., & Molenaar, I.W. (1997). A multidimensional item response model: Constrained latent class analysis using the Gibbs sampler and posterior predictive checks. *Psychometrika, 62,* 171–189.

Holland, P.W. (1981). When are item response models consistent with observed data? *Psychometrika, 46,* 79–92.

Holland, P.W. (1990). The Dutch identity: A new tool for the study of item response models. *Psychometrika, 55,* 5–18.

Holland, P.W., & Rosenbaum, P.R. (1986). Conditional association and unidimensionality in monotone latent trait models. *The Annals of Statistics, 14,* 1523–1543.

Huguenard, B.R., Lerch, F.J., Junker, B.W., Patz, R.J., & Kass, R.E. (1997). Working memory failure in phone-based interaction. *ACM Transactions on Computer-Human Interaction, 4,* 67–102.

Huynh, H. (1994). A new proof for monotone likelihood ratio for the sum of independent Bernoulli random variables. *Psychometrika, 59,* 77–79.

Jaakkola, T.S., & Jordan, M.I. (2000). Bayesian parameter estimation via variational methods. *Statistics and Computing, 10,* 25–37.

Janssen, R., & De Boeck, P. (1997). Psychometric modeling of componentially designed synonym tasks. *Applied Psychological Measurement, 21,* 37–50.

Johnson, E.G., Mislevy, R.J., & Thomas, N. (1994). Theoretical background and philosophy of NAEP scaling procedures. In E.G. Johnson, J. Mazzeo, & D.L. Kline (Eds.), *Technical Report of the NAEP 1992 Trial State Assessment Program in Reading* (pp. 133–146). Washington, DC: Office of Educational Research and Improvement, U.S. Department of Education.

Junker, B.W. (1993). Conditional association, essential independence and monotone unidimensional item response models. *The Annals of Statistics, 21,* 1359–1378.

Junker, B.W. (1998). Some remarks on Scheiblechner's treatment of ISOP models. *Psychometrika, 63,* 73–85.

Junker, B.W., & Ellis, J.L. (1997). A characterization of monotone unidimensional latent variable models. *The Annals of Statistics, 25,* 1327–1343.

Junker, B.W., Koedinger, K.R., & Trottini, M. (2000, July). *Finding improvements in student models for intelligent tutoring systems via variable selection for a linear logistic test model.* Paper presented at the Annual North American Meeting of the Psychometric Society, Vancouver, Canada.

Junker, B.W., & Sijtsma, K. (2000). Latent and manifest monotonicity in item response models. *Applied Psychological Measurement, 24,* 65–81.

Kass, R.E., Tierney, L., & Kadane, J.B. (1990). The validity of posterior expansions based on Laplace's method. In S. Geisser, J.S. Hodges, S.J. Press, & A. Zellner (Eds.), *Bayesian and likelihood methods in statistics and econometrics: Essays in honor of George A. Barnard* (pp. 473–488). New York: North-Holland.

Kelderman, H., & Rijkes, C.P.M. (1994). Loglinear multidimensional IRT models for polytomously scored items. *Psychometrika, 59,* 149–176.

Lee, Y., & Nelder, J.A. (1996). Hierarchical generalized linear models (with discussion). *Journal of the Royal Statistical Society, Series B, 58,* 619–678.

Linacre, J.M. (1989). *Many-faceted Rasch measurement.* Chicago: Mesa Press.

Liu, C., & Rubin, D.B. (1998). Maximum likelihood estimation of factor analysis using the ECME algorithm with complete and incomplete data. *Statistica Sinica, 8,* 729–747.

Loevinger, J. (1948). The technique of homogeneous tests compared with some aspects of "scale analysis" and factor analysis. *Psychological Bulletin, 45,* 507–530.

Lord, F.M. (1952). A theory of test scores. *Psychometric Monographs, 7.*

Maris, E. (1995). Psychometric latent response models. *Psychometrika, 60,* 523–547.

Maris, E., De Boeck, P., & Van Mechelen, I. (1996). Probability matrix decomposition models. *Psychometrika, 61,* 7–29.

Masters, G.N. (1982). A Rasch model for partial credit scoring. *Psychometrika, 47,* 149–174.

McCullagh, P., & Nelder, J.A. (1989). *Generalized linear models* (2nd ed.). New York: Chapman and Hall.

Meijer, R.R. (1996). Person-fit research: An introduction. (Guest editor's introduction to the Special issue: Person-fit research: Theory and applications.) *Applied Measurement in Education, 9,* 3–8.

Meijer, R.R., Sijtsma, K., & Smid, N.G. (1990). Theoretical and empirical comparison of the Mokken and the Rasch approach to IRT. *Applied Psychological Measurement, 14,* 283–298.

Mellenbergh, G.J. (1995). Conceptual notes on models for discrete polytomous item responses. *Applied Psychological Measurement, 19,* 91–100.

Meredith, W. (1965). Some results based on a general stochastic model for mental tests. *Psychometrika, 30,* 419–440.

Mislevy, R.J. (1985). Estimation of latent group effects. *Journal of the American Statistical Association, 80,* 993–997.

Mislevy, R.J. (1996). Test theory reconceived. *Journal of Educational Measurement, 33,* 379–416.

Mislevy, R.J., & Sheehan, K.M. (1989). The role of collateral information about examinees in item parameter estimation *Psychometrika, 54,* 661–679.

Mokken, R.J. (1971). *A theory and procedure of scale analysis: With applications in political research.* The Hague: Mouton.

Mokken, R.J. (1997). Nonparametric models for dichotomous items. In W.J. van der Linden & R.K. Hambleton (Eds.), *Handbook of modern item response theory* (pp. 351–368). New York: Springer-Verlag.

Molenaar, I.W. (1991). A weighted Loevinger H-coefficient extending Mokken scaling to multicategory items. *Kwantitatieve Methoden, 37,* 97–117.

Molenaar, I.W. (1997). Nonparametric methods for polytomous responses. In W.J. van der Linden & R.K. Hambleton (Eds.), *Handbook of modern psychometrics* (pp. 369–380). New York: Springer-Verlag.

Molenaar, I.W., & Sijtsma, K. (2000). User's manual MSP5 for Windows: A program for Mokken scale analysis for polytomous items. Version 5.0 [Software manual]. Groningen: ProGAMMA.

Muraki, E., & Carlson, J.E. (1995). Full-information factor analysis for polytomous item responses. *Applied Psychological Measurement, 19,* 73–90.

Nichols, P., & Sugrue, B. (1999). The lack of fidelity between cognitively complex constructs and conventional test development practice. *Educational Measurement: Issues and Practice, 18,* 18–29.

Patz, R.J., & Junker, B.W. (1999a). A straightforward approach to Markov chain Monte Carlo methods for item response models. *Journal of Educational and Behavioral Statistics, 24,* 146–178.

Patz, R.J., & Junker, B.W. (1999b). Applications and extensions of MCMC in IRT: Multiple item types, missing data, and rated responses. *Journal of Educational and Behavioral Statistics, 24,* 342–366.

Patz, R.J., Junker, B.W., & Johnson, M.S. (2000). *The hierarchical rater model for rated test items and its application to large-scale educational assessment data.* (CMU Statistics Technical Report #712). Carnegie Mellon University, Department of Statistics, Pittsburgh, PA.

Patz, R.J., Junker, B.W., Lerch, F.J., & Huguenard, B.R. (1996). *Analyzing small psychological experiments with item response models* (CMU Statistics Department Technical Report #644). Carnegie Mellon University, Department of Statistics, Pittsburgh, PA.

Post, W.J. (1992). *Nonparametric unfolding models. A latent structure approach.* Leiden: DSWO Press.

Post, W.J., & Snijders, T.A.B. (1993). Nonparametric unfolding models for dichotomous data. *Methodika, 7,* 130–156.

Ramsay, J.O. (1991). Kernel smoothing approaches to nonparametric item characteristic curve estimation. *Psychometrika, 56,* 611–630.

Ramsay, J.O. (1995). A similarity-based smoothing approach to nondimensional item analysis. *Psychometrika, 60,* 323–339.

Ramsay, J.O. (1996). A geometrical approach to item response theory. *Behaviormetrika, 23,* 3–17.

Rasch, G. (1960). *Probabilistic models for some intelligence and attainment tests.* Copenhagen: The Danish Institute of Educational Research. (Expanded edition, 1980. Chicago: The University of Chicago Press.)

Reckase, M.D. (1985). The difficulty of test items that measure more than one ability. *Applied Psychological Measurement, 9,* 401–412.

Resnick, L.B., & Resnick, D.P. (1992). Assessing the thinking curriculum: New tools for educational reform. In B.R. Gifford & M.C. O'Connor (Eds.), *Changing assessments: Alternative views of aptitude, achievement, and instruction* (pp. 37–75). Norwell, MA: Kluwer.

Rigdon, S.E., & Tsutakawa, R.K. (1983). Parameter estimation in latent trait models. *Psychometrika, 48,* 567–574.

Robertson, T., Wright, F.T., & Dykstra, R.L. (1988). *Order restricted statistical inference.* New York: Wiley.

Rosenbaum, P.R. (1984). Testing the conditional independence and monotonicity assumptions of item response theory. *Psychometrika, 49,* 425–435.

Rosenbaum, P.R. (1987a). Probability inequalities for latent scales. *British Journal of Mathematical and Statistical Psychology, 40,* 157–168.

Rosenbaum, P.R. (1987b). Comparing item characteristic curves. *Psychometrika, 52,* 217–233.

Roussos, L. (1994). *Summary and review of cognitive diagnosis models.* Unpublished manuscript, Law School Admissions Council, Newton, PA.

Samejima, F. (1969). Estimation of latent ability using a response pattern of graded scores. *Psychometrika Monograph, 17.*

Samejima, F. (1995). Acceleration model in the heterogeneous case of the general graded response model. *Psychometrika, 60,* 549–572.

Samejima, F. (1997). Departure from normal assumptions: A promise for future psychometrics with substantive mathematical modeling. *Psychometrika, 62,* 471–493.

Scheiblechner, H. (1972). Das Lernen und Lösen komplexer Denkaufgaben [The learning and solving of complex reasoning items]. *Zeitschrift für Experimentelle und Angewandte Psychologie, 3,* 456–506.

Scheiblechner, H. (1995). Isotonic ordinal probabilistic models (ISOP). *Psychometrika, 60,* 281–304.

Seltman, H. (1999). *Hidden stochastic models for biological rhythm data.* Unpublished Ph.D. dissertation, Carnegie Mellon University, Department of Statistics, Pittsburgh, PA.

Shute, V.J., & Psotka, J. (1996). Intelligent tutoring systems: Past, Present and Future. In D. Jonassen (Ed.), *Handbook of Research on Educational Communications and Technology* (pp. 570–600). New York: Macmillan Press.

Sijtsma, K. (1998). Methodology review: Nonparametric IRT approaches to the analysis of dichotomous item scores. *Applied Psychological Measurement, 22,* 3–32.

Sijtsma, K., & Hemker, B.T. (1998). Nonparametric polytomous IRT models for invariant item ordering, with results for parametric models. *Psychometrika, 63,* 183–200.

Sijtsma, K., & Junker, B.W. (1996). A survey of theory and methods of invariant item ordering. *British Journal of Mathematical and Statistical Psychology, 49,* 79–105.

Sijtsma, K., & Junker, B.W. (1997). Invariant item ordering of transitive reasoning tasks. In J. Rost & R. Langeheine (Eds.), *Applications of latent trait and latent class models in the social sciences* (pp. 97–107). Münster: Waxmann Verlag.

Sijtsma, K., & Van der Ark, L.A. (2001). Progress in NIRT analysis of polytomous item scores: Dilemmas and practical solutions. In A. Boomsma, M.A.J. van Duijn, & T.A.B. Snijders (Eds.), *Essays on item response theory* (pp. 297–318). New York: Springer-Verlag.

Snijders, T.A.B. (2001). Two-level nonparametric scaling for dichotomous data. In A. Boomsma, M.A.J. van Duijn, & T.A.B. Snijders (Eds.), *Essays on item response theory* (pp. 319–338). New York: Springer-Verlag.

Snijders, T.A.B., & Nowicki, K. (1997). Estimation and prediction for stochastic block models for graphs with latent block structure. *Journal of Classification, 14* 75–100.

Stegelmann, W. (1983). Expanding the Rasch model to a general model having more than one dimension. *Psychometrika, 48,* 259–267.

Stout, W.F. (1987). A nonparametric approach for assessing latent trait unidimensionality. *Psychometrika, 52,* 589–617.

Stout, W.F. (1990). A new item response theory modeling approach with applications to unidimensionality assessment and ability estimation. *Psychometrika, 55,* 293–325.

Stout, W.F., Habing, B., Douglas, J., Kim, H.R., Roussos, L., & Zhang, J. (1996). Conditional covariance-based nonparametric multidimensionality assessment. *Applied Psychological Measurement, 20,* 331–354.

Tanner, M.A. (1996). *Tools for statistical inference: Methods for the exploration of posterior distributions and likelihood functions* (3rd ed.). New York: Springer-Verlag.

Tatsuoka, K.K. (1990). Toward an integration of item response theory and cognitive error diagnosis. In N. Fredriksen, R. Glaser, A. Lesgold, & M.G. Shafto (Eds.), *Diagnostic monitoring of skill and knowledge acquisition* (pp. 453–488). Hillsdale, NJ: Erlbaum.

Ter Hofstede, F., Steenkamp, J.-B.E.M., & Wedel, M. (1999). *Identifying spatial segments in international markets.* Manuscript submitted for publication.

Thissen, D., & Steinberg, L. (1986). A taxonomy of item response models. *Psychometrika, 51,* 567-577.

Tutz, G. (1990). Sequential item response models with an ordered response. *British Journal of Mathematical and Statistical Psychology, 43,* 39–56.

Van der Ark, L.A. (1999, July). *A reference card for the relations between IRT models for polytomous items and some relevant properties.* Paper presented at the European Meeting of the Psychometric Society in Lüneburg, Germany.

Van der Linden, W.J., & Hambleton, R.K. (Eds.). (1997). *Handbook of modern item response theory.* New York: Springer-Verlag.

VanLehn, K., & Niu, Z. (in press). Bayesian student modeling, user interfaces and feedback: A sensitivity analysis. *International Journal of Artificial Intelligence in Education.*

VanLehn, K., Niu, Z., Siler, S., & Gertner, A. (1998). Student modeling from conventional test data: A Bayesian approach without priors. In B.P. Goetl, H.M. Halff, C.L. Redfield, & V.J. Shute (Eds.), *Proceedings of the Intelligent Tutoring Systems Fourth International Conference, ITS 98* (pp. 434–443). Berlin: Springer-Verlag.

Verhelst, N.D., & Verstralen, H.H.F.M. (1993). A stochastic unfolding model derived from the partial credit model. *Kwantitatieve Methoden, 42,* 73–92.

Verhelst, N.D., & Verstralen, H.H.F.M. (2001). An IRT model for multiple raters. In A. Boomsma, M.A.J. van Duijn, & T.A.B. Snijders (Eds.), *Essays on item response theory* (pp. 89–108). New York: Springer-Verlag.

Wilson, M.R., & Hoskens, M. (1999). *The rater bundle model.* Paper presented at the Annual Meeting of the American Educational Research Association, Montreal, Canada.

Wilson, D., Wood, R.L., & Gibbons, R. (1983). TESTFACT: Test scoring and item factor analysis [Computer software]. Chicago: Scientific Software.

Wu, M.L., Adams, R.J., & Wilson, M.R. (1997). ConQuest: Generalized item response modeling software [Software manual]. Melbourne: Australian Council for Educational Research.

Yamamoto, K., & Gitomer, D.H. (1993). Application of a HYBRID model to a test of cognitive skill representation. In N. Fredriksen & R.J. Mislevy (Eds.), *Test theory for a new generation of tests* (pp. 275–295). Hillsdale, NJ: Erlbaum.

Yuan, A., & Clarke, B. (1999). *Manifest characterization and testing for two latent traits.* Manuscript submitted for publication.

Zhang, J., & Stout, W.F. (1999). The theoretical DETECT index of dimensionality and its application to approximate simple structure. *Psychometrika, 64,* 213–249.

Zimowski, M.F., Muraki, E., Mislevy, R.J., & Bock, R.D. (1997). BILOG-MG: Multiple-group IRT analysis and test maintenance for binary items [Computer software]. Chicago: Scientific Software.

Zwick, R. (1992). The National Assessment of Educational Progress [Special issue]. *Journal of Educational Measurement, 17,* 93–94.

15

Reversibility Revisited and Other Comparisons of Three Types of Polytomous IRT Models

Bas T. Hemker[1]

ABSTRACT Molenaar (1983) was the first to distinguish and to compare three different types of polytomous item response theory (IRT) models: cumulative, continuation, and partial credit models. In his research, and later in research by others, the three types of models were compared on various criteria. One of the criteria introduced in IRT modeling is the property of reversibility. In this chapter, the results of a number of comparison studies are summarized. Next, the property of reversibility is further investigated. Finally, the relationship between the three types of models and psychological reality is discussed.

1 Introduction

In many tests, items are used that have more than two ordered score categories. Such items are called polytomous items. To describe these items, the polytomous item scores can be split up into dichotomies while maintaining the ordinal nature of the categories. These dichotomies are called item steps. In item response theory (IRT) three different definitions of an item step are dominant: the cumulative item step, the conditional item step, and the partial credit item step. Each definition is based on a different way to split up the polytomous item score. For all three definitions, the probability of passing an item step can be written as a function of the unidimensional latent trait θ. Such a function is called an item step response function (ISRF). Consequently, there are three definitions of an ISRF, and each definition is related to a type of polytomous IRT models.

All three types of polytomous IRT models are defined by the assumptions of unidimensionality, local independence, and by restrictions on the functional form of the ISRFs. For example, it can be assumed that the

[1]National Institute for Educational Measurement (CITO), P.O. Box 1034, 6801 MG Arnhem, The Netherlands; e-mail: *bas.hemker@cito.nl*

functional form of the ISRFs is given by a logistic function (Molenaar, 1983) or by a nonparametric function (Molenaar, 1986). The latter function only assumes that the probability of passing an item step increases in the unidimensional latent trait. Molenaar (1983) was the first to compare polytomous IRT models from these three classes of models. Later, the comparison of the three types of polytomous IRT models was continued by Mellenbergh (1995), Hemker (1996), Van Engelenburg (1997), and Akkermans (1998).

In this chapter an overview of the three types of models is given. Studies comparing these three types on a number of properties are summarized. The property of reversibility is further investigated. If a model has this property, it means that after reversing the scoring direction, the model still holds. This property is discussed separately from the summary of comparison studies, because the proofs for the nonparametric models that were given earlier (Hemker, 1996) are improved, and new results for the parametric models are presented. Finally, the relation between the three types of models and "psychological reality" (Samejima, 1995) is discussed; that is, the question is raised as to what typical items or scoring rules the models can be applied. In the discussion of the chapter conclusions are drawn, and some suggestions for further research are given.

2 Three Types of Models for Polytomous Items

The three types of IRT models for polytomous items all assume unidimensionality (UD) and local independence (LI). UD means that it is assumed that all items measure the same latent trait. LI means that for each subject a response to one item is not influenced by the response to any other item in the test. Let the item score on item I_i be denoted X_i, $i = 1, \ldots, k$. The maximum score on item I_i is m_i; thus $X_i = 0, \ldots, m_i$. Let $\pi_{ih}(\theta)$ denote $P(X_i = h \mid \theta)$, $h = 0, \ldots, m_i$, and $G(\theta)$ denote the cumulative distribution function of the latent trait in the population considered. Then UD and LI imply

$$P(X_1 = x_1, \ldots, X_k = x_k) = \int_\theta \prod_{i=1}^{k} \pi_{ix_i}(\theta)\, dG(\theta) \,. \tag{15.1}$$

Equation 15.1 does not restrict the distributions of $\mathbf{X} = (X_1, \ldots, X_k)'$ as long as there are no restrictions on the probabilities $\pi_{ih}(\theta)$, the distribution $G(\theta)$, or both (Holland & Rosenbaum, 1986, Theorem 15.2; see also Suppes & Zanotti, 1981).

The probability that the score on item I_i falls in category h, $\pi_{ih}(\theta)$, is the *category characteristic curve* (CCC) of this category h of item I_i. The sum of the $m_i + 1$ CCCs for item I_i over all categories $h = 0, \ldots, m_i$ is equal to 1. Therefore, restrictions with respect to their specification can be imposed on m_i of the CCCs. The $(m_i + 1)$st CCC is then also specified.

However, instead of restricting the CCCs, restrictions are placed on m_i functions which are related to the CCCs but are much simpler to model: the ISRFs. These ISRFs split up the response variable in (dichotomous) item steps while maintaining the ordinal nature of the response. Let $Y_{ih}(\theta)$ denote the probability of passing an item step h on item I_i as a function of θ. The probability of failing this step is given by $1 - Y_{ih}(\theta)$. Note that in the context of the ISRFs, h runs from 1 to m_i (cf. Fischer, 1995, p. 294).

The $Y_{ih}(\theta)$ is a general notation of the ISRF. Specific functions are also needed because there are three ways to split the response in dichotomies, which implies that there are three different definitions of an item step (Molenaar, 1983; Mellenbergh, 1995), which are known as the cumulative, the conditional, and the partial credit item steps. If a specific definition of an item step is assumed, $Y_{ih}(\theta)$ is replaced by $C_{ih}(\theta)$ if there is a cumulative item step, by $M_{ih}(\theta)$ for a conditional item step, and by $A_{ih}(\theta)$ for a partial credit item step. The three specific definitions of the ISRFs for the three definitions of the item steps are given next.

The ISRF of a *cumulative item step* is given by

$$C_{ih}(\theta) = P(X_i \geq h \mid \theta) \ . \tag{15.2}$$

If the ISRFs are defined by (15.2) for all i and h, it is assumed that item steps have a strict order. If one step is failed, all the following item steps are also failed; if one step is passed, all previous steps are passed. This implies that all possible scores say something about all item steps. The persons who have passed item step h of item I_i have an item score at least equal to h. Persons with a lower item score failed the hth item step; if a person has an item score equal to h, he or she passed the first h item steps and failed the next $m_i - h$ item steps. These types of models are called cumulative models (CUM; Molenaar, 1983; see also Samejima, 1969). They are also known as cumulative probability models (Mellenbergh, 1995), difference models (Thissen & Steinberg, 1986), Thurstone models (Andrich, 1995a), or gradation models (Van Engelenburg, 1997, Chap. 2).

The ISRF of a *conditional item step* is given by

$$M_{ih}(\theta) = \frac{P(X_i \geq h \mid \theta)}{P(X_i \geq h - 1 \mid \theta)} \ . \tag{15.3}$$

The definition of an ISRF by (15.3) for all i and h also assumes a strict order of the item steps. However, in contrast with (15.2), failing an item step defined by (15.3) implies that the following item steps are never considered, that is, neither failed nor passed. Passing an item step implies that the following item step is considered. Thus, item step h is taken by all persons with a score equal to or larger than h, and not taken by all persons with a score equal to $h - 1$; the only persons who fail this item step are those having an item score equal to $h - 1$. Persons with lower item scores did not try item step h, and thus did not fail it. This means that only the scores equal to or larger that $h-1$ say something about item step h, which implies that the number of scores that have implications on the item step depends on the item step. These types of models are called conditional models (CON; Molenaar, 1983). They are also known as continuation ratio models (Mellenbergh, 1995), or sequence models (Van Engelenburg, 1997, Chap. 2).

The ISRF in the case of a *partial credit item step* is given by

$$A_{ih}(\theta) = \frac{\pi_{ih}(\theta)}{\pi_{ih}(\theta) + \pi_{i,h-1}(\theta)} . \tag{15.4}$$

The definition of an ISRF by (15.4) for all i and h implies that the only persons who passed item step h are the ones with score h, and the only ones who failed the step are the ones with score $h - 1$. Only the scores h and $h - 1$ have a direct influence on the ISRF. Indirectly, however, all previous and subsequent steps influence each step, which complicates the interpretation of the difficulty of the item step (Molenaar, 1983; Verhelst & Verstralen, 1997). These types of models are called partial credit models (PCM; Molenaar, 1983). They are also known as adjacent category models (Mellenbergh, 1995), Rasch models (Andrich, 1995a), or partition models (Van Engelenburg, 1997, Chap. 2).

Polytomous models are not only defined by the definition of the ISRF, but also by their functional form. These forms can be parametric or nonparametric, yielding parametric and nonparametric models, respectively.

2.1 Parametric Models

The parametric form that is used in many polytomous IRT models is the logistic function. In the simplest case each step is described by only one parameter β_{ih}, which can be interpreted as a location parameter. This leads to the function

$$Y_{ih}(\theta) = \frac{\exp(\theta - \beta_{ih})}{1 + \exp(\theta - \beta_{ih})} .$$

This equation is often given in terms of a logit:

$$\text{logit}\,[Y_{ih}(\theta)] = \ln\left[\frac{Y_{ih}(\theta)}{1 - Y_{ih}(\theta)}\right] = \theta - \beta_{ih} . \tag{15.5}$$

Recall that $Y_{ih}(\theta)$ denotes an ISRF in general. However, if the ISRF has a specific definition, three different parametric models are obtained by substituting $Y_{ih}(\theta)$ in (15.5) by $C_{ih}(\theta)$, $M_{ih}(\theta)$, or $A_{ih}(\theta)$. This yields the (one-parameter) graded response model (1p-GRM; Samejima, 1969, 1972), sequential Rasch model (SRM; Tutz, 1990), and the partial credit model (1p-PCM; Masters, 1982), respectively. Note that, although the forms of the ISRFs are identical, the interpretation of β_{ih} differs over the three definitions of ISRFs (Molenaar, 1983; Mellenbergh, 1995).

Molenaar (1983) showed for all three types of models and the form of (15.5) that "when logits for one model are straight lines, most of them are curves for the other models" (p. 8). This implies that the parametric models for polytomous items are not hierarchically nested, because if one of the three types of parametric models holds, the other two types cannot hold.

Equation 15.5 can be extended by adding a parameter α_i ($\alpha_i > 0$) to the logit such that

$$\text{logit}\,[Y_{ih}(\theta)] = \alpha_i(\theta - \beta_{ih}) \;. \tag{15.6}$$

This α_i is often referred to as the item discrimination parameter. Models with the form of (15.6) are called two-parameter models with α varying over items, abbreviated as 2p(i)-models.

Analogous to models with the form of (15.5), models with the form of (15.6) are not hierarchically ordered. Thissen and Steinberg (1986) have proved this if (15.2) and (15.3) both have the form of (15.6); Mellenbergh (1995) proved this result for (15.2) and (15.4), and (15.3) or (15.4). For (15.3) or (15.4) the same result was found with α also varying over item steps [2p(ih)-models]. This means that for $\text{logit}\,[Y_{ih}(\theta)] = \alpha_{ih}(\theta - \beta_{ih})$, and all its special cases, (15.3) and (15.4) are not hierarchically nested. It was noted (cf. Mellenbergh, 1995) that similarly to β_{ih}, the interpretation of α_i and α_{ih} differs over the three different types of models.

2.2 Nonparametric Models

If a nonparametric definition of ISRFs is assumed, it means that

$$Y_{ih}(\theta) \;\text{ is nondecreasing in } \theta, \;\text{ for all } i \text{ and } h\;. \tag{15.7}$$

Molenaar (1986) was the first to compare the three models defined by (15.2), (15.3), and (15.4) given (15.7).

The nonparametric graded response model (np-GRM; Hemker, Sijtsma, Molenaar, & Junker, 1996, 1997) is obtained if the ISRFs are defined by (15.2) and have the form of (15.7); that is, $C_{ih}(\theta)$ is nondecreasing in θ. This model is identical to the Mokken (1971) model of monotone homogeneity (MH) for polytomous items (Molenaar, 1982, 1986, 1991, 1997; Hemker, Sijtsma, & Molenaar, 1995). The definition of (15.3) and the form of (15.7) (i.e., $M_{ih}(\theta)$ is nondecreasing in θ) yields the nonparametric sequential model (np-SM; Hemker, 1996). Finally, the definition of (15.4) and the form of (15.7) (i.e., $A_{ih}(\theta)$ is nondecreasing in θ) yields the nonparametric partial credit model (np-PCM; Hemker et al., 1996, 1997). Note that in all these models UD and LI are also assumed.

It was shown (Hemker, 1996) that nondecreasingness of $A_{ih}(\theta)$ in θ, for all i and h, implies nondecreasingness for all $M_{ih}(\theta)$, and that nondecreasingness of $M_{ih}(\theta)$ implies nondecreasingness of all $C_{ih}(\theta)$. Thus, in contrast to the parametric models, the three types of models are hierarchically nested if the ISRFs have a nonparametric form given by (15.7). This means that

$$\text{np-PCM} \Rightarrow \text{np-SM} \Rightarrow \text{np-GRM} .$$

It was also shown (Hemker, 1996) that most parametric models mentioned in this chapter imply the np-PCM. The only exception is the 2p(ih)-sequential model: if α_{ih} varies over item steps, the np-PCM is not implied. Note, however, that the 2p(ih)-sequential model does imply the np-SM.

3 Summary of Comparison Studies

Studies comparing the three types of models on both empirical and theoretical criteria are discussed in this section. Results from studies using empirical and simulated data sets are discussed in Section 3.1, followed by theoretical comparisons on ordering properties and on additivity.

3.1 Comparison with Empirical and Simulated Data

In studies using empirical data it has been investigated which model could be used best in describing the data. First the three one-parameter models were investigated (Molenaar, 1983), and later the three nonparametric models (Molenaar, 1986). While the two studies gave no clearcut conclusion on which model to use, it was suggested (Molenaar, 1983) that "the CUM model is best verified" (p. 9; i.e., fit well), in the investigated data set. A similar result for parametric models for simulated data sets was found by Van Engelenburg (1997, Chap. 1). Another study found a comparable fit when either the 1p-SM or the PCM was applied to the same data set (Verhelst, Glas, & De Vries, 1997). Note that in the case of nonparametric models a preference for the CUM type of model is expected because the np-GRM is the most general model. Thus it is expected that this model fits more data sets.

In a simulation study using the ideal observer index (IOI; Levine, Drasgow, Williams, McCusker, & Thomasson, 1992), Maydeu-Olivares, Drasgow, and Mead (1994) compared the CUM model (15.2) and the PCMs, as defined by (15.4), with a form given by (15.6). Their conclusions were that "both models better fit data generated according to their assumptions than data generated by the alternative model," and that "neither model was found to be uniformly superior" (pp. 253–254). Similar conclusions were drawn in studies in which the CON model with the form of (15.6) was added (Van Engelenburg, 1997; Akkermans, 1998). An additional result (Van Engelenburg, 1997) was that, with regard to the estimation of latent trait values, incorrect models performed as well as correct models.

3.2 Comparison on Ordering Properties

A number of studies were performed comparing the three types of models on their measurement properties. These studies were concerned with the ordering of scores, subjects, or items. The ordering of scores is implied by the property of stochastic ordering of the manifest score (SOM). SOM implies that the expected score is nondecreasing in θ. This was investigated for item scores (SOM of X_i) as well as for the total score (SOM of X_+). SOM of X_+ means that for any two respondents θ_a and θ_b, with $\theta_a < \theta_b$,

$$P(X_+ \geq x_+ \mid \theta_a) \leq P(X_+ \geq x_+ \mid \theta_b) \ .$$

The ordering of subjects is implied by stochastic ordering on the latent trait (SOL): the higher the score of a subject, the higher the expected latent trait value for this subject. Subjects can be ordered on θ by means of an item score (SOL by X_i) or by the total score (SOL by X_+). SOL by X_+ means that for any constant values c, and for all $0 \leq s < t \leq \sum_i m_i$,

$$P(\theta > c \mid X_+ = s) \leq P(\theta > c \mid X_+ = t) \ .$$

The ordering of items is implied by invariant item ordering (IIO; Sijtsma & Junker, 1996). IIO means that the ordering of the items is the same for all values of θ, and thus for all persons. Define the conditional expectation of the item score on item I_i as $E(X_i \mid \theta)$. Then IIO means that for a set of k items the items can be ordered and numbered accordingly such that

$$E(X_1 \mid \theta) \leq E(X_2 \mid \theta) \leq \ldots \leq E(X_k \mid \theta) \ .$$

SOM and SOL are implied by the more technical property of the monotone likelihood ratio (MLR; Lehmann, 1959, Sect. 3.3). Note that for practical use SOL is more relevant than SOM, because it allows an ordering of persons on an attribute that cannot be observed but that can be measured by means of something observable, namely, the item score or total score.

The proofs that show which model implies which property can be found elsewhere (Hemker et al., 1996, 1997). Here we note that all models discussed in this chapter have SOM of X_i and SOM of X_+; the least restrictive of the models discussed here with the properties of MLR of X_i and SOL by X_i is the np-PCM; and the least restrictive of the model is the 1p-PCM with the property of MLR of X_+ and SOL by X_+.

None of the models presented here imply IIO (Sijtsma & Hemker, 1998). However, for all three types of models ISRFs with a restricted form of (15.5) imply IIO. The restriction of (15.5) concerns β_{ih}, which is split up in an item parameter β_i that does not vary over steps, and an item step parameter τ_h that does not vary over items. Thus, one obtains the form

$$\text{logit}\,[Y_{ih}(\theta)] = \theta - (\beta_i + \tau_h) \ . \tag{15.8}$$

This so-called rating scale restriction does not have implications for the form of an individual ISRF, but it does for a group of ISRFs. Note that this restriction can only be applied if all items have the same number of answer categories, and that it does not depend on the definition of the ISRFs. The three models with the form of (15.8) are the (restricted version of the) rating scale GRM (RS-GRM; Muraki, 1990; CUM model), the sequential rating scale model (RS-SM; Tutz, 1990; CON model) and the rating scale model (RS-PCM; Andrich, 1978; PCM type of model). Nonparametric models with the property of IIO are defined by restrictions on the relations between the ISRFs, not on the exact form of the ISRFs (Sijtsma & Hemker, 1998). These nonparametric models are not further discussed here.

3.3 Comparison on Additivity

Another property on which the three types of models have been compared is the property of additivity. In the case of an arbitrary number of categories $m+1$ ($m > 1$), two or more adjacent categories can be joined after the data have been collected. For example, polytomous data can be dichotomized. A model has the property of additivity if the model holds for the original data and for the data after adjacent categories are joined.

It has been shown that if the $Y_{ih}(\theta)$s are nondecreasing in θ, that is, have the form of (15.7), for the original number of answer categories, the $Y_{ih}(\theta)$s are nondecreasing after joining adjacent categories, irrespective of their definition (Hemker, 1996). Thus additivity holds for the np-PCM, np-SM, and np-GRM. If the ISRFs have parametric forms of (15.5), (15.6), or (15.8), the CUM definitions (15.2) possess additivity (Samejima, 1995; Roskam, 1995). Additivity does not hold for these forms under the CON definition (15.3) (Samejima, 1995), or under the PCM definition (15.4) (Jansen & Roskam, 1986; Roskam, 1995; Andrich, 1995a, 1995b). Note that the property of additivity is regarded here in a theoretical sense. From studies on empirical data it has also been concluded that the additivity property does not hold for the parametric PCMs (Andrich, 1995b).

4 The Three Types of Models and Reversibility

Usually items are designed such that they are indicative of the latent trait they are designed to measure: the higher θ, the higher the probability of passing an item step. This is in concordance with the definition of the nonparametric form (15.7) of the ISRF. However, a test may contain items for which the higher θ implies a lower probability of passing an item step, which means that $Y_{ih}(\theta)$ decreases in θ, for example, an item using the number of mistakes as an item score in a test where a high θ means a high ability. Another example is the item "abortion is murder: totally disagree/disagree/do not know/agree/totally agree" in a test measuring "tolerance to abortion," in which a higher agreement yields a higher score. Such items are so-called counterindicative items. In this section it is investigated whether it is possible, within the restrictions of the various models, to "reverse" the data of a counterindicative item, in such a way that it resembles an indicative item.

A counterindicative item score denoted X_i^* has to be reversed to an indicative item score X_i after the data are collected to fit into the models presented here. This reversal of item scores means that $X_i = m_i - X_i^*$. Thus, after reversal of the item score,

$$P(X_i^* = g \mid \theta) = P(X_i = m_i - g \mid \theta) \;, \tag{15.9}$$

for all $g = 0, \ldots, m_i$. For notational convenience assume that all items have the same number of answer categories. The proofs given next also hold for m varying over items, because in the proofs it is sufficient to consider only one item.

If reversibility holds, the reversal of the item scores implies that failing an item step on a counterindicative item is the same as having a success on the same item step after the scores are reversed. It also implies that having a success on an item step of a counterindicative item is the same as failing this step after the scores have been reversed. Reversibility, which has also been referred to as "palindromic invariance" (McCullagh, 1978), also has implications for the order of the item steps. If reversibility holds, a difficult counterindicative item step implies that after reversal of scores the same item step is an easy item step. For example, if many subjects take the step from "totally disagree" to "disagree," its reversed step, which is from "disagree" to "totally disagree," is taken by a few subjects. If reversibility holds, the only change in order of the item steps is the direction, not the order itself (Hemker, 1996).

Reversibility is also referred to as scale reversibility (Mellenbergh, 1997), because this property can also be defined in terms of the reversal of scales. A counterindicative item can be defined as an item measuring the exact opposite of θ, namely, θ^*, with $\theta^* = -\theta$. If a function is nonincreasing in the indicative scale θ, it is nondecreasing in the counterindicative scale θ^*. Reversing the scale also implies reversing the difficulties while maintaining their order. If a model has the property of reversibility, the counterindicative item ISRFs described by a function of θ^* have the same form as the ISRFs after the scores and scale are reversed to θ.

Definition 15.1 Reversibility holds if the same model holds before and after the reversal of item scores, which means that:

(i) the definition of an ISRF before and after the reversal of item scores is the same;

(ii) the form of an ISRF as a function of θ^* before the reversal of item scores is identical to the form of an ISRF as a function of θ after reversal.

Next, some notation is introduced. Let $Y^*_{ig}(\theta)$ denote the ISRF of an counterindicative item. For notational convenience, let $m-g$ be denoted h. If reversibility holds, success on a counterindicative item step is failure on the same step after the scores have been reversed:

$$Y^*_{ig}(\theta) = 1 - Y_{i,m-g}(\theta) = 1 - Y_{ih}(\theta) \ . \tag{15.10}$$

It is important to note that for notational convenience item step g in a counterindicative item is the step *from* $X^*_i = g$ to $X^*_i = g+1$, whereas for indicative items, step h is the step *to* $X_i = h$ from $X_i = h-1$.

Equation 15.10 implies that $\text{logit}\,[Y^*_{ig}(\theta)]$ is identical to $-\text{logit}\,[Y_{ih}(\theta)]$. Thus, if reversibility holds

$$\text{logit}\,[Y^*_{ig}(\theta)] + \text{logit}\,[Y_{ih}(\theta)] = 0 \ . \tag{15.11}$$

If reversibility holds for all g, (15.11) is equivalent to the first necessary condition (i) in the definition of reversibility: $Y^*_{ig}(\theta)$ and $Y_{ih}(\theta)$ are defined by the same model. If (15.11) holds, the forms of $Y^*_{ig}(\theta)$ and $Y_{ih}(\theta)$ are investigated to see whether the second condition is also satisfied.

Next, proofs are given that show which of the models from the three classes have the property of reversibility (Theorem 15.1) and which models do not have this property (Theorem 15.2). In these proofs five lemmas are used. Lemmas 15.1 through 15.3 are concerned with the first necessary condition (i) in Definition 15.1, Lemmas 15.4 and 15.5 with condition (ii). Results for nonparametric models have been obtained by Hemker (1996), but in contrast to those proofs it is not assumed here that the first derivative of an ISRF exists for all θ.

Lemma 15.1 Equation 15.11 holds for $Y^*_{ig}(\theta)$ and $Y_{ih}(\theta)$ defined by (15.2).

Proof $Y^*_{ig}(\theta)$ defined by (15.2) is identical to $C^*_{ig}(\theta)$. Note that $C^*_{ig}(\theta)$ means $P(X^*_i > g \mid \theta)$, and not $P(X^*_i \geq g|\theta)$, because the ISRF g of a counterindicative item is the step from $X^*_i = g$ to $X^*_i = g+1$. Note that $P(X^*_i > g \mid \theta) = \sum_{k>h} \pi^*_{ik}(\theta)$. From (15.9) it follows that this is identical to $1 - \sum_{k\geq h} \pi_{ik}(\theta) = 1 - P(X_i \geq h \mid \theta) = 1 - C_{ih}(\theta)$. Note that $C_{ih}(\theta)$ is the step from $X_i = h-1$ to $X_i = h$. Because $Y^*_{ig}(\theta) = 1 - C_{ih}(\theta)$, (15.11) holds if $Y^*_{ig}(\theta)$ and $Y_{ih}(\theta)$ are defined by (15.2). □

Lemma 15.2 Equation 15.11 holds for $Y^*_{ig}(\theta)$ and $Y_{ih}(\theta)$ defined by (15.4).

Proof The proof is analogous to the proof of Lemma 15.1. $Y^*_{ig}(\theta)$ defined by (15.4) is identical to $A^*_{ig}(\theta)$. Note that $A^*_{ig}(\theta) = \pi^*_{i,g+1}(\theta)/[\pi^*_{ig}(\theta) +$

$\pi^*_{i,g+1}(\theta)]$, because the ISRF g of a counterindicative item is the step from $X^*_i = g$ to $X^*_i = g+1$. By applying (15.9) to $A^*_{ig}(\theta)$ it is shown that $A^*_{ig}(\theta)$ is identical to $\pi_{i,m-g-1}(\theta)/[\pi_{i,m-g}(\theta)+\pi_{i,m-g-1}(\theta)] = \pi_{i,h-1}(\theta)/[\pi_{ih}(\theta)+\pi_{i,h-1}(\theta)]$. This is identical to $1-A_{ih}(\theta)$. Thus, $A^*_{ig}(\theta) = 1-A_{ih}(\theta)$, which means that (15.11) holds if $Y^*_{ig}(\theta)$ and $Y_{ih}(\theta)$ are defined by (15.4). □

Lemma 15.3 Equation 15.11 does not hold for $Y^*_{ig}(\theta)$ and $Y_{ih}(\theta)$ defined by (15.3) if $g = 0$.

Proof For the first ISRF of an item, (15.3) and (15.2) are always identical, and for the last ISRF (15.3) and (15.4) are always identical (Hemker, 1996, Chap. 6, p. 8). Thus, the first counterindicative ISRF (i.e., from $X^*_i = 0$ to 1) is $M^*_{i0}(\theta) = P(X^*_i \geq 1 \mid \theta)$, and the last indicative ISRF (i.e., from $X_i = m-1$ to m) is $M_{im}(\theta) = \pi_{im}(\theta)/[\pi_{i,m-1}(\theta) + \pi_{im}(\theta)]$. Note that $M_{i,m-1}(\theta)$ is the reverse ISRF of $M^*_{i0}(\theta)$, because $h = m$ if $g = 0$ ($h = m-g$). The logit of the ISRF for $g = 0$ is $\text{logit}\,[M^*_{i0}(\theta)] = -\ln[P(X_i \geq m \mid \theta)/P(X_i < m \mid \theta)]$ (see Lemma 15.1). The logit of the ISRF for $h = m-0$ is $M_{im}(\theta) = \ln[\pi_{im}\theta)/\pi_{i,m-1}(\theta)]$ (see Lemma 15.2). The sum of these two logits is $\ln[\pi_{im}(\theta)/\pi_{i,m-1}(\theta)] - \ln[\pi_{im}(\theta)/P(X_i < m \mid \theta)]$. Because $\pi_{i,m-1}(\theta)$ and $P(X_i < m \mid \theta)$ are not identical, (15.11) does not hold if the ISRFs are defined by (15.3). □

In Lemma 15.3 $g = 0$, but note that (15.11) also does not hold for any other value of g if the conditional definition of the item step is used. For $g = m-1$ ($h = m-g = 1$), the proof is analogous to the proof of Lemma 15.3. For $0 < g < m-1$, it can be shown that $\text{logit}\,[M^*_{ig}(\theta)] + \text{logit}\,[M_{ih}(\theta)] = \ln[P(X_i < h \mid \theta)/P(X_i = h \mid \theta)] + \ln[P(X_i \geq h \mid \theta)/P(X_i = h-1 \mid \theta)]$ which is obviously not identical to 0 for all θ. However, Lemma 15.3 is given in this form because it is sufficient for the proof of Theorem 15.2.

Lemma 15.4 If Equation 15.11 holds, (15.7) for $Y^*_{ig}(\theta^*)$ implies (15.7) for $Y_{ih}(\theta)$.

Proof Equation 15.7 for $Y^*_{ig}(\theta^*)$ means that for counterindicative items the ISRFs are nondecreasing in θ^*. Note that $-\theta^* = \theta$. Therefore, $Y^*_{ih}(\theta)$, and thus $\text{logit}[Y^*_{ig}(\theta)]$, is nonincreasing in θ. This implies that the function $-\text{logit}[P(Y_{ih}(\theta)]$ is nonincreasing in θ, if (15.11) holds, that is, if $\text{logit}[Y^*_{ig}(\theta)] = -\text{logit}[Y_{ih}(\theta)]$. This means that $\text{logit}[Y_{ih}(\theta)]$ is nondecreasing in θ, and thus $Y_{ih}(\theta)$ is nondecreasing in θ, which is identical to (15.7) for $Y_{ih}(\theta)$. □

Lemma 15.5 If Equation 15.11 holds, (15.6) for $Y^*_{ig}(\theta^*)$ implies (15.6) for $Y_{ih}(\theta)$.

Proof For parametric models the item step difficulty of a counterindicative item step g (from $X_i = g$ to $X_i = g + 1$) is denoted β^*_{ig}. Thus, (15.6) for $Y^*_{ig}(\theta^*)$ means that $\text{logit}[Y^*_{ig}(\theta^*)] = \alpha_i(\theta^* - \beta^*_{ig})$. If (15.11) holds for the ISRF before and after the reversal of the item scores or of the scale, the order of item step difficulties is reversed; only the direction, or, for example, the labels "easy" and "hard" are reversed. Thus, the β_{ih}s can be obtained by multiplying β^*_{ig} by -1, for all i and g ($g = 0, \ldots, m-1$): $\beta_{ih} = \beta_{i,m-g} = -\beta^*_{ig}$. Note that reversal of the score, or of the scale, also means that $\theta = -\theta^*$.

Therefore, if (15.11) holds, $\text{logit}\,[Y^*_{ig}(\theta^*)]$ can be written in terms of indicative item parameters: $\alpha_i(\theta^* - \beta^*_{ig}) = \alpha_i[-\theta - (-\beta_{ih})] = -\alpha_i(\theta - \beta_{ih})$. From (15.11) it follows that $-\alpha_i(\theta - \beta_{ih})$ is equal to $-\text{logit}\,[Y_{ih}(\theta)]$, and therefore $\text{logit}\,[Y_{ih}(\theta)] = \alpha_i(\theta - \beta_{ih})$, which implies (15.6) for $Y_{ih}(\theta)$. □

Note that the proof of Lemma 15.5 for equations other than (15.6) that have linear forms of the logit, such as (15.8), (15.5), and (15.6) with α_{ih} instead of α_i, is analogous to the proof of Lemma 15.5. These proofs are not given here.

Theorem 15.1 The following models have the property of reversibility.
a. np-PCM;
b. RS-PCM, 1p-PCM, 2p(*i*)-PCM and the 2p(*ih*)-PCM;
c. np-GRM; and
d. RS-GRM, 1p-GRM, and 2p(*i*)-GRM.

Proof The first condition for reversibility is that before and after the reversal of item scores the definition of an ISRF is identical. This is true for the PCM type of models (*a* and *b*), which are defined by (15.2) (Lemma 15.1), and for the CUM type of models (*c* and *d*), which are defined by (15.4) (Lemma 15.2). Equation 15.11 holds for both definitions.

The second condition is that the form of an ISRF as a function of θ^* before the reversal of item scores is identical to the form of an ISRF as a function of θ after reversal. Lemma 15.4 shows that this is true for the nonparametric models (*a* and *c*), and Lemma 15.5 shows that this is true for the 2p(*i*)-PCM and 2p(*i*)-GRM. For the other parametric models in *b* and *d*, the logits of the ISRFs are linear functions, so proofs analogous to Lemma 15.5 can be given for all these models. This means that both conditions for reversibility, as given by the definition, hold for *a*, *b*, *c*, and *d*, which proves Theorem 15.1. □

Theorem 15.2 The CON type of models does not have the property of reversibility.

Proof The CON type of models, such as np-SM, RS-SM, 1p-SM, 2p(*i*)-SM, and the 2p(*ih*)-SM, are defined by (15.3). Lemma 15.3 yields that the first necessary condition for reversibility is not satisfied for models defined by (15.3), irrespective of their form. Therefore reversibility does not hold. □

The hierarchical relation of the three nonparametric models implies that np-GRM is the most general model, and the np-PCM is the most restrictive of the three. It has just been shown that both models have the property of reversibility. The np-SM, however, which can be considered the "middle model" because it is implied by the np-PCM and it implies the np-GRM, does not have the reversibility property. This may seem strange, but can be explained as follows.

The nonparametric models are defined by a nondecreasing form of their least restrictive ISRF definition. It was shown that (15.11) holds for $C_{ih}(\theta)$ and $A_{ih}(\theta)$, but not for $M_{ih}(\theta)$. The np-GRM is defined by nondecreasingness of $C_{ih}(\theta)$, which implies nondecreasingness of $M_{ih}(\theta)$ and of $A_{ih}(\theta)$. Even if nondecreasingness of $M_{ih}(\theta)$ is assumed by the np-GRM, reversibility holds because the focus is on $C_{ih}(\theta)$. On the other hand, even if nondecreasingness of $A_{ih}(\theta)$ is assumed by the np-SM, the focus is on $M_{ih}(\theta)$, and therefore it is concluded that reversibility does not hold for this model.

5 Comparison on Psychological Agreement

A theoretical comparison of the three types of models on a more philosophical level considers their "psychological agreement with reality" (Samejima, 1995), that is, to what kind of typical items or scoring rules the models can be applied. Recently, two classifications of the three types were introduced. They are very similar, but have a different emphasis: the first focuses on the classification of models by means of task features or item features (Van Engelenburg, 1997), whereas the second procedure focuses on the scoring procedures or scoring rules (Akkermans, 1998). Note that one could argue that the item features are different if the scoring rules are different, and vice versa, and that there is a direct relation between the two.

Both classifications are based on several questions. The first question concerns the step process: are the item steps taken simultaneously or separately? The steps are taken simultaneously if the examinee gives one overall judgment on a scale (task feature), or if the examinee is evaluated in one overall judgment (scoring rule). This process or this type of item is described by CUM models.

In the case where the steps within one item are taken separately the second classification criterion is the continuation rule, with the related question: is the scoring or the task continued after a step is failed, or is the scoring or task stopped after a step is failed? If the subject tries all steps and the score is the sum of the scores on these separate (binary) tasks, it is called parallel scoring. This type of process or item is described by PCM models.

If the scoring is sequential, that is, if the stop-if-failed rule is applied, there is a third question: is the ordering of the steps completely fixed, or not? Only if a fixed ordering is assumed, may the CON models describe the item (or the process). If there is no fixed ordering of steps, there is no specific type of model related to the process. Note that if reversibility holds, the steps can be reversed, which means that the order of the steps is not completely fixed. Thus, the result that reversibility does not hold for CON models is in accordance with the assumption of a fixed ordering of item steps for this type of model.

All three types of processes can be illustrated by an item from a test that is used to measure to ability to speak Dutch, the "NT2 profieltoets" ["Dutch as a foreign language profile test" (CITO, 1999)]. The item is presented to the examinee by an examiner. It consists of a case followed by a question and reads (translated from Dutch):

> You work at an office. You have to fax a letter for your boss. You have no experience with the fax machine. You know a colleague who knows how to fax. What do you ask your colleague?

The verbal reply of the examinee is judged on content, grammar, and pronunciation by the examiner. The item can be scored in various ways. A first possibility is one overall judgment by the examiner, taking into account the three criteria, rating the response as "very bad," "bad," "good" or "very good," yielding a score of 0, 1, 2, or 3, respectively. Following the classification, the item is best described by a CUM model if this scoring procedure is applied.

A second possibility is that the score is the number of criteria that is performed well on the task. Thus, a sentence in which content and grammar are good, but pronunciation is bad, yields a score of 2. Applying the classification, this scoring procedure is described by PCM-type models.

A third possibility is a sequential scoring procedure. First the reply is judged on content. If the response is incorrect, the first step is failed and the examiner gives an item score of 0. The two other criteria are not considered. If the examinee's response is correct, the response is judged on grammar. If the performance on this criterion is incorrect, the examiner gives an item score of 1, whereas if it is performed correctly, the response is judged on pronunciation. If the pronunciation is bad, an item score of 2 is given, and a good performance on pronunciation yields the maximum item score of 3. This type of scoring or item is described by the CON model.

This third approach is sensible if it is expected that the examinee knows only a few sentences, and these sentences are trained until grammar and pronunciation are perfect. The examinee may give one of these sentences as a reply on all questions, even if it is not a sensible answer. In a partial credit modeling process the score on each item would be $X_i = 2$ for the correct grammar and pronunciation. Only content is scored as incorrect. This score of $X_i = 2$ on an item is too high if a person has no true speaking ability, and is only trained to the test.

Also, it may be sensible not to judge the pronunciation performance of people who failed in grammar, if the latter is considered to be more important than pronunciation. In the PCM the item score obtained for performing well on grammar and badly on pronunciation is equal to performing well on pronunciation and badly on grammar.

Of course, scoring procedures may be applied that are not described by one of the three models. The actual scoring of the item in the "NT2 profieltoets" is a combination of sequential and partial credit scoring. First the content is judged sequentially with the stop-if-fail rule, and if this step is passed both grammar and pronunciation are judged in a partial credit scoring.

It may be argued that not only the definition of the ISRFs should reflect the scoring process, but also that the form of the ISRFs should agree with psychological reality. For polytomous items with ordered item scores nondecreasingness of the ISRFs (15.7) is sensible for most processes. The specific parametric form of the logistic function is arbitrary. Other nondecreasing parametric functions could be chosen equally well without challenging the idea which is in agreement with psychological reality that a subject with a higher latent trait value has a higher probability of taking an item step.

5.1 *Interpretation of Parameters*

One reason to relate models to psychological reality is to give an interpretation of the model parameters in terms of scoring rules or task features. Fitting the wrong model to a process yields fundamental conceptual difficulties (Akkermans, 1998). For example, fitting the 1p-PCM to a sequential process yields certain interpretation problems. First, the odds of obtaining a score k rather than $k + 1$ depends on only one parameter (Molenaar, 1983), whereas in the sequential procedure it is expected that more parameters play a role. Second, each probability of obtaining a particular item score depends on all parameters (Andrich, 1995c), whereas in the sequential process one would expect that the lower scores do not depend on the parameters of the higher steps. Another example can be given if the CON model is used in the case of graded scoring or parallel scoring. The lack of reversibility of the CON models may lead to problems in the interpretation of the parameters when applied to scoring processes that should be described by the CUM and PCM types of models.

It seems that the importance of selecting the right model to describe a particular scoring process lies in the possibility of interpreting the parameters. Even if the right model is used, however, the interpretation of parameters is cumbersome. In the relatively simple case with only parameters β_{ih}, it is argued that this parameter cannot be interpreted as the difficulty parameter (e.g., Sijtsma & Hemker, 1998). It seems impossible to interpret β_{ih} in terms of a simple item step characteristic irrespective of the definition of the item step. Note also that the interpretation of β_{ih} depends on the relative position of this step to the other steps. This implies that it may be impossible to give a good interpretation of β_{ih} that does not depend on the other steps. Therefore, it does not seem to be possible to make generalizations about step h apart from item I_i and apart from the specific order of this step in this item. Even if this would be possible, the interpretation is blurred if a parameter α_i is added (e.g., Sijtsma & Hemker, in press). More research on the interpretation of the parameters in various polytomous models is needed to clarify these problems.

Another problem of interpreting model parameters, related to specific scoring procedures or task features, is that not all scoring procedures are described by one of the three models, for example, when the first step uses a stop-if-fail rule and the next steps use a partial scoring rule.

Related to this problem is that different but equally valid solution strategies may be used to solve a task. These different solution strategies yield different steps, or even different numbers of steps. For example, an item that is often used to illustrate CON-models (Akkermans, 1998; for a similar item see also Tutz, 1995),

$$\sqrt{\frac{7.5}{0.3} - 16}\,,$$

is assumed to have three consecutive solution steps which would yield a score of 3 points: (7.5/0.3), 25 − 16, and taking the square root of 9. However, the solution may also be done in four steps: (16 × 0.3), 7.5 − 4.8, 2.7/0.3, and taking the square root of 9. This second process is somewhat more cumbersome than the first, but the two solutions are equally valid. If, however, the scoring rules are set by the first solution process, but the second approach is used by the subject, this subject obtains a score of 0: in the CON models the scoring is stopped because the first step (7.5/0.3) was not taken, although the equally valid other first step (16 × 0.3) is performed correctly. Another problem in this case is that the final step, which is exactly the same in both approaches, namely, taking the root of 9, may have different parameter values.

It is obvious from these problems that the interpretation of the parameters is troublesome. If interpretation of the parameters of the correct model is almost as hard as interpreting the wrong model, or if the right model does not exist, a model could be selected on other criteria and the underlying process may not be relevant. This is still a matter of discussion.

6 Discussion

In this chapter a summary was given of various studies that compared three types of models: cumulative, conditional, and partial credit models, which differ in their definition of ISRFs. Models with the same ISRF definition may differ in the form of the ISRFs. They can be parametrically defined with a different number of parameters, or they can be nonparametrically defined as nondecreasing functions.

Studies that compared the three types of models on fit with empirical data (Molenaar, 1983; Van Engelenburg, 1997) gave no clearcut results as to which model could best be used. Comparison on the basis of simulated data showed that all models could "recognize" data equally well that were produced by their "own" model. However, with respect to the estimation of the latent trait values, incorrect models (i.e., models fit on data not produced by that model) performed as well as correct models. Therefore, it can be argued that using an incorrect model will not necessarily lead to severe problems. It is also argued that if the wrong model is fit to a specific task or scoring procedure, the interpretation of parameters may be troublesome. It seems, however, that even if the correct model is selected the interpretation of the parameters is almost as hard.

If model selection based on parameter interpretation is troublesome, other criteria come into focus where underlying processes are not relevant. Then, the availability of observable properties to test the goodness-of-fit, and the ease of application (Molenaar, 1995, p. 13) become more important. Probably one of the most important criteria is whether a model has the measurement properties needed to solve the problem at hand.

An overview of models showing their specific properties may be of help in model selection. Such an overview should also reveal which properties are implied by the models. In many cases a property may hold, even it is not *implied* by the model. That is because some of the counterexamples that were used to show that a property is not implied by a model, are extreme cases that do not often occur. Further research is needed to investigate how often a model lacks a certain property empirically. Different results are expected for different types of models as preliminary results show that in most cases the ordering of persons by their total score under the 2p-PCM is much better when the 2p-GRM holds (see also Van der Ark, 2000).

It is obvious from the results in this chapter that more research is needed on the comparison of models with respect to the measurement properties. Also more attention should be given to the usefulness, feasibility, and variability of interpretation of the parameters of the item step difficulties, because the problem of misinterpretation of the parameters can be huge in the case of polytomous items. An important question that remains to be answered is whether it is possible to obtain useful interpretations of the parameters for the three types of models.

References

Akkermans, W. (1998). *Studies on statistical models for polytomous items.* Unpublished doctoral dissertation, Twente University, The Netherlands.

Andrich, D. (1978). A rating scale formulation for ordered response categories. *Psychometrika, 43*, 561–573.

Andrich, D. (1995a). Models for measurement, precision, and the nondichotomization of graded responses. *Psychometrika, 60*, 7–26.

Andrich, D. (1995b). Further remarks on nondichotomization of graded responses. *Psychometrika, 60*, 37–46.

Andrich, D. (1995c). Distinctive and incompatible properties of two common classes of IRT models for graded responses. *Applied Psychological Measurement, 19*, 101-119.

CITO (1999). *Nederlands als tweede taal profieltoets* [Dutch as a foreign language profile test]. Arnhem: CITO.

Fischer, G.H. (1995). The derivation of polytomous Rasch models. In G.H. Fischer & I.W. Molenaar (Eds.), *Rasch models: Foundations, recent developments, and applications* (pp. 293–305). New York: Springer-Verlag.

Hemker, B.T. (1996). *Unidimensional IRT models for polytomous items, with results for Mokken scale analysis.* Unpublished doctoral dissertation, Utrecht University, The Netherlands.

Hemker, B.T., Sijtsma, K., & Molenaar, I.W. (1995). Theoretical results and item selection from a multidimensional item bank in the Mokken IRT model for polytomous items. *Applied Psychological Measurement, 19*, 337–352.

Hemker, B.T., Sijtsma, K., Molenaar, I.W., & Junker, B.W. (1996). Polytomous IRT models and monotone likelihood ratio of the total score. *Psychometrika, 61*, 679–693.

Hemker, B.T., Sijtsma, K., Molenaar, I.W., & Junker, B.W. (1997). Stochastic ordering using the latent trait and the sum score in polytomous IRT models. *Psychometrika, 63*, 331–347.

Holland, P.W., & Rosenbaum, P.R. (1986). Conditional association and unidimensionality in monotone latent variable models. *The Annals of Statistics, 14*, 1523–1543.

Jansen, P.G.W., & Roskam, E.E. (1986). Latent trait models and dichotomization of graded responses. *Psychometrika, 51*, 69–92.

Lehmann, E.L. (1959). *Testing statistical hypotheses.* New York: Wiley.

Levine, M. V., Drasgow, F., Williams, B., McCusker, C., & Thomasson, G.L. (1992). Measuring the difference between two models. *Applied Psychological Measurement, 16*, 261–278.

Masters, G.N. (1982). A Rasch model for partial credit scoring. *Psychometrika, 47*, 149–174.

Maydeu-Olivares, A., Drasgow, F., & Mead, A.D. (1994). Distinguishing among parametric item response models for polychotomous ordered data. *Applied Psychological Measurement, 18*, 245–256.

McCullagh, P. (1978). A class of parametric models for the analysis of squared contingency tables with ordered categories. *Biometrika, 65*, 413–418.

Mellenbergh, G.J. (1995). Conceptual notes on models for discrete polytomous item responses. *Applied Psychological Measurement, 19*, 91–100.

Mellenbergh, G.J. (1997). *Polytomous item response models.* Unpublished manuscript.

Mokken, R.J. (1971). *A theory and procedure of scale analysis: With applications in political research.* The Hague: Mouton.

Molenaar, I.W. (1982). Mokken scaling revisited. *Kwantitatieve Methoden, 8*, 145–164.

Molenaar, I.W. (1983). *Item steps* (Heymans Bulletin HB–83–630–EX). Groningen: University of Groningen, Vakgroep Statistiek en Meettheorie.

Molenaar, I.W. (1986). Een vingeroefening in item response theorie voor drie geordende antwoordcategorieën [An exercise in item response theory for three ordered response categories]. In G.F. Pikkemaat & J.J.A. Moors (Eds.), *Liber Amicorum Jaap Muilwijk* (pp. 39–57). Groningen: Econometrisch Instituut.

Molenaar, I.W. (1991). A weighted Loevinger H-coefficient extending Mokken scaling to multicategory items. *Kwantitatieve Methoden, 37*, 97–117.

Molenaar, I.W. (1995). Some background for item response theory and the Rasch model. In G.H. Fischer & I.W. Molenaar (Eds.), *Rasch models: Foundations, recent developments, and applications* (pp. 3–14). New York: Springer-Verlag.

Molenaar, I.W. (1997). Nonparametric models for polytomous responses. In W.J. van der Linden & R.K. Hambleton (Eds.), *Handbook of modern item response theory.* New York: Springer-Verlag.

Muraki, E. (1990). Fitting a polytomous item response model to Likert-type data. *Applied Psychological Measurement, 14*, 59-71.

Roskam, E.E. (1995). Graded responses and joining categories: A rejoinder to Andrich' "Models for measurement, precision, and the nondichotomization of graded responses." *Psychometrika, 60*, 27–35.

Samejima, F. (1969). Estimation of latent trait ability using a response pattern of graded scores. *Psychometrika Monograph, 17.*

Samejima, F. (1972). A general model for free-response data. *Psychometrika Monograph, 18.*

Samejima, F. (1995). Acceleration model in the heterogeneous case of the general graded response model. *Psychometrika, 60*, 549–572.

Sijtsma, K., & Hemker, B.T. (1998). Nonparametric item response theory models for invariant item ordering, with results for parametric models. *Psychometrika, 63*, 183–200.

Sijtsma, K., & Hemker, B.T. (in press). A taxonomy of IRT models for ordering persons and items using simple sum scores. *Journal of Educational and Behavioral Statistics.*

Sijtsma, K., & Junker, B.W. (1996). A survey of theory and methods of invariant item ordering. *British Journal of Mathematical and Statistical Psychology, 49*, 79–105.

Suppes, P., & Zanotti, M. (1981). When are probabilistic explanations possible? *Synthese, 48*, 191–199.

Thissen, D., & Steinberg, L. (1986). A taxonomy of item response models. *Psychometrika, 51*, 567–577.

Tutz, G. (1990). Sequential item response models with an ordered response. *British Journal of Mathematical and Statistical Psychology, 43*, 39–55.

Tutz, G. (1997). Sequential models for ordered responses. In W.J. van der Linden & R.K. Hambleton (Eds.), *Handbook of modern item response theory* (pp. 139–152). New York: Springer-Verlag.

Van der Ark, L.A. (2000). *Practical consequences of stochastic ordering of the latent trait under various polytomous IRT models.* Manuscript in preparation.

Van Engelenburg, G. (1997). *On psychometric models for polytomous items with ordered categories within the framework of item response theory.* Unpublished doctoral dissertation, University of Amsterdam.

Verhelst, N.D., Glas, C.A.W., & De Vries, H.H. (1997). A steps model to analyze partial credit. In W.J. van der Linden & R.K. Hambleton (Eds.), *Handbook of modern item response theory* (pp. 123–138). New York: Springer-Verlag.

Verhelst, N.D., & Verstralen, H.H.F.M. (1997). *Modelling sums of binary responses by the partial credit model* (Measurement and Research Department Reports 97-7). Arnhem: CITO.

16

Progress in NIRT Analysis of Polytomous Item Scores: Dilemmas and Practical Solutions

Klaas Sijtsma and L. Andries van der Ark[1]

ABSTRACT This chapter discusses several open problems in nonparametric polytomous item response theory: (1) theoretically, the latent trait θ is not stochastically ordered by the observed total score X_+; (2) the models do not imply an invariant item ordering; and (3) the regression of an item score on the total score X_+ or on the restscore R is not a monotone nondecreasing function and, as a result, it cannot be used for investigating the monotonicity of the item step response function. Tentative solutions for these problems are discussed. The computer program MSP for nonparametric IRT analysis is based on models that neither imply the stochastic ordering property nor an invariant item ordering. Also, MSP uses item restscore regression for investigating item step response functions. It is discussed whether computer programs may be based (temporarily) on models that lack desirable properties and use methods that are not (yet) supported by sound psychometric theory.

1 Mokken Scale Analysis for Polytomous Item Scores

Nonparametric item response theory (NIRT) for polytomous ordered item scores is characterized by the combination of practical methods for the analysis of empirical questionnaire data and complex theoretical developments, which have many pitfalls and several unsolved problems. The combination of unfinished and sometimes seemingly unsolvable theoretical problems and the need for practical solutions for data analysis problems poses an interesting challenge to the applied statistician. We focus on the nonparametric methods for scaling persons and items, initially proposed for dichotomous

---

[1]Both authors are at the Department MTO, Tilburg University, P.O. Box 90153, 5000 LE Tilburg, The Netherlands; e-mail: *k.sijtsma@kub.nl* and *a.vdark@kub.nl*

item scores by Mokken (1971) and extended to polytomous ordered item scores by Molenaar (1982, 1986, 1997). In Section 1 we start by defining the two NIRT models that are central in the work of Mokken and Molenaar and then discuss the scaling procedures built into the computer program MSP5 for Windows (abbreviated throughout as MSP; Molenaar & Sijtsma, 2000). In Section 2 we discuss the state of the art for the theory underlying the two NIRT models and provide suggestions and results for open problems. In Section 3 we discuss the methodological issue of whether it is wise, desirable, or avoidable to implement as yet imperfect data analysis methods into a computer program.

1.1 Definition of the Models

We assume that a test or questionnaire consists of k items, each of which has $m+1$ ordered answer categories, reflecting the ordering on an underlying latent trait θ, such as an attitude or a personality trait. Scores are assigned to answer categories such that the ordering of the scores corresponds with the hypothesized ordering of the answer categories on θ. Let the items be indexed I_i, $i = 1, \ldots, k$, and let the random variable giving the score on item I_i be denoted X_i, with realizations $x_i = 0, \ldots, m$. We assume that the k items measure the same θ. An observable summary score based on the k item scores is used to describe the respondent's position on θ. Usually, this summary score is the unweighted sum of k item scores, denoted X_+, and is defined as

$$X_+ = \sum_{i=1}^{k} X_i \ .$$

In addition to unidimensional measurement, we assume local independence of the item scores. Let $\mathbf{X} = (X_1, \ldots, X_k)'$ and $\mathbf{x} = (x_1, \ldots, x_k)'$. Furthermore, $P(X_i = x \mid \theta)$ denotes the conditional probability that a score of x has been obtained on item I_i. The assumption of local independence can be expressed as

$$P(\mathbf{X} = \mathbf{x} \mid \theta) = \prod_{i=1}^{k} P(X_i = x_i \mid \theta) \ .$$

Thus, θ alone explains the relationships between items. An implication of local independence is that the covariances between items conditional on θ equal 0. By integrating over the distribution of θ, denoted $G(\theta)$, the unconditional multivariate distribution of the k item scores is obtained,

$$P(\mathbf{X} = \mathbf{x}) = \int_{\theta} \prod_{i=1}^{k} P(X_i = x_i \mid \theta) \, dG(\theta) \ .$$

Suppes and Zanotti (1981) showed that, unless further restrictions are placed upon the conditional probabilities $P(X_i = x_i \mid \theta)$, the distribution of θ, or both, the multivariate distribution of the k item scores is not restricted. Usually, item response models further restrict the conditional probabilities by introducing additional assumptions.

With respect to the assumptions restricting the conditional probabilities, first we consider three possibilities (see also Agresti, 1990; Akkermans, 1998; Hemker & Sijtsma, 1998; Mellenbergh, 1995; Molenaar, 1983; Van Engelenburg, 1997) and then we concentrate on the model that is relevant in the present NIRT context. In our discussion, we use both $P(X_i = x_i \mid \theta)$ and $P(X_i \geq x_i \mid \theta)$. These probabilities are related by two simple equations:

$$P(X_i \geq x \mid \theta) = \sum_{n=x}^{m} P(X_i = n \mid \theta) \ ,$$

and

$$P(X_i = x \mid \theta) = P(X_i \geq x \mid \theta) - P(X_i \geq x+1 \mid \theta) \ . \tag{16.1}$$

To further restrict the conditional response probabilities, the *cumulative probability* definition assumes that $P(X_i \geq x \mid \theta)$ is a monotone nondecreasing function of θ,

$$P(X_i \geq x \mid \theta) \text{ nondecreasing in } \theta, \text{ for all } x, i, \tag{16.2}$$

with $x = 0$ representing an uninteresting case; that is, $P(X_i \geq 0 \mid \theta) = 1$.

The *adjacent category* definition assumes that the conditional probability of having a score of x given that either a score of $x-1$ or a score of x has been observed, and given θ, is a monotone nondecreasing function of θ,

$$P(X_i = x \mid X_i = x-1 \vee x; \theta) \text{ nondecreasing in } \theta, \text{ for all } x, i, \tag{16.3}$$

where

$$P(X_i = x \mid X_i = x-1 \vee x; \theta) = \frac{P(X_i = x \mid \theta)}{P(X_i = x-1 \mid \theta) + P(X_i = x \mid \theta)} \ ,$$

with $x = 0$ producing a probability of 1.

Finally, the *continuation ratio* definition assumes that the conditional probability of having at least an item score of x, given that we know that at least a score of $x-1$ was obtained, is a nondecreasing function of θ,

$$P(X_i \geq x \mid X_i \geq x-1; \theta) \text{ nondecreasing in } \theta, \text{ for all } x, i, \tag{16.4}$$

where

$$P(X_i \geq x \mid X_i \geq x-1; \theta) = \frac{P(X_i \geq x \mid \theta)}{P(X_i \geq x-1 \mid \theta)} \,,$$

with $x = 0$ again producing a probability of 1.

By choosing parametric functions for the conditional probabilities, parametric item response models for polytomous ordered item scores are obtained (Mellenbergh, 1995; Molenaar, 1983). For example, Samejima's (1969) graded response model is readily seen to be a parametric special case of (16.2),

$$P(X_i \geq x \mid \theta) = \frac{\exp[\alpha_i(\theta - \beta_{ix})]}{1 + \exp[\alpha_i(\theta - \beta_{ix})]} \,,$$

where α_i denotes a nonnegative slope parameter and β_{ix} denotes a location parameter. Hemker and Sijtsma (1998) defined three NIRT models based on the assumptions of unidimensionality, local independence, and the order restrictions in (16.2), (16.3), and (16.4), respectively. The hierarchical relationships between the three classes of nonparametric and parametric item response models for polytomous items were investigated by Hemker, Sijtsma, Molenaar, and Junker (1996, 1997) and Hemker, Van der Ark, and Sijtsma (2000). Van Engelenburg (1997) investigated the psychological response processes underlying each type of model and Akkermans (1998) studied the item scoring rules for these models.

Molenaar (1986) considered the cumulative probability definition to be more plausible than the adjacent category definition, because the conditioning only on $x - 1$ and x, thus isolating these two scores from their natural item context, is relatively unrealistic. Also, he dismissed the continuation ratio definition because conclusions based on it depend on the direction of the item scale whereas for most attitude items reversal of the scale direction should not lead to conclusions other than those based on the original scale direction (also, see Hemker, 2001; Hemker & Sijtsma, 1998).

Molenaar (1982, 1986, 1997) defined his model of *monotone homogeneity* (MH model) by assuming unidimensionality, local independence, and monotonicity in the sense of the cumulative probability definition, henceforth simply called monotonicity. In the MH model a higher θ implies a higher expected item score X_i, and also a higher expected unweighted total score X_+ (Hemker et al., 1997). In practice, however, the total score X_+ is used for ordering respondents on θ, so the inference uses X_+ to say something about θ. It has been shown (Hemker et al., 1997) that inference about the ordering on θ based on X_+ is not implied by the MH model for polytomous item scores. We thus run into the first theoretical problem of polytomous NIRT modeling, to which we return in Section 2.

In this chapter, the conditional probability $P(X_i \geq x \mid \theta)$ is called the item step response function (ISRF), because it describes the conditional

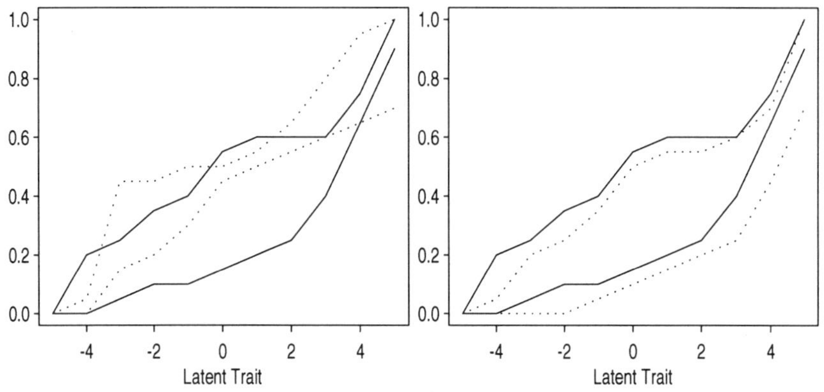

FIGURE 16.1. Two items with two ISRFs each (solid for one item, dotted for the other), satisfying monotonicity in the left panel, and double monotonicity in the right panel.

probability of taking the imaginary step x between the answer categories $x-1$ and x. Let Y_{xi} denote the binary variable indicating whether (score 1) or not (score 0) step x has been taken. Then we may write $P(X_i \geq x \mid \theta) = P(Y_{xi} = 1 \mid \theta)$. In the MH model the ISRFs are nondecreasing functions by definition.

Molenaar (1982) also defined the *double monotonicity* (DM) model which, like the MH model, is based on unidimensionality, local independence, and monotonicity, and adds the fourth assumption that the $k \times m$ ISRFs do not intersect. That is, for any pair of ISRFs, say, g of item I_i and h of item I_j, of which we know for one θ_0 that $P(X_i \geq g \mid \theta_0) < P(X_j \geq h \mid \theta_0)$, nonintersection means that

$$P(X_i \geq g \mid \theta) \leq P(X_j \geq h \mid \theta) , \quad \text{for all } \theta, g, h;\ i \neq j.$$

It may be noted that the ISRFs of the same item do not intersect by definition. Nonintersection of $k \times m$ ISRFs means that the ISRFs have the same ordering, with the exception of possible ties, for each value of θ. The DM model thus implies an ordering of item steps, which is valid for all θs and thus within all possible subpopulations from the population of interest. Monotonicity and double monotonicity are illustrated in Figure 16.1. The left panel of Figure 16.1 depicts the ISRFs of two items, indicated by a solid line for one item and a dotted line for the other, that satisfy monotonicity but not double monotonicity. The right panel of Figure 16.1 depicts the ISRFs of two items that satisfy both monotonicity and double monotonicity.

1.2 Scaling Procedure

Several researchers (e.g., Hemker, Sijtsma, & Molenaar, 1995; Molenaar, 1982, 1986, 1991; Molenaar & Sijtsma, 1988, 2000; Sijtsma, Debets, & Molenaar, 1990) discussed methods for determining the fit of the MH and DM models to polytomous item scores. These methods were implemented in three consecutive versions of the computer program MSP (Debets & Brouwer, 1989; Molenaar, Debets, Sijtsma, & Hemker, 1994; Molenaar & Sijtsma, 2000). The following methods have been implemented.

Scalability coefficients for pairs of items (denoted H_{ij}), for individual items with respect to the other $k-1$ items in the questionnaire (denoted H_i), and for a set of k items as a whole (denoted H) (Molenaar, 1991). These scalability coefficients can be written as normed (sums of) covariances between pairs of items. H_{ij} is defined as the ratio of the covariance of items I_i and I_j, $\text{Cov}(X_i, X_j)$, and the maximum covariance given the marginals of the bivariate cross-classification table of scores on items I_i and I_j, $\text{Cov}(X_i, X_j)_{\max}$; that is,

$$H_{ij} = \frac{\text{Cov}(X_i, X_j)}{\text{Cov}(X_i, X_j)_{\max}} .$$

The item scalability coefficient H_i is defined as

$$H_i = \frac{\sum\limits_{j \neq i} \text{Cov}(X_i, X_j)}{\sum\limits_{j \neq i} \text{Cov}(X_i, X_j)_{\max}} ,$$

and the scalability coefficient H for k items is defined as

$$H = \frac{\sum\limits_{i=1}^{k-1} \sum\limits_{j=i+1}^{k} \text{Cov}(X_i, X_j)}{\sum\limits_{i=1}^{k-1} \sum\limits_{j=i+1}^{k} \text{Cov}(X_i, X_j)_{\max}} .$$

Hemker et al. (1995) have shown that given the MH model,

$$0 \leq H_{ij} \leq 1, \quad \text{for all } i, j;\ i \neq j ,$$

$$0 \leq H_i \leq 1, \quad \text{for all } i,$$

and

$$0 \leq H \leq 1 .$$

Under the MH model, H-values of 0 mean that for at least $k-1$ items the ISRFs are constant functions of θ (Hemker et al., 1995). Higher H-values mean that the slope of the ISRFs tends to be steeper, which implies that the item steps and the items discriminate better among values of θ. Obviously, high H-values are desirable for a questionnaire and its items. When constructing questionnaires, we require both

$$H_{ij} > 0, \quad \text{for all } i, j;\ i \neq j\ , \tag{16.5}$$

and

$$H_i \geq c,\ c > 0, \quad \text{for all } i. \tag{16.6}$$

Because $H \geq \min(H_i)$, (16.6) implies that $H \geq c$. The usual choice is $c = 0.3$. Other values have been discussed by Hemker et al. (1995).

An **automated item selection procedure** of which the H-coefficients are the core. The item selection procedure is a bottom-up algorithm, which selects items from an item pool into subsets that satisfy (16.5) and (16.6). The item selection starts with a kernel of items, usually the two items with the highest H_{ij} value which is significantly greater than 0 (using a standard normal test; Molenaar & Sijtsma, 2000), and adds items from the remaining items one-by-one, in each selection step maximizing H, until no more items can be selected that satisfy (16.5) and (16.6). If there are items left after the first scale has been selected, the algorithm tries to select a second scale from the remaining items, a third, a fourth, and so on, until no more scales can be formed or no more items remain.

Descriptive methods, graphical displays, and statistical tests for investigating whether the data support the assumption that single ISRFs are nondecreasing functions and whether several ISRFs are nonintersecting. Among the methods for investigating intersection of the ISRFs are two that directly address the estimation of two ISRFs, say, g and h of the items I_i and I_j, respectively, through the regression of an item step score on the total score on the $k-2$ items excluding items I_i and I_j. Another method is an extension of the $P(++)$ and $P(--)$ matrices which Mokken (1971, pp. 132–133) introduced for investigating the nonintersection of the item response functions (IRFs) of dichotomously scored items. For dichotomous items, a method for investigating intersection of k ISRFs based on H-coefficients is available (Sijtsma & Meijer, 1992).

Comparative scalability results for several relevant subgroups from a group, for example, based on country, age, social-economic status, religion, or gender. This enables the comparison of groups on important scalability criteria, the assessment of the usefulness of a scale in different groups, and a first evaluation of differential item functioning (DIF) between groups.

A method for estimating the **reliability** of the total score X_+, which assumes that the ISRFs do not intersect (Molenaar & Sijtsma, 1988). Thus, before interpreting the reliability estimate it should be checked whether the DM model fits.

2 Three Open Theoretical Problems in NIRT

For each of the three problems discussed here, we first provide the theory, next we discuss current research aimed at solving the problem and, finally, we discuss the method for investigating the problem in real data.

2.1 Ordering Persons on the Latent Scale

Theory. Typical of NIRT is that ISRFs are subject to order restrictions but ISRFs are not parametrically defined. This means that the estimation of the latent person parameter θ by means of maximum likelihood methods, well known from parametric models, is not feasible here. Thus, NIRT necessarily resorts to observable person summary scores, such as the much used X_+. In an item response theory (IRT) context, one would expect the following order relationship between X_+ and θ: for higher values of X_+ we expect the conditional expected θ to increase or remain equal. That is, for two fixed values of X_+, say, s and t, we expect that

$$s < t \Rightarrow E(\theta \mid X_+ = s) \leq E(\theta \mid X_+ = t) . \tag{16.7}$$

We call this property *ordering of the expected latent trait* (OEL). OEL is implied by two better known properties. Hemker et al. (1996) investigated the monotone likelihood ratio (MLR) of X_+ given θ. MLR means that under the assumptions of unidimensionality, local independence, and monotonicity of the ISRFs, the ratio

$$\frac{P(X_+ = t \mid \theta)}{P(X_+ = s \mid \theta)}$$

is nondecreasing in θ whenever $s \leq t$. MLR implies OEL. They showed that none of the well-known polytomous IRT models from the class of cumulative probability models, including the MH and DM models, implies MLR. Of the popular models from the class of adjacent category models, only the partial credit model (PCM; Masters, 1982) and special cases of this model imply MLR. Hemker et al. (2000) showed that none of the known continuation ratio IRT models implies MLR.

Hemker et al. (1997) investigated whether θ was stochastically ordered by X_+. This property is called stochastic ordering of the latent trait (SOL). MLR implies SOL, but SOL does not imply MLR (Lehmann, 1959, p. 74). SOL means that, for any fixed value of θ, say, θ_0,

$$P(\theta \geq \theta_0 \mid X_+ = s) \leq P(\theta \geq \theta_0 \mid X_+ = t) , \quad \text{for all } s < t . \tag{16.8}$$

It may be noted that (16.8) expresses that with an increasing total score the cumulative distribution of θ is uniformly higher for s than for t (also, see Sijtsma & Hemker, in press). OEL (16.7) expresses the relationship between the means of these distributions. Hemker et al. (1997) showed that none of the well-known cumulative probability models (including the MH and the DM models) implies SOL, and from the class of adjacent category models of the well-known models only the partial credit model (Masters, 1982) and its special cases imply SOL. For continuation ratio models, Hemker et al. (2000) showed that none of the known models implies SOL. Finally, it may be noted that for *dichotomous* items, the MH model and, by implication, the DM model imply MLR (see Grayson, 1988), and thus SOL and OEL.

Present research. Current research addresses two topics. The first is whether the MH and DM models *theoretically* imply the weaker OEL property (16.7). With an example we show that they do not, so the second topic is whether the MH model implies OEL in most *practical* situations.

The next example provides an answer to the first question. We use a numerical example with one item ($k = 1$) and three ordered answer categories ($m = 2$), and a discrete θ with two values θ_1 and θ_2, both with probability 0.5 and $\theta_1 < \theta_2$, to show that the DM model does not imply OEL (16.7); consequently, the more general MH model does not imply OEL either.

Example. *The DM model does not imply OEL.* Define $P(X \geq x \mid \theta)$ as

x	0	1	2
$P(X \geq x \mid \theta_1)$	1.0	.88	.40
$P(X \geq x \mid \theta_2)$	1.0	.94	.88

Next, $P(X = x \mid \theta)$ is computed using (16.1) as

x	0	1	2
$P(X = x \mid \theta_1)$	.12	.48	.40
$P(X = x \mid \theta_2)$	.06	.06	.88

$P(\theta \mid X = x)$ is obtained by $P(X = x \mid \theta)P(\theta)/P(X = x)$ and results in

x	0	1	2
$P(\theta_1 \mid X = x)$	.67	.89	.31
$P(\theta_2 \mid X = x)$	.33	.11	.69

If we let $\theta_1 = 0$ and $\theta_2 = 1$ then $E(\theta \mid X) = P(\theta_2 \mid X = x)$, and the last table shows that $E(\theta \mid X)$ is decreasing between $X = 0$ and $X = 1$. It may be readily checked that for all other choices of θ_1 and θ_2 with $\theta_1 < \theta_2$, $E(\theta \mid X)$ decreases between $X = 0$ and $X = 1$.

Next, we used a simulation study to investigate whether the MH model implies OEL in practical situations. The design of the simulation study was the following. We used a parametric definition of an ISRF within the context of the class of continuation ratio models (16.4), which was loosely inspired by Samejima's (1995) acceleration model, and which allowed us to simulate with great flexibility ISRFs within the context of the cumulative probability model (16.2), by using

$$P(X_i \geq x \mid \theta) = \prod_{n=1}^{x} \frac{P(X_i \geq n \mid \theta)}{P(X_i \geq n-1 \mid \theta)} . \tag{16.9}$$

Our definition of the continuation ratio ISRF uses five item parameters for describing the lower and upper asymptotes of the ISRFs (γ_L and γ_U, respectively), the location (β_{ix}) and the slope (α_{ix}) of the ISRF, and the acceleration parameter (ξ_{ix}), which pushes down ($\xi_{ix} > 1$) or lifts up ($0 < \xi_{ix} < 1$) the entire ISRF. The continuation ratio ISRF is now defined as

$$\frac{P(X_i \geq x \mid \theta)}{P(X_i \geq x-1 \mid \theta)} = \gamma_L + (\gamma_U - \gamma_L) \left\{ \frac{\exp[\alpha_{ix}(\theta - \beta_{ix})]}{1 + \exp[\alpha_{ix}(\theta - \beta_{ix})]} \right\}^{\xi_{ix}} . \tag{16.10}$$

Inserting (16.10) into (16.9) does not produce an ISRF with a parameter structure that has a sensible interpretation in the context of parametric IRT, but the result is a monotone increasing function of θ which, therefore, can be used for the investigation of OEL in the context of Molenaar's MH model for polytomous item scores.

In each cell of our simulation design we simulated 1000 tests, and evaluated for each test whether $E(\theta \mid X_+)$ is nondecreasing in X_+. If $E(\theta \mid X_+)$ decreased for some increasing values of X_+, we determined the probability that two randomly selected simulees are incorrectly ordered by X_+ given their ordering on θ. For example, a simulated test may yield

x_+	$P(X_+ = x_+)$	$E(\theta \mid X_+ = x_+)$	Nondecreasing
0	.08	−.83	
1	.21	.60	Yes
2	.42	**1.42**	Yes
3	.21	**1.41**	No
4	.08	1.74	Yes

For this test we say that $E(\theta \mid X_+)$ is not nondecreasing in X_+ and that the probability of incorrectly ordering two randomly selected simulees is $2 \times .42 \times .21 = .1764$.

In our simulation study, five items had $m + 1$ ordered categories whose ISRFs were determined by (16.9) and (16.10). Parameters α^*_{ix} were inde-

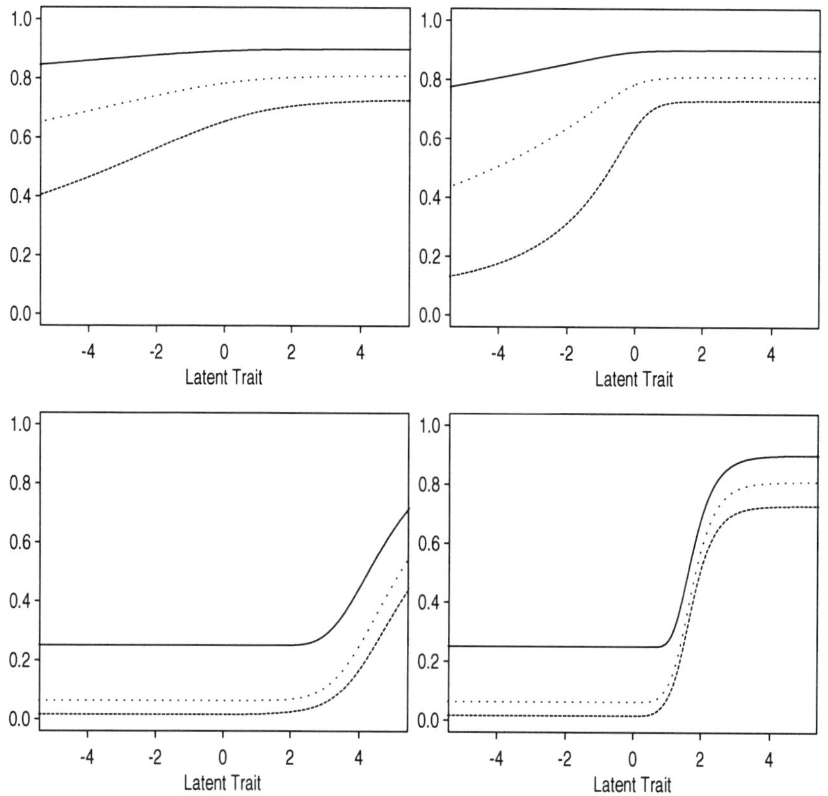

FIGURE 16.2. Typical ISRFs (simulation study, $m = 3$), each set based on $\gamma_L = .25$ and $\gamma_U = .9$, and also $\mu_{\alpha^*} = 0$, $\mu_{\xi^*} = -3$ (top left), $\mu_{\alpha^*} = 1$, $\mu_{\xi^*} = -3$ (top right), $\mu_{\alpha^*} = 0$, $\mu_{\xi^*} = 3$ (bottom left), $\mu_{\alpha^*} = 1$, $\mu_{\xi^*} = 3$ (bottom right).

pendent draws from $\mathcal{N}(\mu_{\alpha^*}, .2)$. To avoid negative values, the slope parameters in (16.10) were defined as $\alpha_{ix} = \exp(\alpha^*_{ix})$. The location parameters in (16.10), β_{ix}, were independent draws from $\mathcal{N}(0, .5)$. Parameters ξ^*_{ix} were independent draws from $\mathcal{N}(\mu_{\xi^*}, 1)$ and the acceleration parameters in (16.10) were defined as $\xi_{ix} = \exp(\xi^*_{ix})$. The latent trait θ was assumed to have a standard normal distribution. Different types of tests ($k = 5$) were generated by varying m, μ_{α^*}, μ_{ξ^*}, and two combinations of γ_L and γ_U. In the first combination $\gamma_L = 1/(m + 1)$ and γ_U was distributed uniformly, $\mathcal{U}(.9, 1)$; in the second combination $\gamma_L = 0$ and $\gamma_U = 1$. The complete simulation design had order 3 (number of answer categories, $m+1$) $\times 2$ (mean μ_{α^*} parameter) $\times 2$ (mean μ_{ξ^*} parameter) $\times 2$ (combination of γ_L and γ_U). It may be noted that within the class of continuation ratio models, α_{ix} can be interpreted as the slope parameter, and ξ_{ix} as the acceleration parameter, but the α^*_{ix} and ξ^*_{ix} parameters do not have a meaningful

TABLE 16.1. Number of times 1000 tests yield $E(\theta \mid X_+)$ nondecreasing in X_+.

γ_L/γ_U		$\frac{1}{m+1}/\sim\mathcal{U}(.9,1)$		0/1	
μ_{α^*}	$m+1$	$\mu_{\xi^*}=-3$	$\mu_{\xi^*}=3$	$\mu_{\xi^*}=-3$	$\mu_{\xi^*}=3$
0	3	530	791	629	977
	4	316	793	383	887
	5	245	702	209	770
1	3	698	784	747	835
	4	546	674	448	486
	5	423	514	228	231

TABLE 16.2. Probability of correctly ordering two randomly selected simulees.

γ_L/γ_U		$\frac{1}{m+1}/\sim\mathcal{U}(.9,1)$		0/1	
μ_{α^*}	$m+1$	$\mu_{\xi^*}=-3$	$\mu_{\xi^*}=3$	$\mu_{\xi^*}=-3$	$\mu_{\xi^*}=3$
0	3	.9792	.9875	.9884	1.0000
	4	.9765	.9683	.9807	1.0000
	5	.9843	.9596	.9767	1.0000
1	3	.9911	.9912	.9921	.9999
	4	.9908	.9833	.9886	.9998
	5	.9932	.9765	.9874	.9997

interpretation due to the logarithmic transformation. Within the class of cumulative probability models, of which the MH model is a special case, none of the parameters α_{ix}, ξ_{ix}, α^*_{ix}, and ξ^*_{ix} can be interpreted meaningfully due to the transformation expressed by (16.9). The only objective of different parameter choices is to obtain differently shaped ISRFs. Four examples of ISRFs resulting from inserting (16.10) into (16.9) are displayed in Figure 16.2.

The results of the simulation study are summarized in Tables 16.1 and 16.2. Table 16.1 shows for each cell in the design that often $E(\theta \mid X_+)$ was decreasing in some adjacent values of X_+. Models with $\mu_{\alpha^*}=0$, $\mu_{\xi^*}=3$, $\gamma_L=0$, and $\gamma_U=1$ produced the highest frequency of tests in which OEL held, probably because the slopes of the ISRFs were more similar than those in other design cells. Table 16.1 also shows that results became worse as $m+1$ increased.

The results in Table 16.2 show a much brighter picture. The model with $\mu_{\alpha^*}=0$, $m+1=5$, $\mu_{\xi^*}=3$, and $\gamma_L=0.2$, and $\gamma_U\sim\mathcal{U}(.9,1)$, yielded the lowest probability of ordering two randomly drawn simulees correctly, which was still as high as .9596, which is the mean of 1000 probabilities, each based on two probabilities $P(X_+ = x_+)$ calculated using (16.10). Models with $\mu_{\xi^*} = 3$, and $\gamma_L = 0$ and $\gamma_U = 1$, produced the highest probabilities, which were all at least .999. Overall, no serious violations of OEL were found. Moreover, a visual inspection of the $E(\theta \mid X_+)$s suggested that if $E(\theta \mid X_+ = s) > E(\theta \mid X_+ = t)$ when $s < t$, the numerical difference between these expected values was very small. For example, a typical string of $E(\theta \mid X_+ = x_+)$ for $x_+ = 0, \ldots, 20$ was

$$
\begin{array}{rrrrrrr}
-.83, & .90, & 1.43, & 1.73, & 1.95, & 2.24, & 2.55, \\
2.74, & 2.92, & 3.12, & 3.32, & 3.46, & 3.67, & \mathbf{3.93}, \\
\mathbf{3.88}, & 4.18, & 4.52, & 4.93, & 4.94, & 4.95, & 4.97,
\end{array}
$$

where the bold values indicate a reversal of the expected order. Results for $k = 10$ and results for slope parameters in (16.10), which are the exponential of independent draws from $\mathcal{N}(\mu_{\alpha^*}, \sigma^2_{\alpha^*} = 1)$, were also investigated for several cells of the design. These results were similar to the results for $k = 5$ and $\sigma^2_{\alpha^*} = .2$ and, therefore, are not presented here.

Based on these first results for the cumulative probability models, we tentatively conclude that the probability of incorrectly ordering persons on θ by means of X_+ is sufficiently small to continue the practice of measuring on the X_+ scale. Throughout this simulation study we assumed that θ had a standard normal distribution. It may be noted that OEL is not invariant for monotone transformations of θ (T.A.B. Snijders, personal communication, 2000) and ordering results may be different for nonlinear transformations of θ. This simulation study was only a first step in the investigation of OEL for polytomous IRT models. A more comprehensive study including investigation of SOL, for all three classes of polytomous IRT models, is presently being performed.

Method. MSP provides the user with methods for analyzing item response data (both dichotomous and polytomous) under the MH and DM models. When a model fits polytomous data the researcher is advised (Molenaar & Sijtsma, 2000, p. 18) to consider X_+ as a proxy for ordering respondents on θ. Our simulation results tentatively suggest that, in practice, the use of X_+ does not lead to serious errors when ordering respondents on θ. For dichotomous items, MLR and SOL hold; thus, X_+ can be used safely for ordering persons on θ.

2.2 Ordering Items by Their Difficulty

Theory. Parametric IRT models for polytomous item scores have location parameters on the θ scale for each answer category. Verhelst and Verstralen (1991) argued that these location parameters cannot be interpreted as difficulty parameters. For example, in the PCM (Masters, 1982) β_{ix} gives the

θ value at which the response curves of the adjacent answer categories $x-1$ and x intersect, but not the difficulty of answering in either one of these categories. Sijtsma and Junker (1996) gave a definition of item difficulty for parametric and nonparametric IRT models for dichotomous items, which was used also for polytomous items by Sijtsma and Hemker (1998; based on Chang & Mazzeo, 1994). The difficulty of an item with $m+1$ ordered answer categories is defined as $E(X_i \mid \theta)$, for $i = 1, \ldots, k$. By definition k items have an invariant item ordering (IIO) if they can be ordered (and numbered accordingly) such that

$$E(X_1 \mid \theta) \leq \ldots \leq E(X_k \mid \theta) , \quad \text{for all } \theta. \tag{16.11}$$

For dichotomous items, Rosenbaum (1987) said that items satisfying (16.11) have a *latent scale.*

An IIO can be important in applications that assume that respondents or subgroups of respondents have the same item ordering. For example, starting and stopping rules in intelligence testing assume that for all respondents the same items are easy and the same items are difficult. More specifically, the use of such rules assumes the same item ordering of the k items across θ, with the exception of possible ties; see (16.11). Another example is DIF which, for dichotomous items, assumes that the IRF $P(X_i = 1 \mid \theta)$ is identical for different subgroups from the population of interest. In a NIRT context, one could require an IIO, both for dichotomous and polytomous items (note that for dichotomous items with 0/1 scoring $E(X_i|\theta) = P(X_i = 1|\theta)$). In a particular population an IIO is defined at the θ level; therefore, an IIO implies the same item ordering by mean scores for subgroups, for example, boys and girls, different ethnic groups, and different age groups. One way to study DIF in a nonparametric context is by checking the item ordering at the subgroup level. Other applications of IIO were discussed by Sijtsma and Junker (1996); see also Hemker (2001).

Only a few polytomous IRT models imply an IIO. Sijtsma and Hemker (1998) found that of the popular IRT models from the classes of cumulative probability models and adjacent category models, only the rating scale model (Andrich, 1978) implies an IIO. The rating scale model is a special case of the PCM with linear restrictions on the location parameters. Hemker et al. (2000) found that within the class of continuation ratio models only the sequential rating scale model (Tutz, 1990) implies an IIO.

Sijtsma and Hemker (1998) gave a sufficient condition for IIO. First, unidimensionality, local independence, monotonicity of the ISRFs, and nonintersection of the ISRFs is assumed; that is, the DM model is assumed. Second, it is assumed that if for a given item score x_0,

$$P(X_1 \geq x_0 \mid \theta) < P(X_2 \geq x_0 \mid \theta) < \ldots < P(X_k \geq x_0 \mid \theta) , \text{ for all } \theta, \tag{16.12}$$

then for each item score x,

$$P(X_1 \geq x \mid \theta) \leq P(X_2 \geq x \mid \theta) \leq \ldots \leq P(X_k \geq x \mid \theta) , \quad \text{for all } \theta. \tag{16.13}$$

TABLE 16.3. Numerical example illustrating two methods of investigating IIO.

Restgroup Order	Subgroup Item Means	First Pooling	Second Pooling
1	0.40	0.40	0.40
2	**1.00**	0.80	0.80
3	**0.80**	0.80	0.80
4	**0.60**	0.80	0.80
5	1.20	1.20	1.20
6	2.00	2.00	2.00
7	2.20	**2.20**	2.07
8	**2.40**	**2.00**	2.07
9	**1.60**	**2.00**	2.07

Again, the item numbering follows the item ordering. In words, if we know the ordering of conditional probabilities for one item score x_0, then it follows that the ordering for the other item scores is the same with the exception of possible ties. Sijtsma and Hemker (1998) called the DM model with the additional restrictions (16.12) and (16.13) the strong DM model and noted that this is a highly restrictive model.

Present research. Here, we tentatively suggest a method for investigating whether a set of k polytomously scored items has an IIO, based on a first attempt published by Verweij, Sijtsma, and Koops (1999). First, the total group of interest is split into the relevant subgroups. Second, within each subgroup w, $w = 1, \ldots, W$, the item means are estimated using the sample means $\overline{X}_{iw}$, $i = 1, \ldots, k$. Third, for all w, the item means are also estimated within the total group with subgroup w excluded, say, the restgroup. The restgroup means are denoted $\overline{X}_{i(-w)}$, $i = 1, \ldots, k$. Fourth, for each subgroup the items are placed in the ordering suggested by the corresponding restgroup item means $\overline{X}_{i(-w)}$ and it is checked whether there are reversals in the subgroup item order compared to the corresponding restgroup item order, which thus serves as the criterion. It may be noted that the item ordering may vary across restgroups, and that other ordering criteria may be preferred. Fifth, when reversals of the criterion order are found within the subgroup, the string of items involved is marked as in Table 16.3, second column, where two strings are printed in boldface (the example involves nine items). We suggest certain procedures for evaluating reversals.

The first procedure investigates whether reversals within a string are significantly different from tied item means using the signed rank test. The second procedure searches for the nearest ordering that matches the restgroup ordering according to the least-squares criterion. The principle is illustrated in the third and fourth columns of Table 16.3, and is known as the pool-adjacent-violators algorithm (Barlow, Bartholomew, Bremner, & Brunk, 1972, pp. 13–18). The idea is to replace strings of decreasing sample means (second column, boldface) by their mean (third column), and to repeat this routine until no more violations remain (fourth column). Sijtsma (1988, p. 46) suggested using a discrepancy measure based on the mean residual sum of squares as a descriptive statistic. These methods and variations are currently being subjected to more elaborate research.

Method. MSP provides the opportunity to investigate the scalability of a set of items within several groups, defined by means of a grouping variable. The program does not contain methods that explicitly investigate whether a set of items has an IIO. Because the MH and DM models do not imply an IIO, and because researchers often like to know whether item difficulty orderings are the same for different relevant subgroups, the inclusion of a method for investigating the item ordering in relevant subgroups would be an important new feature of MSP, or any other IRT program for models not implying an IIO.

2.3 Estimation of Item Step Response Functions

Theory. The monotonicity assumption says that $P(X_i \geq x \mid \theta)$ is nondecreasing in θ, for all x and all i. For dichotomous items, Junker (1993) showed that the MH model does *not* imply that the observable probability $P(X_i = 1 \mid X_+)$ is nondecreasing in X_+. This means that this probability cannot be used for investigating the monotonicity property of IRFs. Junker (1993) also showed that if we condition on a summary score that does not contain X_i, for example, the restscore

$$R = X_+ - X_i \ ,$$

the MH model implies manifest monotonicity (MM),

$$P(X_i = 1 \mid R) \text{ nondecreasing in } R.$$

Thus, MM can be used for investigating the monotonicity property of IRFs.

For polytomous items, Junker and Sijtsma (2000) discussed an example (due to B.T. Hemker, personal communication, 1997) which shows that the MH model does not imply MM; thus, in general,

$$P(X_i \geq x \mid R) \text{ NOT nondecreasing in } R.$$

Alternatively, Junker (1996) suggested conditioning on a restscore D based on the $k - 1$ item scores all dichotomized at the same x, yielding dichotomized item scores D_j and $D = \sum_j D_j$ $(j \neq i)$, such that for consecutive fixed values $x = 1, \ldots, m$,

$$D_j = \begin{cases} 0 \,, & \text{if } X_j < x \,, \\ 1 \,, & \text{otherwise} \,, \end{cases} \tag{16.14}$$

and then investigating whether

$$P(X_i \geq x \mid D) \,, \quad \text{for all } x, \tag{16.15}$$

is nondecreasing in restscore D. Hemker (1996, Chap. 6) proved that if all k items are dichotomized as in (16.14)—thus including X_i (in 16.15), where $X_i \geq x$ is recoded as $X_i = 1$—the MH model still holds. This means that (16.15) has the property of MM and provides a valid means for investigating nondecreasingness of the ISRFs. Junker (1996) noted that different items can also be dichotomized at different values of x, but we refrain from this possibility. Finally, let $\mathbf{X}_{D_x}$ be the binary $N \times k$ data matrix resulting from dichotomization at item score x (16.14) and let $\mathbf{X}_P$ be the polytomous data matrix. Then, it is readily checked that

$$\mathbf{X}_P = \sum_{x=1}^{m} \mathbf{X}_{D_x} \,.$$

Thus, the m dichotomizations together contain all information from the polytomous data, which would not be the case when fewer dichotomizations were considered.

Future research. We suggest three lines of research. The first is to investigate the practical usefulness of the voluminous results produced by the method proposed by Junker (1996), based on the dichotomization of polytomous item response data. It may be noted that for each individual ISRF m estimates have to be investigated (one for each dichotomization of the $k-1$ items constituting the restscore D). The question then becomes how to combine all sample results into one conclusion, in particular, when for some but not for all ISRFs violations are found.

The second research topic is to take the polytomous data as the starting point and to investigate other conditioning variables based on observable data that are (1) substitutes for X_+ or R; (2) useful estimates of θ; and (3) when inserted in (16.15) produce MM. Candidate conditioning variables that are sum scores from the class of unweighted total scores X_+^* based on $1 \leq S \leq k-1$ items can be excluded beforehand, because the subset of S items can be defined to constitute a new test. We are currently investigating alternative and viable item summaries.

The third research topic is the investigation of the practical usefulness of $P(X_i \geq x \mid R)$ as an estimate of the ISRF by means of simulated polytomous item scores. This research is currently taking place, but no definitive results can be reported as yet.

Method. MSP makes ample use of $P(X_i \geq x|R)$ for estimating the ISRFs. These estimates are used for checking whether the ISRFs are monotone nondecreasing and whether ISRFs are nonintersecting. Future research has to (1) show whether for practical use these methods give valid information about the ISRFs in the sense that errors are negligible; and (2) indicate whether methods presently implemented should be replaced by theoretically sound methods as discussed above.

3 Discussion

3.1 Use Another Polytomous IRT Model?

Hemker et al. (1997) showed that the PCM (and special cases) is the only known polytomous IRT model that implies SOL, and Van der Ark (2000) showed that the PCM also is the only model that implies OEL. This means that when the researcher requires SOL or the weaker OEL as a measurement property, the PCM (or special cases) is the only option. Because the PCM is a restrictive model, assuming equal discrimination for all k items, its use in practical data analysis will often lead to the rejection of many items from the test. This is a sacrifice many researchers may not be prepared to make. Thus, the PCM is a limited option as a practical alternative for polytomous IRT models that do not imply SOL or OEL. This conclusion underlines the need for robustness studies that show to what degree SOL and OEL are violated under several models in practical testing situations. The first robustness results presented here tentatively support the use of X_+ for ordering respondents under the MH model for polytomous items.

Sijtsma and Hemker (1998) showed that only highly restrictive polytomous IRT models, such as their strong DM model and the parametric rating scale model (Andrich, 1978) imply an IIO. The application of any of these models to data probably leads to the rejection of many items. The restrictiveness of these models supports the need for methods that can be used for investigating IIO without the need for simultaneously fitting a particular IRT model. An example of such a method may be one that compares the item ordering by mean scores in each relevant subgroup with an overall item ordering found from the data or hypothesized on the basis of a priori knowledge.

Our simulation study gave some positive first results with respect to the practical robustness of models against failure of SOL or OEL. Of course, future research may show that there exist nontrivial situations in which SOL or OEL is not guaranteed. In those cases, new models have to be developed, for example, incorporating SOL as an assumption, and observable consequences from such models have to be studied, which can be used for investigating model-data fit. Based on our present knowledge we expect that methods for investigating IIO will be developed and applied successfully and will become part of future versions of MSP. Also, we expect that alternative methods for investigating properties of ISRFs as the one proposed by Junker (1996; see Equation 16.15) will be further developed.

3.2 Progress Through Success and Failure

The situation described in this chapter, in which several practical data analysis methods are supported by sound psychometric theory while others are only partly supported by theory, is typical of scientific research. New methods are developed and initially believed to be correct until someone shows that they suffer from flaws or someone else proposes a better method.

The availability of user-friendly software provides the opportunity for practical researchers to analyze their test and questionnaire data by means of advanced methods. For psychometricians, a user-friendly program is the only way in which they can effectively promote their newly developed methods among researchers. One could argue that programs be made available only after their theoretical basis has been established completely, but we have already seen that scientific knowledge develops rather through successes and apparent successes, which after a while may be found to suffer from flaws (or worse, turn out to be complete failures) and then are improved or replaced by better methods.

Much of the progress in psychometric theory is stimulated by the application of programs such as MSP to empirical data from various kinds of applications. For example, after a developmental psychologist had analyzed his polytomous data with MSP and OPLM (Verhelst, 1992), he asked whether the fit of the DM model and the generalized PCM implied that the items had equal ordering by mean scores for different age groups. This reasonable question could not be answered by most of the polytomous IRT models, and this fact then became the basis for the IIO research (Sijtsma & Junker, 1996; Sijtsma & Hemker, 1998). Although, looking back, their origins were less clear, without doubt the MLR, SOL, and item restscore regression research was also stimulated by practical data analysis problems and questions asked by researchers using MSP and other programs. It is this mutual cross-fertilization between the theory and practice of psychometrics that stimulates the progress in the development of polytomous IRT modeling.

References

Agresti, A. (1990). *Categorical data analysis.* New York: Wiley.

Akkermans, L.M.W. (1998). *Studies on statistical models for polytomously scored test items.* Unpublished doctoral dissertation, University of Twente, Enschede, The Netherlands.

Andrich, D. (1978). A rating scale formulation for ordered response categories. *Psychometrika, 43*, 561–573.

Barlow, R.E., Bartholomew, D.J., Bremner, J.M., & Brunk, H.D. (1972). *Statistical inference under order restrictions.* New York: Wiley.

Chang, H., & Mazzeo, J. (1994). The unique correspondence of the item response function and item category response functions in polytomously scored item response models. *Psychometrika, 59*, 391–404.

Debets, P., & Brouwer, E. (1989). MSP: A program for Mokken scale analysis for polytomous items [Software manual]. Groningen: ProGAMMA.

Grayson, D.A. (1988). Two-group classification in latent trait theory: Scores with monotone likelihood ratio. *Psychometrika, 53*, 383–392.

Hemker, B.T. (1996). *Unidimensional IRT models for polytomous items, with results for Mokken scale analysis.* Unpublished doctoral dissertation, Utrecht University, The Netherlands.

Hemker, B.T. (2001). Reversibility revisited and other comparisons of three types of polytomous IRT models. In A. Boomsma, M.A.J. van Duijn, & T.A.B. Snijders (Eds.), *Essays on item response theory* (pp. 277–296). New York: Springer-Verlag.

Hemker, B.T., & Sijtsma, K. (1998). *A comparison of three general types of unidimensional IRT models for polytomous items.* Manuscript submitted for publication.

Hemker, B.T., Sijtsma, K., & Molenaar, I.W. (1995). Selection of unidimensional scales from a multidimensional item bank in the polytomous Mokken IRT model. *Applied Psychological Measurement, 19*, 337–352.

Hemker, B.T., Sijtsma, K., Molenaar, I.W., & Junker, B.W. (1996). Polytomous IRT models and monotone likelihood ratio of the total score. *Psychometrika, 61*, 679–693.

Hemker, B.T., Sijtsma, K., Molenaar, I.W., & Junker, B.W. (1997). Stochastic ordering using the latent trait and the sum score in polytomous IRT models. *Psychometrika, 62*, 331–347.

Hemker, B.T., Van der Ark, L.A., & Sijtsma, K. (2000). *On measurement properties of sequential IRT models.* Manuscript submitted for publication.

Junker, B.W. (1993). Conditional association, essential independence and monotone unidimensional item response models. *The Annals of Statistics, 21*, 1359–1378.

Junker, B.W. (1996). *Exploring monotonicity in polytomous item response data.* Paper presented at the Annual Meeting of the National Council on Measurement in Education, New York.

Junker, B.W., & Sijtsma, K. (2000). Latent and manifest monotonicity in item response models. *Applied Psychological Measurement, 24*, 65–81.

Lehmann, E.L. (1959). *Testing statistical hypotheses.* New York: Wiley.

Masters, G.N. (1982). A Rasch model for partial credit scoring. *Psychometrika, 47*, 149–174.

Mellenbergh, G.J. (1995). Conceptual notes on models for discrete polytomous item responses. *Applied Psychological Measurement, 19*, 91–100.

Mokken, R.J. (1971). *A theory and procedure of scale analysis: With applications in political research.* The Hague: Mouton.

Molenaar, I.W. (1982). Mokken scaling revisited. *Kwantitatieve Methoden, 8*, 145–164.

Molenaar, I.W. (1983). *Item steps* (Heymans Bulletin HB–83–630–EX). Groningen: University of Groningen, Vakgroep Statistiek en Meettheorie.

Molenaar, I.W. (1986). Een vingeroefening in item response theorie voor drie geordende antwoordcategorieën [An exercise in item response theory for three ordered answer categories]. In G.F. Pikkemaat & J.J.A. Moors (Eds.), *Liber amicorum Jaap Muilwijk* (pp. 39–57). Groningen: Econometrisch Instituut.

Molenaar, I.W. (1991). A weighted Loevinger H-coefficient extending Mokken scaling to multicategory items. *Kwantitatieve Methoden, 37*, 97–117.

Molenaar, I.W. (1997). Nonparametric models for polytomous responses. In W.J. van der Linden & R.K. Hambleton (Eds.), *Handbook of modern item response theory* (pp. 369–380). New York: Springer-Verlag.

Molenaar, I.W., Debets, P., Sijtsma, K., & Hemker, B.T. (1994). User's manual MSP [Software manual]. Groningen: ProGAMMA.

Molenaar, I.W., & Sijtsma, K. (1988). Mokken's approach to reliability estimation extended to multicategory items. *Kwantitatieve Methoden, 28*, 115–126.

Molenaar, I.W., & Sijtsma, K. (2000). User's manual MSP5 for Windows: A program for Mokken scale analysis for polytomous items. Version 5.0 [Software manual]. Groningen: ProGAMMA.

Rosenbaum, P.R. (1987). Probability inequalities for latent scales. *British Journal of Mathematical and Statistical Psychology, 40*, 157–168.

Samejima, F. (1969). Estimation of latent ability using a response pattern of graded scores. *Psychometrika Monograph, 17.*

Samejima, F. (1995). Acceleration model in the heterogeneous case of the general graded response model. *Psychometrika, 60*, 549–572.

Sijtsma, K. (1988). *Contributions to Mokken's nonparametric item response theory.* Amsterdam: Free University Press.

Sijtsma, K., Debets, P., & Molenaar, I.W. (1990). Mokken scale analysis for polychotomous items: Theory, a computer program and an application. *Quality & Quantity, 24*, 173–188.

Sijtsma, K., & Hemker, B.T. (1998). Nonparametric polytomous IRT models for invariant item ordering, with results for parametric models. *Psychometrika, 63*, 183–200.

Sijtsma, K., & Hemker, B.T. (in press). A taxonomy of IRT models for ordering persons and items using simple sum scores. *Journal of Educational and Behavioral Statistics.*

Sijtsma, K., & Junker, B.W. (1996). A survey of theory and methods of invariant item ordering. *British Journal of Mathematical and Statistical Psychology, 49*, 79–105.

Sijtsma, K., & Meijer, R.R. (1992). A method for investigating the intersection of item response functions in Mokken's nonparametric IRT model. *Applied Psychological Measurement, 16*, 149–157.

Suppes, P., & Zanotti, M. (1981). When are probabilistic explanations possible? *Synthese, 48*, 191–199.

Tutz, G. (1990). Sequential item response models with an ordered response. *British Journal of Mathematical and Statistical Psychology, 43*, 39–55.

Van der Ark, L.A. (2000). *Practical consequences of stochastic ordering of the latent trait under various polytomous IRT models.* Manuscript in preparation.

Van Engelenburg, G. (1997). *On psychometric models for polytomous items with ordered categories within the framework of item response theory.* Unpublished doctoral dissertation, University of Amsterdam.

Verhelst, N.D. (1992). *Het eenparameter logistisch model* (OPLM) [The one-parameter logistic model (OPLM)] (OPD Memorandum 92–3). Arnhem: CITO.

Verhelst, N.D., & Verstralen, H.H.F.M. (1991). *The partial credit model with non-sequential solution strategies.* Arnhem: CITO.

Verweij, A.C., Sijtsma, K., & Koops, W. (1999). An ordinal scale for transitive reasoning by means of a deductive strategy. *International Journal of Behavioral Development, 23*, 241–264.

17

Two-Level Nonparametric Scaling for Dichotomous Data

Tom A.B. Snijders[1]

ABSTRACT This chapter considers a two-level design where the objects to be scaled are the higher-level units. Nested within each object are lower-level units, called subjects, and a set of dichotomous items is administered to each subject. The subjects are regarded as strictly parallel tests for the objects. Examples are the scaling of teachers on the basis of their pupils' responses, or of neighborhoods on the basis of responses by inhabitants. A two-level version of the nonparametric scaling method first proposed by Mokken (1971) is elaborated upon. The probabilities of positive responses to the items are assumed to be increasing functions of the value on a latent trait. The latent trait value for each subject is composed of an object-dependent value and a subject-dependent deviation from this value. The consistency of responses within, but also between objects is expressed by two-level versions of Loevinger's H-coefficients. The availability of parallel tests is used to calculate a reliability coefficient.

1 Introduction

In item response theory we are accustomed to situations where the subjects who provide the responses are also the objects that are to be measured. Formulated more generally, it is usual that for each individual to be scaled, exactly one set of item responses is available. There are situations, however, where the roles of measured object and responding subject are played by distinct entities, as is demonstrated by the following examples.

- Pupils are asked questions about their teachers, with the purpose of assessing the teachers (as perceived by the pupils) on some dimension.

- Employees are asked to respond to a questionnaire about the department where they work, in order to scale the departments with respect to working climate.

[1]Department of Statistics and Measurement Theory, University of Groningen, Grote Kruisstraat 2/1, 9712 TS Groningen, The Netherlands; e-mail: *t.a.b.snijders@ppsw.rug.nl*

- Social settings such as neighborhoods are assessed by a survey of multiple informants, for example, inhabitants. The answers given by the informants are aggregated to provide an assessment of the social setting (this is called *ecometrics* by Raudenbush & Sampson, 1999).

- In a study of personal networks, for each respondent (ego) in a survey, a list is made of the other persons (alters) to whom the respondent is related according to certain criteria. A number of characteristics of the alters are collected in order to characterize ego's social environment (Jansson & Spreen, 1998).

The *objects* to be measured in these examples are, respectively, teachers, departments, social settings, and respondents. For each object a set of multiple measurements is available: measurements corresponding to pupils, employees, informants, and alters. The units corresponding to each measurement are in general called *subjects*, although this word may be unexpected in the fourth example. Each of the multiple measurements is supposed to yield responses to the same set of k *items*. This defines a three-level nesting structure: responses denoted X_{vji} to items I_i are nested within subjects S_{vj}, which in their turn are nested within objects O_v. Equivalently, this can be regarded as a multivariate two-level nesting structure, where subjects are nested within objects and each subject provides a vector of responses $\mathbf{X}_{vj} = (X_{vj1}, \ldots, X_{vjk})'$.

The purpose of the method of analysis presented below is to give scale values to the objects, while the subjects are regarded as parallel tests of these objects. It is assumed that the item responses X_{vji} are dichotomous, scored as 1 (positive, correct) and 0 (negative, false). The statistical approach elaborated here is a nonparametric latent trait model in the sense that a one-dimensional latent trait θ is assumed to govern (stochastically) the item responses, and only qualitative assumptions are made about how the response probabilities depend on the latent trait. These assumptions do not restrict these probabilities to a particular mathematical function. The following questions are treated in this chapter.

1. What, in this two-level design, is a suitable nonparametric definition of the relation between latent trait and observed responses?

2. How can it be checked empirically whether the items are sufficiently strongly associated to be regarded as indicators of a one-dimensional latent variable; and whether the contribution of the objects to the item responses, relative to the contribution by the subjects, is large enough for measuring the objects?

3. How can the objects be scored on this latent dimension?

4. What is the reliability of the resulting scale?

These questions are answered by elaborating a two-level version of the nonparametric scaling method first proposed by Mokken (1971). Introductions to this technique are given by, for example, Mokken and Lewis (1982), Mokken (1997), and Sijtsma (1998). In the discussion section an alternative parametric model is mentioned.

The Mokken scaling model is an attractive nonparametric model for unidimensional cumulative scaling. This model applies to the standard single-level design, where there is no distinction between subjects and objects. It can be summarized as follows. Subjects, indicated by S_v, respond to dichotomously scored items indicated by I_i. The response (scored 0 or 1) of subject S_v to item I_i is denoted X_{vi}. It is assumed that a latent trait θ exists for which each subject has a value, denoted θ_v, determining the probability distribution of S_v's vector of responses. It is further assumed that for each item I_i, there exists a nondecreasing function $f_i(\cdot)$ such that the probability of a positive response, given the latent trait value θ_v, is

$$P\,(X_{vi} = 1 \mid \theta_v) = f_i(\theta_v)\ .$$

These functions $f_i(\cdot)$ are called item characteristic curves (ICCs). Furthermore, the Mokken model assumes local stochastic independence: conditional on the value of the latent trait, the responses of the subject to different items are outcomes of independent random variables.

Important questions on the application of this scaling method are the scalability of a given set of items, the selection of a scalable subset from a given set of items, and the attribution of scores to subjects. Procedures for data analysis according to this model are described in the literature mentioned above, and implemented in the program MSP (Molenaar & Sijtsma, 2000).

2 A Two-Level Model for Nonparametric Scaling of Dichotomous Data

In the two-level design assumed here, X_{vji} denotes the response of subject S_{vj} to item I_i with regard to object O_v. The set of items scored is assumed to be the same for each object–subject combination, while the number of subjects for object O_v, denoted n_v, may be object-dependent.

The model elaborated for this situation is a cumulative scaling model governed by a unidimensional latent trait θ. Assume that each object O_v has a value θ_v on the latent trait, and each subject S_{vj} nested within object O_v has a deviation δ_{vj} from this value. This yields the value $\theta_v + \delta_{vj}$ on the latent trait for the combination of subject S_{vj} and object O_v. The deviation δ_{vj} may be considered as a (random) subject effect together with an object $\times$ subject interaction. Conditional on these latent trait values, we assume stochastic independence of the responses to the different items and by the different subjects. We also assume the existence of nondecreasing ICCs $f_i(\cdot)$ that are specific to the items but do not depend on object or subject. The probabilities of positive responses are defined by the item ICCs evaluated in the combined latent trait values:

$$P(X_{vji} = 1 \mid \theta_v, \delta_{vj}) = f_i(\theta_v + \delta_{vj}) \ . \tag{17.1}$$

Furthermore, we assume that the δ_{vj} are independent and identically distributed random variables. The responses by the different subjects therefore are independent and identically distributed, conditional on θ_v. It is not a restriction to assume also that δ_{vj} has expected value 0.

In terms of multilevel analysis (see, e.g., Bryk & Raudenbush, 1992; Goldstein, 1995; Snijders & Bosker, 1999), this may be regarded as a three-level model for binary data where the random parts at levels two and three are composed of random intercepts without random slopes. The nonparametric approach, however, is unusual in multilevel analysis.

The scaling method must provide scale values for the objects. The influence of the subjects on these scale values is a nuisance from the point of view of measuring the objects. When $\text{Var}(\theta_v)$ is large compared to $\text{Var}(\delta_{vj})$, then the differences between the subjects hardly matter compared to the differences between the objects. This is desirable from the point of view of scaling the objects. When, on the contrary, $\text{Var}(\theta_v)$ is small compared to $\text{Var}(\delta_{vj})$, then the probabilities of the responses to the items are determined by the subjects more strongly than by the objects, leading to a comparatively unreliable scale (unless the number of subjects per object is large).

It follows from model (17.1) that the marginal probability of a positive response for a randomly drawn subject to object O_v, given the latent trait value θ_v, is given by

$$\phi_i(\theta_v) = P(X_{vji} = 1 \mid \theta_v) = E_\delta(X_{vji} \mid \theta_v) = E_\delta\left[f_i(\theta_v + \delta_{vj})\right] , \tag{17.2}$$

where the symbol E_δ denotes that the expected value is taken with respect to the distribution of δ_{vj}. In analogy to classical testing theory, the function $\phi_i(\cdot)$ can be regarded as the "true score" of object O_v on item I_i, as it is the average of X_{vji} over the probability distribution corresponding to the contribution δ_{vj} of the randomly drawn subject S_{vj}. Since $f_i(\cdot)$ is nondecreasing, the function $\phi_i(\cdot)$ is nondecreasing, too.

Definition (17.2) is similar to Lewis's (2001) definition of the expected response function. In (17.2), the random subject contribution is integrated out; in the expected response function, the uncertainty about the item parameter is integrated out. If the item parameter is a location parameter (which is the most frequent case), then (17.2) and the expected response function are identical from a mathematical point of view.

The function $\phi_i(\cdot)$ will be flatter than the original item characteristic curve $f_i(\cdot)$, because of the process of averaging with respect to the distribution of δ_{vj}. For small values of $\text{Var}(\delta_{vj})$ this can be seen as follows. Assume that $f_i(\cdot)$ is twice continuously differentiable. Then a Taylor expansion shows that, for small δ,

$$f_i(\theta_v + \delta) \approx f_i(\theta_v) + \delta f_i'(\theta_v) + \frac{1}{2}\delta^2 f_i''(\theta_v) \ .$$

Since $E(\delta_{vj}) = 0$, definition (17.2) implies that, for small values of $\text{Var}(\delta_{vj})$,

$$\phi_i(\theta_v) \approx f_i(\theta_v) + \frac{1}{2}\text{Var}(\delta_{vj}) f_i''(\theta_v) \ .$$

This shows that if the function $f_i(\theta)$ is convex for small θ and concave for large θ (which is the usual situation), then for small variances of δ_{vj}, the function $\phi_i(\theta)$ will be larger than $f_i(\theta)$ for small θ, and smaller than $f_i(\theta)$ for large θ; and the difference between the two functions will be largest where the second derivative of $f_i(\theta)$ is largest.

In Section 3, definitions for scalability coefficients are proposed, in order to express the degree to which a set of items forms a cumulative scale, and to assess the relative influence of δ_{vj} as compared to θ_v. Section 4 is devoted to the interpretation of these coefficients, and Section 5 to their estimation. Sections 6 and 7 present a definition of scale values for the objects and a measure for their reliability. A simulated and a real example are presented in Sections 8 and 9.

3 Scalability Coefficients

3.1 *Scalability Coefficients in a Single-Level Design*

In the usual single-level form of Mokken scaling, scalability coefficients are used to express whether the items indeed form a cumulative scale. For these scalability coefficients it is assumed that objects have been drawn at random from some population, or that—more loosely—the θ_v values of the objects can be considered to be representative for some population. This assumption is needed because the scalability coefficients are defined relative to the *marginal* joint probability distribution of the item scores, that is, the distribution of $(X_{v1}, X_{v2}, \ldots, X_{vk})$ for a randomly drawn object; the coefficients depend jointly on the items and the population of objects.

To define the scalability coefficients, denote the marginal probability of a positive response to item I_i by $P_i = P(X_{vi} = 1)$. Loevinger's H-coefficients of scalability are defined (see, e.g., Mokken, 1971; Sijtsma, 1998; Sijtsma & Van der Ark, 2001) for each item pair $(I_i, I_{i'})$ as the ratio of the covariance between item scores X_{vi} and $X_{vi'}$ to their maximum possible covariance, given the marginal probabilities P_i and $P_{i'}$. This covariance is equal to $P(X_{vi} = X_{vi'} = 1) - P_i P_{i'}$. The maximum possible value of the term

$P(X_{vi} = X_{vi'} = 1)$, given the marginal probabilities, is $\min\{P_i, P_{i'}\}$. This implies that the scalability coefficient for an item pair is given by

$$H_{ii'} = \frac{P(X_{vi} = X_{vi'} = 1) \; - \; P_i \, P_{i'}}{\min\{P_i, P_{i'}\} - P_i \, P_{i'}} \; .$$

In the case of a perfect Guttman scale (i.e., when the responses are deterministic so that the ICCs assume only the values 0 and 1), $H_{ii'}$ equals 1; in the case of independent (unrelated) items this coefficient equals 0. When both ICCs f_i and $f_{i'}$ are nondecreasing, $H_{ii'}$ must be nonnegative.

A similar coefficient can be defined for each item as the sum of the covariances between this item and all other items, divided by the sum of the maximum possible covariances, given the marginal probabilities. This is expressed by

$$H_i = \frac{\sum\limits_{i \neq i'} \left[P(X_{vi} = X_{vi'} = 1) \; - \; P_i \, P_{i'} \right]}{\sum\limits_{i \neq i'} \left[\min\{P_i, P_{i'}\} - P_i \, P_{i'} \right]} \; .$$

Similarly, the H-coefficient for the whole scale is based on the similar ratio for the sum of all covariances:

$$H = \frac{\sum\limits_{i < i'} \left[P(X_{vi} = X_{vi'} = 1) \; - \; P_i \, P_{i'} \right]}{\sum\limits_{i < i'} \left[\min\{P_i, P_{i'}\} - P_i \, P_{i'} \right]} \; .$$

The scalability coefficients are completely determined by the ICCs $f_i(\cdot)$ together with the probability distribution in the object population of the latent trait values θ_v. The H-coefficients are higher when ICCs are steeper and when the object distribution has greater dispersion. High values are indications of good scales. As a rule of thumb, H-values for items are required to be at least 0.30, and scales are considered to be strong when these values are at least 0.50.

3.2 Scalability Coefficients in a Two-Level Design

These notions can be adapted to the two-level design studied in this chapter. We wish to know if the association between the items is strong enough, and if the set of responses X_{vji} $(i = 1, \ldots, k; \, j = 1, \ldots, n_v)$ is adequate for measuring a given object O_v. It is to be expected that responses given by the same subject are more consistent (i.e., more strongly associated) than responses given by different subjects. Accordingly, we distinguish *within-subject scalability coefficients* from *between-subject scalability coefficients*, where the former are expected to be higher than the latter. The marginal "popularity" $P(X_{vji} = 1)$ of item I_i for a randomly chosen object O_v and subject S_{vj} is denoted again by P_i.

The within-subject scalability coefficient for item pairs is defined as

$$H_{ii'}^{W} = \frac{P(X_{vji} = X_{vji'} = 1) \; - \; P_i \, P_{i'}}{\min\{P_i, P_{i'}\} - P_i \, P_{i'}} \; , \tag{17.3}$$

that is, as the usual Loevinger coefficient applied to the responses of a single subject S_{vj}. The between-subject scalability coefficient is defined as

$$H_{ii'}^{B} = \frac{P(X_{vji} = X_{vj'i'} = 1) \; - \; P_i \, P_{i'}}{\min\{P_i, P_{i'}\} - P_i \, P_{i'}} \; , \qquad j \neq j' \; , \tag{17.4}$$

which is the Loevinger coefficient applied to the response given by a random subject S_{vj} to item I_i and by a different subject $S_{vj'}$ to item $I_{i'}$.

Analogous scalability coefficients for the items are the within-subject scalability coefficient

$$H_{i}^{W} = \frac{\sum_{i \neq i'} \left[P(X_{vji} = X_{vji'} = 1) \; - \; P_i \, P_{i'} \right]}{\sum_{i \neq i'} \left[\min\{P_i, P_{i'}\} - P_i \, P_{i'} \right]} \tag{17.5}$$

and the between-subject scalability coefficient

$$H_{i}^{B} = \frac{\sum_{i \neq i'} \left[P(X_{vji} = X_{vj'i'} = 1) \; - \; P_i \, P_{i'} \right]}{\sum_{i \neq i'} \left[\min\{P_i, P_{i'}\} - P_i \, P_{i'} \right]} \; , \qquad j \neq j' \; . \tag{17.6}$$

For the entire scale, the within-subject scalability coefficient is defined by

$$H^{W} = \frac{\sum_{i < i'} \left[P(X_{vji} = X_{vji'} = 1) \; - \; P_i \, P_{i'} \right]}{\sum_{i < i'} \left[\min\{P_i, P_{i'}\} - P_i \, P_{i'} \right]} \tag{17.7}$$

and the between-subject scalability coefficient by

$$H^{B} = \frac{\sum_{i < i'} \left[P(X_{vji} = X_{vj'i'} = 1) \; - \; P_i \, P_{i'} \right]}{\sum_{i \neq i'} \left[\min\{P_i, P_{i'}\} - P_i \, P_{i'} \right]} \; , \qquad j \neq j' \; . \tag{17.8}$$

The *within-subject* scalability coefficients refer to the situation where the two levels are disaggregated by considering every object–subject combination as a single case, unrelated to the other cases. They are based on the correlations between item responses to I_i and $I_{i'}$ by the same subject. The probabilities in the definitions of the within-subject scalability coefficients are determined by the ICCs $f_i(\cdot)$ and the distribution of the latent parameter values $\theta_v + \delta_{vj}$, for example,

$$P(X_{vji} = X_{vji'} = 1) = E_{\theta} E_{\delta} \left[f_i(\theta_v + \delta_{vj}) \, f_{i'}(\theta_v + \delta_{vj}) \right] . \tag{17.9}$$

The *between-subject* scalability coefficients, on the other hand, are based on the correlations between item responses by different subjects. This implies that the probabilities in the definitions of these coefficients are determined by the ICCs $\phi_i(\cdot)$ given in (17.2) and the distribution of the objects' latent trait values θ_v, as is expressed in

$$P(X_{vji} = X_{vj'i'} = 1) = E_\theta \left[\phi_i(\theta_v) \, \phi_{i'}(\theta_v) \right] . \tag{17.10}$$

In the Appendix it is proven that the model assumptions imply that within-subject scalability coefficients are greater than the corresponding between-subject scalability coefficients, unless the deviations δ_{vj} are constant, in which case they are equal. The proof rests on the properties that the ICCs $\phi_i(\cdot)$ are flatter than $f_i(\cdot)$, and the distribution of $\theta_v + \delta_{vj}$ is more dispersed than that of θ_v.

Thus, the assumption of nondecreasing ICCs implies that $0 \leq H^B \leq H^W$. If the variances of θ_v and δ_{vj} are positive and the ICCs are strictly increasing, then $0 < H^B < H^W$. In the extreme situation that $\mathrm{Var}(\delta_{vj}) = 0$, it holds that $0 < H^B = H^W$. In the opposite extreme situation that $\mathrm{Var}(\theta_v) = 0$, we have $0 = H^B < H^W$, and scaling of objects makes no sense.

4 Interpretation of Scalability Coefficients

To assess the quality of the scale as a measurement instrument for the objects, the homogeneity of the item set and the reliability of measurement must be considered. This section is about the assessment of homogeneity; the reliability is discussed in Section 7.

In the two-level situation, where the measurement operates via the subjects' responses, two aspects of homogeneity must be distinguished. The first aspect is the homogeneity of the items for measuring the object–subject combinations with associated trait values $\theta_v + \delta_{vj}$. The second is the importance of the objects as compared to the subjects in the variation of the latent trait. These two questions are discussed in this section.

For the investigation of the quality of the scale as a unidimensional cumulative scale for object–subject combinations, the within-subject scalability coefficients are the most relevant. Their interpretation is similar to the interpretation of scalability coefficients in the single-level Mokken scaling model. For the two-level design, however, there is more information about each object because there are multiple subjects per object. The limited simulation-based experience presented below suggests that, when the number of subjects per object is reasonably high, meaningful scales can be obtained when within-subject scalability coefficients are at least 0.2. Further experience is necessary to corroborate or criticize this proposal.

The degree to which the responses to the items are determined by the objects' values on the latent trait θ_v, rather than by the subjects' deviations

δ_{vj}, could be expressed in a parametric model by the intraclass correlation coefficient for the latent trait $\text{Var}(\theta_v)/[\text{Var}(\theta_v) + \text{Var}(\delta_{vj})]$. Our nonparametric model, however, is invariant under increasing transformations of the latent variable θ, so that these variances are not identified and the intraclass correlation cannot be used. Instead, we can compare between-subject scalability coefficients to within-subject scalability coefficients.

At the end of the preceding section it was mentioned that the model assumptions imply that between-subject scalability coefficients are always smaller than or equal to the corresponding within-subject scalability coefficients, and they are equal if $\text{Var}(\delta_{vj}) = 0$, that is, if the subjects are of no consequence for the responses. This implies that the ratio H^B/H^W can be used as an indication of variability between objects to total variability of the item responses. When H^B is almost as large as H^W, then the responses for the object are hardly affected by the particular subject (the influence of the random deviation δ_{vj} is small); when, on the contrary, H^B is much smaller than H^W, then the responses for the object are strongly affected by the particular subject (the influence of δ_{vj} is large). Similar interpretations can be attached to the ratios $H^B_{ii'}/H^W_{ii'}$ and H^B_i/H^W_i. Relatively small values of the between-subject coefficients imply that one needs a large number of subjects (or a large number of items) for a reliable estimation of object scores. Very small values of the between-subject coefficients suggest that the object parameter θ_v is not a very relevant parameter in the sense that its influence on the responses to the items is small compared to the influence of the subjects, expressed by the deviations δ_{vj}.

5 Estimation of Scalability Coefficients

The various scalability coefficients can be estimated by taking the defining formulae and substituting relative frequencies for the probabilities. Recall that the number of subjects presented to object O_v is n_v; define the total number of subjects presented by $n_+ = \sum_v n_v$ and the total number of objects by N.

Sometimes it may be suspected that the numbers n_v could be associated stochastically with the latent trait values θ_v and δ_{vj}. For example, objects with higher latent trait values could tend to have larger (or smaller) values of n_v. The estimators proposed in this section are not affected by such a stochastic dependence.

First consider the probability $P_i = P(X_{vji} = 1)$, which is the probability of a positive response to item I_i given by a randomly chosen subject with respect to a randomly chosen object. This probability can be expressed by

$$P(X_{vji} = 1) \;=\; E_\theta E_\delta \left[P(X_{vji} = 1 \mid \theta_v,\, \delta_{vj}) \right] = E_\theta \left[\phi_i(\theta_v) \right] .$$

The "true item score" $\phi_i(\theta_v)$ for object O_v is estimated unbiasedly by the average score on item I_i for this object,

$$\overline{X}_{v.i} = \frac{1}{n_v} \sum_{j=1}^{n_v} X_{vji} \ . \tag{17.11}$$

Hence an unbiased estimator for P_i is given by the average of $\overline{X}_{v.i}$ over all objects,

$$\hat{P}_i = \frac{1}{N} \sum_{v=1}^{N} \overline{X}_{v.i} \ . \tag{17.12}$$

The unbiasedness of (17.11) for estimating $\phi_i(\theta_v)$ implies that (17.12) is an unbiased estimator for P_i even if in the population of objects there is a stochastic association between n_v and θ_v.

Estimators for the probabilities of joint positive responses to two items are defined in an analogous way by

$$\hat{P}^W_{ii'}(1,1) = \hat{P}(X_{vji} = X_{vji'} = 1) = \frac{1}{N} \sum_v \frac{1}{n_v} \sum_j X_{vji} X_{vji'} \ , \tag{17.13}$$

$$\begin{aligned} \hat{P}^B_{ii'}(1,0) &= \hat{P}(X_{vji} = X_{vj'i'} = 1) \ , \qquad j \neq j' \ , \\ &= \frac{1}{N} \sum_v \frac{1}{n_v(n_v - 1)} \sum_{j \neq j'} X_{vji} X_{vj'i'} \ . \end{aligned} \tag{17.14}$$

Substitution of these estimators leads to the following estimators of the scalability coefficients,

$$\hat{H}^W_{ii'} = \frac{\hat{P}^W_{ii'}(1,1) - \hat{P}_i \hat{P}_{i'}}{\min\{\hat{P}_i, \hat{P}_{i'}\} - \hat{P}_i \hat{P}_{i'}} \ , \qquad i \neq i' \ , \tag{17.15}$$

$$\hat{H}^W_i = \frac{\sum_{i \neq i'} [\hat{P}^W_{ii'}(1,1) - \hat{P}_i \hat{P}_{i'}]}{\sum_{i \neq i'} [\min\{\hat{P}_i, \hat{P}_{i'}\} - \hat{P}_i \hat{P}_{i'}]} \ , \qquad i \neq i' \ , \tag{17.16}$$

$$\hat{H}^W = \frac{\sum_{i < i'} [\, \hat{P}^W_{ii'}(1,1) - \hat{P}_i \hat{P}_{i'} \,]}{\sum_{i < i'} [\, \min\{\hat{P}_i, \hat{P}_{i'}\} - \hat{P}_i \hat{P}_{i'} \,]} \ , \tag{17.17}$$

and similarly for $\hat{H}^B_{ii'}$, $\hat{H}^B_i$, and $\hat{H}^B$.

6 Object Scores

In this nonparametric framework the latent parameter θ_v itself is not identifiable, because no assumptions are made with respect to the shape of the

ICCs. Therefore, θ_v itself cannot be estimated. Instead, the object scores presented in this section estimate a monotone function of θ_v.

An intuitively acceptable object score is defined by the average,

$$\overline{X}_{v..} = \frac{1}{n_v k} \sum_{j=1}^{n_v} \sum_{i=1}^{k} X_{vji} \; . \tag{17.18}$$

The expected value of $\overline{X}_{v..}$,

$$\mu(\theta_v) = \frac{1}{k} \sum_{i=1}^{k} \phi_i(\theta_v) \; , \tag{17.19}$$

is a monotone nondecreasing function of θ_v. Thus, the object score $\overline{X}_{v..}$ is an unbiased estimator for a monotone function of the latent trait value θ_v. Other possibilities for defining object scores, as well as further properties of the stochastic relation between object scores and latent trait values, are topics for further research.

7 Reliability

For the single-level Mokken scaling method there are several methods for estimating reliability, based on the matrix of joint positive responses to pairs of items, the so-called P matrix. A review and comparison of these methods is given by Meijer, Sijtsma, and Molenaar (1995). A nice feature of the two-level design is the availability of independent within-object replications. These replications imply that a test–retest correlation coefficient can be estimated from the intraclass correlation coefficient, where classes are defined by the objects. In this section, reliability is defined by the test–retest correlation coefficient for the objects, and an estimator for this reliability coefficient is given. Also an estimator for the standard error of estimation of $\overline{X}_{v..}$ is given.

The basic parameters required for the definition of the reliability coefficient are based on the average scores for object–subject combinations,

$$\overline{X}_{vj.} = \frac{1}{k} \sum_{i} X_{vji} \; .$$

Define σ_0^2 as the variance of average scores for randomly drawn object–subject combinations,

$$\sigma_0^2 = \mathrm{Var}(\overline{X}_{vj.}) \; ,$$

and ρ as their within-object between-subject correlation,

$$\rho = \mathrm{Cor}(\overline{X}_{vj.}, \overline{X}_{vj'.}) = \frac{1}{\sigma_0^2} \mathrm{Cov}(\overline{X}_{vj.}, \overline{X}_{vj'.}) \; , \qquad j \neq j' \; .$$

Then ρ can be regarded as an intraclass correlation coefficient and is called the *intraobject correlation coefficient.*

The intraobject correlation ρ is the test–retest reliability for the parallel tests corresponding to the different single subjects. Another test–retest reliability coefficient is defined as the correlation between two independent scores $\overline{X}_{v..}$ for the same object O_v, obtained from two independent sets each consisting of n_v independent subjects. The latter coefficient is a function of ρ (see, e.g., Snijders & Bosker, 1999, Sect. 3.5),

$$\rho(\overline{X}_{v..}, \overline{X}'_{v..}) = \frac{n_v \rho}{(n_v - 1)\rho + 1} . \tag{17.20}$$

This coefficient is important for assessing the quality of the scale.

To define estimators for σ_0^2 and ρ, denote the average object score by

$$\overline{X}_{...} = \frac{1}{N} \sum_{v=1}^{N} \overline{X}_{v..} .$$

(Note that $\overline{X}_{...}$ is not the unweighted mean of all X_{vji} values, but the mean of the object scores $\overline{X}_{v..}$; these two means are the same if all the n_v are equal.) Furthermore, denote between-object and within-object variances, respectively, by

$$S_B^2 = \frac{1}{N-1} \sum_{v=1}^{N} (\overline{X}_{v..} - \overline{X}_{...})^2 ,$$

$$S_W^2 = \frac{1}{N} \sum_{v=1}^{N} \frac{1}{n_v - 1} \sum_{j=1}^{n_v} (\overline{X}_{vj.} - \overline{X}_{v..})^2 .$$

Define the harmonic mean $\tilde{n}$ of the n_v by

$$\frac{1}{\tilde{n}} = \frac{1}{N} \sum_{v=1}^{N} \frac{1}{n_v} .$$

With these definitions it can be proved that

$$E(S_W^2) = \sigma_0^2(1 - \rho) \quad \text{and} \quad E(S_B^2) = \frac{\sigma_0^2}{\tilde{n}} \left[1 + (\tilde{n} - 1)\rho\right] .$$

This yields the estimators

$$\hat{\sigma}_0^2 = \frac{\tilde{n} - 1}{\tilde{n}} S_W^2 + S_B^2 , \tag{17.21}$$

$$\hat{\rho} = \frac{1}{\hat{\sigma}_0^2} \left(S_B^2 - \frac{1}{\tilde{n}} S_W^2\right) . \tag{17.22}$$

The reliability coefficient (17.20) now can be estimated by

$$\hat{\rho}(\overline{X}_{v..}, \overline{X}'_{v..}) = \frac{n_v \hat{\rho}}{(n_v - 1)\hat{\rho} + 1} . \tag{17.23}$$

8 Examples for Simulated Data

The presentation of some results for simulated data may give a better understanding of the various parameters introduced and the numerical values these may assume. Three data sets were generated for $N = 500$ objects, each confronted with $n_v = 10$ subjects for which $k = 6$ items were scored. The item ICCs of the items satisfied the Rasch model (see, e.g., Fischer & Molenaar, 1995):

$$f_i(\theta) = \frac{\exp(\theta - \beta_i)}{1 + \exp(\theta - \beta_i)},$$

where β_i is the difficulty parameter of item I_i. The difficulty parameters were chosen as $-2.0, -1.2, -0.4, 0.4, 1.2, 2.0$. The distributions of θ_v and δ_{vj} were normal with mean 0.

The three simulated data sets differed according to the ratio of within-subject to between-subject variances: the variances were $\sigma_\theta^2 = 0.5$, $\sigma_\delta^2 = 0.5$ (simulation 1), $\sigma_\theta^2 = 0.8$, $\sigma_\delta^2 = 0.2$ (simulation 2), and $\sigma_\theta^2 = 0.2$, $\sigma_\delta^2 = 0.8$ (simulation 3). Thus, in all three cases the variance of the combined latent trait values $\theta_v + \delta_{vj}$ is equal to 1. This implies that the population values of the item popularities (marginal probabilities of a positive response) and the within-subject H-coefficients are the same in the three simulations.

For simulation 1, where the variance of object values for the latent trait is equal to the variance of the subject-related deviations, the popularities and scalability coefficients were estimated as presented in Tables 17.1 and 17.2.

We see that H-coefficients are greater when the item popularities are farther apart, as is well known for Mokken scaling. Furthermore, the between-subject $\hat{H}$-coefficients are indeed considerably smaller than the within-subject coefficients. The within-subject coefficients are slightly lower than the value of 0.3 that is considered a reasonable minimum for single-level Mokken scaling. All items have within-subject scalability coefficients higher than 0.22. The ratio $\hat{H}^B/\hat{H}^W$ for the entire scale is 0.443.

Values for the variance of single subject scores, the intraobject correlation coefficient, and the reliability coefficients for various values of n_v are given in Table 17.3. We see that at least 10 to 20 subjects per object are needed in order to get a sufficiently reliable scale.

For the second data set, where the influence of the subjects is much smaller ($\sigma_\theta^2 = 0.8$, $\sigma_\delta^2 = 0.2$), we obtained $\hat{H}^W = 0.260$, $\hat{H}^B = 0.211$,

TABLE 17.1. Estimated item probabilities for simulated data set 1.

i	1	2	3	4	5	6
$\hat{P}_i$	0.158	0.266	0.409	0.577	0.734	0.842

TABLE 17.2. Estimated within-subject and between-subject scalability coefficients for simulated data set 1.

$\hat{H}^W_{ii'}$ i'	1	2	3	4	5	6
i						
1	—	0.138	0.210	0.349	0.411	0.432
2		—	0.155	0.283	0.309	0.452
3			—	0.212	0.290	0.401
4				—	0.221	0.278
5					—	0.182
6						—
i	1	2	3	4	5	6
$\hat{H}^W_i$	0.253	0.227	0.231	0.253	0.256	0.301
Whole Scale $\hat{H}^W$		0.250				

$\hat{H}^B_{ii'}$ i'	1	2	3	4	5	6
i						
1	—	0.067	0.098	0.195	0.164	0.182
2		—	0.078	0.137	0.147	0.188
3			—	0.113	0.104	0.135
4				—	0.086	0.111
5					—	0.075
6						—
i	1	2	3	4	5	6
$\hat{H}^B_i$	0.121	0.108	0.102	0.120	0.103	0.118
Whole Scale $\hat{H}^B$		0.111				

and therefore a high value $\hat{H}^B/\hat{H}^W = 0.809$. Note that the population values for H^W and σ_0^2 are the same in all three simulations. Table 17.3 gives estimated values for σ_0^2, ρ, and the reliability coefficients $\hat{\rho}(\overline{X}_{v..}, \overline{X}'_{v..})$. Much smaller numbers of subjects suffice here for obtaining a reliable scale than for simulation 1.

For the third simulation, where the influence of subjects is much greater, we obtain the reverse results. The scalability coefficients for the entire scale are $\hat{H}^W = 0.258$ and $\hat{H}^B = 0.058$. Table 17.3 gives values for the other parameters and the reliability coefficients $\rho(\overline{X}_{v..}, \overline{X}'_{v..})$. This situation, with $\hat{H}^B/\hat{H}^W = 0.226$, seems to be close to the lower limit of the ratio of H^B/H^W where scaling objects still is feasible. This is also reflected by the low reliabilities.

TABLE 17.3. Estimated parameters and reliability coefficients.

	Data Set 1	Data Set 2	Data Set 3
n_v		$\hat{\rho}(\overline{X}_{v..}, \overline{X}'_{v..})$	
4	0.549	0.708	0.321
8	0.709	0.829	0.487
10	0.753	0.858	0.542
20	0.859	0.924	0.703
30	0.901	0.945	0.781
$\hat{\sigma}_0^2$	0.052	0.052	0.053
$\hat{\rho}$	0.233	0.377	0.106

Some tentative conclusions may be drawn from these simulations and used in provisional rules of thumb. The fact that multiple parallel measurements are available allows the use of scales with lower H-values than in single-level nonparametric scaling without sacrificing reliability. Within-subject scalability coefficients for items and for the whole scale should be greater than 0.2 for a good scale; it may be considered acceptable that for some directly consecutive item pairs, the $\hat{H}^W_{ii'}$-coefficient is between 0.1 and 0.2. The consistency between subjects is satisfactory when between-subject homogeneity coefficients are at least 0.1 (with possibly some exceptions for between-item $\hat{H}^B_{ii'}$-coefficients). For the ratio $\hat{H}^B/\hat{H}^W$, values over 0.3 are reasonable and values over 0.6 are excellent. As indicators for the quality of the scale one should use not only the homogeneity coefficients, but also the reliability coefficient for practical n_v values.

9 Example: Assessment of Teachers by Pupils

In a study by Bosker and Hendriks (1997), pupils in primary schools were asked to respond to a questionnaire about their teacher and classroom. As an example, some questions about classroom climate and order are used. The objects to be scaled are the teachers/classrooms, the responding subjects are the pupils, and the latent trait is order in the classroom.

Six items from the questionnaire were selected on the basis of face value considerations (mainly their unambiguous relation to order in the class). The items are statements with three answer categories (true, somewhat true, not true). They were recoded to values 1, 2, 3, where 1 denotes the most orderly and 3 the most chaotic situation in the classroom. A cross-validatory approach was chosen in which a scale selection phase was distinguished from a scale evaluation phase. In the scale selection phase, data for group-6 pupils (age 9–10 years) were used for the selection of a good subset of items and dichotomization thresholds. In the scale evaluation phase the data for group-7 pupils (age 10–11 years) were used for the evaluation of the resulting scale along the lines of the preceding sections.

In the scale selection phase, items and thresholds were chosen (in a trial and error procedure) so as to yield relatively high between-subject H-coefficients for each item. This resulted in a scale of four items, all dichotomized by contrasting the values 1 and 2 (most and intermediate orderly, coded as $X = 0$) with the value 3 (least orderly, coded as $X = 1$). The number of items is low, but it is shown below that the reliability is satisfactory when the number of subjects is reasonable. The four items, ordered by increasing frequency of the least orderly outcome, are the following.

1. When the teacher tells us something, we listen well (inversely coded).
2. It usually is quiet in the classroom (inversely coded).
3. There often is much noise in the classroom.
4. The teacher often tells us to be quiet.

In the scale evaluation phase of the data analysis, these items were subjected to a scale analysis for the group-7 pupils (whose data were not used in the selection of items and thresholds). There were data for 1530 pupils (subjects) and 77 classrooms/teachers (objects). Results are presented in Tables 17.4 to 17.6. The intraobject correlation $\hat{\rho}$ is 0.26 and the variance of single subject scores $\hat{\sigma}_0^2$ is 0.090.

It can be concluded that this four-item scale is very satisfactory. All within-subject scalability coefficients are over 0.50, so, when regarded as a scale for measuring pupil–classroom combinations, this scale is strong by the classical standards of Mokken scaling. Furthermore, all items have between-subject scalability coefficients over 0.20. The between-to-within-subject ratio of the scalability coefficient for the entire scale $\hat{H}^B/\hat{H}^W$ is 0.37 and these ratios are over 0.35 for all items.

This implies that the consistency of responses within subjects, as well as between subjects within objects, is large enough to use these four questions for assessing classroom climate. The reliability coefficients show that about 10 pupils are sufficient to give a reasonably reliable measurement instrument.

TABLE 17.4. Estimated item probabilities for classroom climate scale.

i	1	2	3	4
$\hat{P}_i$	0.06	0.28	0.35	0.52

TABLE 17.5. Estimated within-subject and between-subject scalability coefficients for classroom climate scale.

$\hat{H}^W_{ii'}$ i' / i	1	2	3	4	$\hat{H}^B_{ii'}$ / i	1	2	3	4
1	—	0.70	0.74	0.68	1	—	0.32	0.32	0.40
2		—	0.56	0.56	2		—	0.18	0.20
3			—	0.56	3			—	0.20
4				—	4				—
i	1	2	3	4	i	1	2	3	4
$\hat{H}^W_i$	0.71	0.58	0.58	0.57	$\hat{H}^B_i$	0.34	0.21	0.20	0.22
Whole Scale $\hat{H}^W$		0.59			$\hat{H}^B$	0.22			

TABLE 17.6. Estimated reliability coefficients for classroom climate scale.

n_v	4	8	10	20	30
$\hat{\rho}(\overline{X}_{v..}, \overline{X}'_{v..})$	0.58	0.74	0.78	0.87	0.91

10 Discussion

A nonparametric method has been presented to scale objects who are known to the researcher only through responses that refer to subjects connected in some way to the objects. Often the subjects are the individuals providing the responses. A test composed of dichotomous items is administered to each object–subject combination. The subjects, nested within the objects, are supposed to be a random sample from a population. This yields a two-level data structure.

The consistency of answer patterns within subjects can be measured by *within-subject* Loevinger's H-coefficients applied to the disaggregated object–subject combinations. To scale the objects on the basis of such data requires, however, that there is enough consistency in the answer patterns also between the subjects within the same objects. This is measured by *between-subject, within-object* Loevinger's H-coefficients.

To investigate how well the objects are scaled by this set of items, these measures have been presented: (1) the ratio H^B/H^W of between-subject to within-subject scalability coefficients, which gives an indication of the extent to which scale values are determined by objects rather than by subjects; and (2) the reliability coefficient of resulting scale. The reliability coefficient depends, of course, on the number of subjects per object.

Scales for objects defined in this way can be used, for example, to investigate relations between the corresponding latent trait and other variables referring to the objects, such as the relation between classroom climate and teacher behavior, or the relation between neighborhood climate and policy instruments applied to the neighborhoods.

A possible alternative to the nonparametric approach presented here is a parametric three-level model for dichotomous data, where the levels are items, subjects, and objects. A standard approach, elaborated for the three-level case by Gibbons and Hedeker (1997) and utilized in the present context by Raudenbush and Sampson (1999), is to assume normal distributions for the latent trait components θ_v and δ_{vj}, and a logistic link function for the dichotomous responses. Although this may be called a standard approach from the point of view of model building, the numerical calculations necessary to estimate the parameters are very complicated. Reviews of parametric models for multilevel dichotomous data (but with a focus on two-level models) are given by Goldstein (1995, Chap. 7) and Snijders and Bosker (1999, Chap. 14). Such an approach yields the rewards connected to the richer parametric structure: the possibility of using covariates for subjects and/or objects, and the availability of standard statistical methods such as maximum likelihood to obtain estimates and standard errors. However, for these models the estimation methods are numerically quite complex and computationally demanding. Moreover, the parametric assumptions will often be questionable: the logistic link function means that one assumes the Rasch model for the object–subject combinations, and the Rasch model is known to be quite a strong assumption for a cumulative unidimensional scale. Advantages of the nonparametric method presented here are the less demanding assumptions and the easy computations.

This chapter only lays out the groundwork for two-level nonparametric scaling. In future research this methodology will have to be elaborated upon, inspired by results obtained in the past 30 years of research on single-level nonparametric scaling as well as by multilevel methodology. It would be worthwhile to develop an inductive search method for finding scalable subsets of items in a larger item set, analogous to the search method of Mokken (1971) implemented in the MSP program (Molenaar & Sijtsma, 2000). The usefulness of the method could be further enhanced by extensions to polytomous items (cf. Molenaar, 1997) and by the development of goodness-of-fit methods. It may be worthwhile to investigate stochastic monotonicity properties (cf. Hemker, Sijtsma, Molenaar, & Junker, 1996, and Sijtsma & Van der Ark, 2001). In addition, other methods for scaling objects could be developed along the lines of posterior means as used in multilevel modeling (cf. Snijders & Bosker, 1999, Sect. 4.7).

All statistics that were proposed in this chapter can be calculated by the PC program TWOMOK which is available from the Web site at URL `http://stat.gamma.rug.nl/snijders/multilevel.htm`.

11 Appendix: Proof that Between-Subject Scalability Coefficients Are Not Larger Than Within-Subject Coefficients

It is proved that

$$P(X_{vji} = X_{vji'} = 1) - P(X_{vji} = X_{vj'i'} = 1) \geq 0 , \qquad j \neq j' . \tag{17.24}$$

This inequality is equivalent to $H^B_{ii'} \leq H^W_{ii'}$, and it implies $H^B_i \leq H^W_i$ and $H^B \leq H^W$. From the expressions analogous to (17.9) and (17.10) we can conclude that the left-hand side of (17.24) equals

$$\begin{aligned} &E_\theta E_\delta \left[f_i(\theta_v + \delta_{vj}) \, f_{i'}(\theta_v + \delta_{vj}) \right] - E_\theta \left[\phi_i(\theta_v) \, \phi_{i'}(\theta_v) \right] \\ &\quad = E_\theta \left\{ E_\delta \left[f_i(\theta_v + \delta_{vj}) \, f_{i'}(\theta_v + \delta_{vj}) \mid \theta_v \right] - \phi_i(\theta_v) \, \phi_{i'}(\theta_v) \right\} . \end{aligned}$$

Because of the definition of ϕ_i, given in (17.2), this is equal to

$$E_\theta \left\{ \mathrm{Cov}_\delta [f_i(\theta_v + \delta_{vj}), \, f_{i'}(\theta_v + \delta_{vj}) \mid \theta_v] \right\} .$$

The conditional covariance, $\mathrm{Cov}_\delta [f_i(\theta_v + \delta_{vj}), \, f_{i'}(\theta_v + \delta_{vj}) \mid \theta_v]$, is a covariance of two nondecreasing functions of the random variable δ_{vj}. Such a covariance is always nonnegative, which proves (17.24). If the ICCs $f_i(\cdot)$ are strictly increasing and $\mathrm{Var}(\delta_{vj}) > 0$, then this covariance is strictly positive, which implies $H^W_{ii'} > H^B_{ii'}$.

In a similar way it can be proved that

$$P(X_{vji} = X_{vj'i'} = 1) - P_i \, P_{i'} = \mathrm{Cov}_\theta \left[\phi_i(\theta_v), \, \phi_{i'}(\theta_v) \right] \geq 0 , \tag{17.25}$$

which implies $H^B_{ii'} \geq 0$.

References

Bosker, R.J., & Hendriks, M.A. (1997). *Betrouwbaarheid, validiteit en bruikbaarheid van een instrumentarium voor zelfevaluatie door basisscholen* [Reliability, validity, and usability of instruments for school self-evaluation in primary education]. Enschede: Universiteit Twente, OCTO.

Bryk, A.S., & Raudenbush, S.W. (1992). *Hierarchical linear models.* Newbury Park, CA: Sage.

Fischer, G.H., & Molenaar, I.W. (1995). *Rasch models: Foundations, recent developments, and applications.* New York: Springer-Verlag.

Gibbons, R.D., & Hedeker, D. (1997). Random effects probit and logistic regression models for three-level data. *Biometrics, 53,* 1527–1537.

Goldstein, H. (1995). *Multilevel statistical models* (2nd ed.). London: Arnold.

Hemker, B.T., Sijtsma, K., Molenaar, I.W., & Junker, B.W. (1996). Polytomous IRT models and monotone likelihood ratio of the total score. *Psychometrika, 61*, 679–693.

Jansson, I., & Spreen, M. (1998). The use of local networks in a study of heroin users: Assessing average local networks. *Bulletin de Méthodologie Sociologique, 59,* 49–61.

Lewis, C. (2001). Expected response functions. In A. Boomsma, M.A.J. van Duijn, & T.A.B. Snijders (Eds.), *Essays on item response theory* (pp. 163–171). New York: Springer-Verlag.

Meijer, R.R., Sijtsma, K., & Molenaar, I.W. (1995). Reliability estimation for single dichotomous items based on Mokken's IRT model. *Applied Psychological Measurement, 19,* 323–335.

Mokken, R.J. (1971). *A theory and procedure of scale analysis: With applications in political research.* The Hague: Mouton.

Mokken, R.J. (1997). Nonparametric models for dichotomous responses. In W.J. van der Linden & R.K. Hambleton (Eds.), *Handbook of modern item response theory* (pp. 351–367). New York: Springer-Verlag.

Mokken, R.J., & Lewis, C. (1982). A nonparametric approach to the analysis of dichotomous item responses. *Applied Psychological Measurement, 6,* 417–430.

Molenaar, I.W. (1997). Nonparametric models for polytomous responses. In W.J. van der Linden & R.K. Hambleton (Eds.), *Handbook of modern item response theory* (pp. 369–380). New York: Springer-Verlag.

Molenaar, I.W., & Sijtsma, K. (2000). User's manual MSP5 for Windows: A program for Mokken scale analysis for polytomous items. Version 5.0 [Software manual]. Groningen: ProGAMMA.

Raudenbush, S.W., & Sampson, R.J. (1999). "Ecometrics": Toward a science of assessing ecological settings, with applications to the systematic social observation of neighborhoods. In M.E. Sobel & M.P. Becker (Eds.), *Sociological methodology 1999,* (pp. 1–41). Boston, MA: Blackwell.

Sijtsma, K. (1998). Methodology review: Nonparametric IRT approaches to the analysis of dichotomous item scores. *Applied Psychological Measurement, 22,* 3–31.

Sijtsma, K., & Van der Ark, L.A. (2001). Progress in NIRT analysis of polytomous item scores: Dilemmas and practical solutions. In A. Boomsma, M.A.J. van Duijn, & T.A.B. Snijders (Eds.), *Essays on item response theory* (pp. 297–318). New York: Springer-Verlag.

Snijders, T.A.B., & Bosker, R.J. (1999). *Multilevel analysis: An introduction to basic and advanced multilevel modeling.* London: Sage.

18

The Circles of Our Minds: A Nonparametric IRT Model for the Circumplex

Robert J. Mokken,[1] Wijbrandt H. van Schuur,[2] and Ard Jan Leeferink[3]

ABSTRACT A nonparametric item response theory (IRT) model for the circumplex is introduced, based on Mokken's (1971) nonparametric IRT model for cumulative scaling, and Van Schuur's (1984) nonparametric IRT model for unfolding. Some examples of circumplex representations in the social sciences are given. Model fit is based first on an extension of Loevinger's coefficient of homogeneity using quadruples as elementary units of analysis. Diagnostics for the probabilistic circumplex are suggested. Assignment of (ordinal) scale values is based on the notion of item steps, as developed by Molenaar (1983). The model presented here is based on dichotomous *pick any/k* data. Suggestions for extension to rank m/k data and to polytomous data are discussed.

1 Introduction

The vast majority of item response theory (IRT) models assume a unidimensional latent trait, in which a subject's response to an item indicates a *dominance* relationship between the subject and the item. A less familiar class of IRT items concerns unfolding models—IRT models in which a subject's response to an item indicates a proximity relationship between the subject and the item, including those of Van Schuur (1984), Van Blokland-Vogelesang (1990), Hoijtink (1991), Post and Snijders (1993), Andrich and Luo (1993), and Roberts and Laughlin (1996). In this chapter we extend

[1]Faculty of Social and Behavioral Sciences, University of Amsterdam. Private address: Franklinstraat 18, 1171 BL Badhoevedorp, The Netherlands; e-mail: *mokken@ccsom.uva.nl*

[2]Department of Sociology, University of Groningen, Grote Rozenstraat 31, 9712 TG Groningen, The Netherlands; e-mail: *h.van.schuur@ppsw.rug.nl*

[3]Department of Policy Research of the City of Groningen, Postbus 1125, 9701 BC Groningen, The Netherlands; e-mail: *j.g.y.leeferink@sozawe.groningen.nl*

these to a new nonparametric IRT model: the circumplex. We introduce it as a generalization of Mokken's nonparametric dominance model and Van Schuur's nonparametric proximity model.

The term *circumplex* was introduced by Guttman (1954), although the idea of a circular scale is older (e.g., Ross, 1938), and related to facet theory (e.g., Levy & Guttman, 1975; Shye, Elizur, & Hoffman, 1994). A circumplex is a structure of items distributed along the circumference of a circle.

An example is the abridged big-five circumplex (De Raad & Hofstee, 1993). This is a circumplex approach to a five-factor model that assumes that all personality traits can be described as pairwise combinations of five factors: extroversion, agreeableness, conscientiousness, emotional stability, and intellect or openness to experience. The 10 pairs of factors are represented as circumplexes, in which specific combinations of positive and negative values on each of the factors form the circumference of a circle. Both items and subjects are represented along the circumference, such that each subject only considers these personality traits as applicable to a specific person when they are close to that person's value on the circle.

Other examples of circumplex representations are Leary (1957), Benjamin (1974), and Kiesler (1983) on interpersonal characteristics, Wiggins (1979, 1980) and Gurtman (1992) on interpersonal problems, Russell (1980), Russell, Lewicka, and Niit (1989), and Larsen and Diener (1992) on emotional judgment, Marcus (1988) on emotional responses to presidential candidates; and Browne (1992) and Tracey and Rounds (1993) on vocational interest.

Empirical evidence for a circumplex structure in data usually comes from an interpretation of the correlation structure of the items, often after using factor analysis. The correlation matrix of a set of items that can be represented as a circumplex can be shown to follow a pattern where items that are close together on the circle correlate highly, the correlations decreasing as the distance between items along the circumference of the circle increases. Correlations are minimal for items that are opposite on the circle.

In this chapter we analyze a set of data as a circular unfolding (CU) scale by a procedure of general nonparametric unfolding (GU). This procedure was derived by Van Schuur (1984), using ideas for nonparametric monotone scale analysis proposed by Mokken (1971, 1997). We develop CU here for dichotomous items only, but with the expectation that the model can be extended to polytomous items, as was possible for the regular unfolding version (Van Schuur, 1993), implemented in MUDFOLD (Van Schuur & Post, 1998). Section 2 introduces the circumplex scale in a deterministic perspective, which enables us to establish criteria for homogeneity and error criteria for the items. Section 3 develops a bottom-up search procedure for finding a circumplex scale, and concludes with an example of scale construction. Section 4 describes forms of parameter estimation, suggesting scoring methods along a (unit) circle and a technique for finding scale values for subjects. Section 5 treats the problem of model fit on the basis of a probabilistic version of the CU model, which allows the application of diagnostics of Molenaar and Sijtsma (2000), and Post and Snijders (1993). Section 6 wraps up the chapter with a conclusion.

2 The Circumplex Scale

These basic assumptions are common to most IRT models.

1. *Unidimensionality.* A single latent continuum is assumed, on which subjects and items are represented each by a single parameter, θ_v and β_i, respectively. General unfolding (GU) is linear in the sense that preference ordering and item picking (affirmation) are related to a latent continuum with two extremes, one in the negative and one in the positive direction. Circular (or circumplicial) unfolding (CU) is related to a circular continuum (a circumplex) with no extremes.
2. *Local independence.* For all subjects with a given value of θ, responses (picking items) are stochastically independent.

Two additional assumptions are specific to the circumplex representation. Assumption 3 gives the deterministic perspective, and Assumption 4 a probabilistic one.

3. *Deterministic.* Subjects only "pick," that is, give a positively keyed response to items that are sufficiently close to their own position on the circumference of the circle. Each item can be described by a second parameter, which defines the interval around the subjects' positions within which they pick. This interval is called the "latitude of acceptance" of the item.
4. *Probabilistic.* The probability that a subject picks an item decreases with increasing distance between subject and item along the circumference of the circle.

2.1 Admissible Response Patterns

We denote our data set as follows. A number of n subjects pick some m items of a total of k items, $m \leq k$. The procedure and example, given in Section 3, is based on the *pick any/k* situation, where m can be any integer between 0 and k. This is the usual case for dominance data of the cumulative scaling type. In *pick any/k* data each of the 2^k response patterns is potentially observable for k items. However, for explanatory purposes we first use the *pick m/k* setup, with constant m.

In this section we investigate conditions on m and k that are necessary for the falsification of the deterministic CU model. For *pick m/k* the following elementary properties, comparing CU and GU can be verified, given the ordering of items, for example, anticlockwise along the circle; from left to right along the linear continuum.

Property 1 For *pick m/k* the total number of different response patterns picks equals $\binom{k}{m}$.

Property 2 The number of perfect patterns is

$$a)\ \text{for CU},\quad k,\quad \text{and}$$

$$b)\ \text{for GU},\quad k-(m-1)=k-m+1\ .$$

Note that for CU the number of perfect patterns does not depend on m: in a deterministic circumplex scale only those response patterns are admissible in which all positively keyed (pick) responses are given to items that are adjacent to each other, and all negatively keyed (nonpick) responses are given to the remaining items, which also are adjacent to each other.

Property 3 The number of error patterns is

$$a)\ \text{for CU},\quad \binom{k}{m}-k,\quad \text{and}$$

$$b)\ \text{for GU},\quad \binom{k}{m}-k+(m-1)\ .$$

Note that $m-1$ error patterns are specific for linear GU only. These *discriminating* patterns are perfect in CU but in error for GU, given the item ordering. From the perspective of GU, they are the patterns where *both* endpoints are picked and the other $m-2$ picked items are adjacent to either one of the endpoints. All other $k-m$ items form a set of zeros with no ones in between. These are the $m-1$ patterns specific for GU. The other $\binom{k}{m}-k$ error patterns are common to both CU and GU. Hence these specific error patterns, involving the extremes, specific for GU and absent in CU, suggest a testing zone for testing CU against GU.

Property 4 To ensure the possibility of falsification for both CU and GU analyses it is required that $k \geq 4$ and $2 \leq m \leq k-2$.

Falsification of a model is defined by the possible occurrence of inadmissible response patterns (error cells), given model and item ordering. From Property 3 we have that for $m = 0, 1$, or k, neither CU nor GU can be falsified. For $m = k-1$ CU is not falsifiable, but GU is if $k \geq 3$. Moreover, CU can only be falsified for $k \geq 4$.

Property 5 The smallest CU falsifying response patterns are those for $m = 2$ and $k = 4$ (quadruples), where the positive responses are not adjacent, that is, the response patterns 1010 and 0101.

TABLE 18.1. Perfect patterns for Guttman scale, Coombs scale, and circumplex.

Guttman Scale						Unfolding Coombs Scale									Circumplex						
A	B	C	D	E	F	A	B	C	D	E	F	G	H	I	A	B	C	D	E	F	G
0	0	0	0	0	0	1	1	1	0	0	0	0	0	0	1	1	1	0	0	0	0
1	0	0	0	0	0	0	1	1	1	0	0	0	0	0	0	1	1	1	0	0	0
1	1	0	0	0	0	0	0	1	1	1	0	0	0	0	0	0	1	1	1	0	0
1	1	1	0	0	0	0	0	0	1	1	1	0	0	0	0	0	0	1	1	1	0
1	1	1	1	0	0	0	0	0	0	1	1	1	0	0	0	0	0	0	1	1	1
1	1	1	1	1	0	0	0	0	0	0	1	1	1	0	1	0	0	0	0	1	1
1	1	1	1	1	1	0	0	0	0	0	0	1	1	1	1	1	0	0	0	0	1

Observe that each of the eight possible responses to a three-item set can be considered as a perfect pattern in a corresponding three-item CU, whereas, for instance, 101 is an error pattern in GU.

As a corollary to Property 4 we have that for procedures based on *pick any/k* the observations with patterns corresponding to $m = 0, 1, k, k-1$, can be disregarded or omitted from the analysis.

The relationships between admissible patterns in the Mokken scale, the unfolding scale, and the circumplex are illustrated in Table 18.1. All seven admissible response patterns of six items are given for the deterministic dominance scale, *pick any/6*, as introduced by Guttman, compared with the admissible response patterns for *pick 3/9* data in the deterministic unfolding scale as introduced by Coombs, and for *pick 3/7* data in the deterministic circumplex scale as given in this chapter.

2.2 Model Violation in Response Patterns with $k > 4$

Property 4 implies that an error pattern in CU can be observed only for sets of at least four items (quadruples). Given the order of all items, the number of errors in an ordered response pattern with more than four items is defined as the number of its quadruples that are CU error patterns, denoted q^o_{ijkl}. This gives a measure of person fit for the circumplex model. The total number of observed quadruple errors in a data set, denoted q^o, for n subjects and k items, is taken as the sum over all subjects of the errors in each response pattern.

As in Van Schuur's (1984) unfolding procedure, errors can be evaluated against the expected number of errors for the null case of marginal independence given the item pick proportions. The evaluation criterion consists of an adaptation of Loevinger's coefficient of homogeneity; that is,

$$H = 1 - \frac{q^o}{q^e}, \tag{18.1}$$

where q^e is the expected number of errors under the null model of statistical independence.

For any ordered quadruple of items $(ABCD)$, collected as *pick any/k* data, the expected number of error patterns is

$$q^e_{ijkl} = n\{p(A)\,[1-p(B)]\,p(C)\,[1-p(D)] + [1-p(A)]\,p(B)\,[1-p(C)]\,p(D)\}\ ,$$

where $p(i)$ is the proportion of positive responses given to item I_i and n is the number of subjects S_v. Thus,

$$H_{ijkl} = 1 - \frac{q^o_{ijkl}}{q^e_{ijkl}}\ . \qquad (18.2)$$

Alternatively, error quadruples can also be counted per item. For item I_i, $i = 1, \ldots, k$, the total number of observed errors q^o_i is taken as the sum over the subjects of the number of error quadruples containing item I_i. Obviously, due to quadruple counting, we have $q^o_i = \frac{1}{4}\sum_{i=1}^{n} q^o_i$. Therefore, an item coefficient of homogeneity H_i is defined as

$$H_i = 1 - \frac{q^o_i}{q^e_i}\ . \qquad (18.3)$$

3 A Search Procedure for a Circumplex Scale

So far we have assumed that the items have already been ordered according to the criteria of the circumplex model. In a confirmatory analysis, a user can then assess the homogeneity of a set of (ordered) items as one circumplex scale by calculating the homogeneity coefficients over ordered quadruples for the whole scale.

However, we usually do not know in which order the items form a possible circumplex scale. We do not even know whether all items can be represented along a circle. The procedure therefore is a bottom-up search procedure, where first a best smallest circumplex scale is either defined or found, which is then extended with more items, one by one, as long as the criteria of a circumplex scale are met.

3.1 Finding the Best Elementary Scale

An elementary CU (sub)scale is a quadruple of items that fits a circumplex. In order to find the best elementary scale in a set of k, its $\binom{k}{4}$ quadruples have to be investigated in their different permutations. Not all permutations need be considered, however. First, in a circumplex scale a permutation $ABCD$ is considered equivalent to its reverse order $DCBA$. It makes no difference if the items are represented clockwise or anticlockwise. And

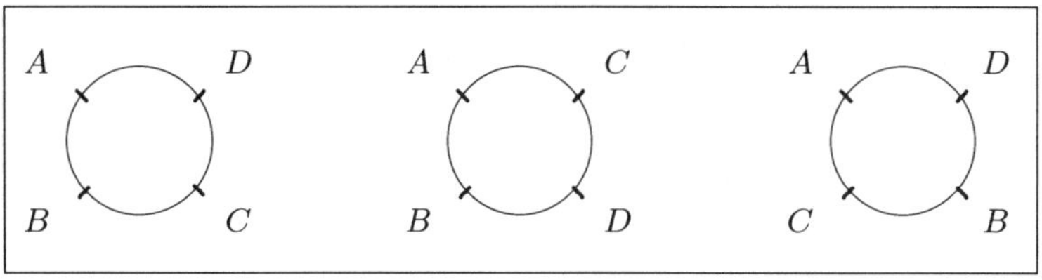

FIGURE 18.1. The three essential permutations for items *ABCD*.

second, since the origin on the circumference of a circle is arbitrary, the circular sequential orderings *ABCD*, *BCDA*, *CDAB*, and *DABC* are equivalent because the circular order of items does not change. These patterns are called *circular equivalent*.

Hence only three essential permutations remain to be analyzed: *ABCD*, *ABDC*, and *ACBD*. Any four items *A*, *B*, *C*, and *D* can form a possible circumplex scale in one of three different arrangements; see Figure 18.1.

According to Property 5, for any ordered quadruple *ABCD* the two error patterns are 1010 and 0101. So for a given four-item (sub)set *ABCD* six response patterns correspond to error patterns in its three essential permutations:

Set	*A*	*B*	*C*	*D*	Error In
a)	1	0	1	0	*A B C D*
	0	1	0	1	*A B C D*
b)	1	0	0	1	*A B D C*
	0	1	1	0	*A B D C*
c)	1	1	0	0	*A C B D*
	0	0	1	1	*A C B D*

For a particular quadruple *ijkl* the corresponding error frequencies can therefore be identified as

(*a*) the frequency of the 1010 and 0101 patterns gives the error frequency for order *ijkl*,

(*b*) the frequency of the 1001 and 0110 patterns gives the error frequency for order *ijlk*, and

(*c*) the frequency of the 1100 and 0011 patterns gives the error frequency for order *ikjl*.

Using these error sums the corresponding observed and expected error frequencies, and three coefficients of homogeneity, H_{ijkl}, H_{ijlk}, and H_{ikjl}, as defined by (18.2), can now be calculated, each corresponding to one of the three different essential ordered quadruples.

To select the best elementary scale the following criteria are used analogous to the MUDFOLD unfolding procedures (Van Schuur & Post, 1998).

1. The H-coefficient of an ordered quadruple of items must be positive in only one of the three possible permutations. The quadruple in this permutation is called a *unique quadruple.*

2. The value of H_{ijkl} should be higher than a user-specified lower bound. In MUDFOLD the default value of the lower bound is 0.30.

3. The best elementary scale is the one for which the sum of the frequencies of the patterns with two adjacent items is the highest. For the ordered quadruple $ijkl$ these frequencies involve the response patterns $ijkl$ (1100), $ijkl$ (0110), $ijkl$ (0011), and $ijkl$ (1001).

3.2 Extending the Elementary Scale

When the best elementary scale has been found, additional items are added to the scale in recursive steps. Consider a circumplex scale consisting of h items, $4 \leq h < k$. Then h positions are possible for the $(h+1)$st item to form an enlarged circumplex scale, where the $(h+1)$st item is an item from the $k-h$ remaining items. When, for instance, the four items $ABCD$ form an elementary scale in this order, then four positions are possible for a fifth Item E to form a circumplex scale: $AEBCD$, $ABECD$, $ABCED$, and $ABCDE$, respectively. Since we are dealing with circularity, the scale in which Item E is placed to the left of A, $EABCD$, is equal to the scale in which item E is placed to the right of Item D, $ABCDE$, and therefore generates no additional information. The $(h+1)$st item that should be added to the scale can be identified by means of the following criteria derived from the MUDFOLD unfolding model of Van Schuur (1984).

1. All quadruples formed by the $(h+1)$st item together with a triple of the previously established h-item scale are considered. For all these $\binom{h}{3}$ new quadruples the H-value should be positive. If, for instance, Item E fits in the position between Items A and B, where the items $ABCD$ form a scale in this order, then the H-values of the four new quadruples, $AEBC$, $AEBD$, $AECD$, and $EBCD$ should be positive.

2. The H_i value of the $(h+1)$st item and the H-value of the whole scale are larger than a user-specified lower bound. Here again the value of 0.30 is suggested.

3. If a candidate $(h+1)$st item is only representable in one position (a unique representation) and satisfies criteria 1 and 2, this item is always preferred over other candidate items that can be represented in more than one position. In general, we compare only those items that are representable by the smallest number of admissible positions.

4. If the candidate item is added to the scale in a preferred position, the H-value of the scale as a whole should be higher than when this item is in any other position, or for any other item in any other position.

5. In attempting to extend the h-item scale to an $(h+1)$-item scale, each of the remaining $k-h$ items will be tried in succession. If an item has at least one negative H_{ijkl} value, or its H_{ijkl} value is below the prespecified lower bound, it cannot be added to the h-item scale. If more than one item satisfies these criteria, then the scale is extended with the item that is representable in the smallest number of possible positions. In the case of a tie, the item that leads to the highest H-value for the scale as a whole is chosen.

After the $(h+1)$st item is selected, the procedure starts again with criterion 1 to add an $(h+2)$nd item to the $(h+1)$-item scale. The procedure ends when none of the remaining items satisfies the criteria, or when no items remain. If four or more residual items are left in the item pool after the procedure is completed, a new procedure for scale construction can be initiated for the remaining items. This implies that more than one circumplex scale can be found, formed by subsets of items from a pool of items. This procedure is called *multiple scaling.*

3.3 An Example of Scale Construction

We now illustrate the search procedure with a subscale of five items from data ($n = 198$) collected by Diener (see Larsen & Diener, 1992). Students gave an assessment on a seven-point scale about the appropriateness of 16 items as descriptive of their current mood. The items were taken from a previous study of Russell (1980) as a good small demonstration sample. They were dichotomized as "appropriate" (values $5-7 \rightsquigarrow 1$) and "inappropriate" (values $1-4 \rightsquigarrow 0$). The five items considered here are successively: A. calm, B. happy, C. sad, D. passive, and E. nervous.

The question to be answered now is whether these five items form a circumplex scale. The five items give rise to five quadruples. Each of these must be examined in terms of the three essential permutations $ijkl$, $ijlk$, and $ikjl$. To that purpose we follow the five steps described in Section 3.2. Table 18.2 shows the number of observed errors, expected errors, and the H-coefficients for all ordered quadruples in each of the three permutations.

TABLE 18.2. Observed errors q^o, expected errors q^e, and H-coefficients in three permutations.

	q^o			q^e			H		
Quadruple	$ijkl$	$ijlk$	$ikjl$	$ijkl$	$ijlk$	$ikjl$	$ijkl$	$ijlk$	$ikjl$
$ABCD$	7	32	51	14.3	23.9	45.0	0.51	−0.34	−0.13
$ABCE$	12	4	96	7.2	8.2	79.2	−0.66	0.49	−0.21
$ABDE$	39	6	50	23.3	14.6	43.2	−1.14	0.59	−0.16
$ACDE$	77	4	4	60.5	10.7	8.4	−0.27	0.63	0.53
$BCDE$	55	15	7	51.5	10.3	9.0	−0.07	−0.46	0.22

Step 1. To determine the best elementary scale we first need to know which of the five constituting quadruples have positive H-values in just one of their three possible permutations, which thus indicates the unique permutation. Four quadruples satisfy this condition. Their unique permutations are given in one of the three columns under H in Table 18.2.

- $ABCD$, with unique permutation $ABCD$ ($ijkl$).
- $ABCE$, with unique permutation $ABEC$ ($ijlk$).
- $ABDE$, with unique permutation $ABED$ ($ijlk$).
- $BCDE$, with unique permutation $BDCE$ ($ikjl$).

Step 2. The second requirement is that the H_{ijkl}-value has to be above a specified lower bound. Taking the value 0.30 for this example, the ordered quadruple $BDCE$ drops out because $H_{BDCE} = 0.22$.

Step 3. The best elementary scale among the remaining three ordered quadruples $ABCD$, $ABEC$, and $ABED$ is then found by inspecting the frequencies of the perfect 2/4 responses (two adjacent items generate positive responses). The frequency of perfect 2/4 patterns for the unique ordered quadruples is equal to the sum of observed errors in the two permutations in which the quadruples have a *negative* H-value. These sums are $32 + 51 = 83$ for $ABCD$, $12 + 96 = 108$ for $ABEC$, and $39 + 50 = 89$ for $ABED$.

Step 4. The sum of perfect 2/4 responses is the highest for the ordered quadruple $ABEC$, so it is selected as the best elementary scale. The value of H_{ABEC} is 0.49.

Step 5. Having selected $ABEC$ as the best elementary scale, we investigate whether the only remaining Item D can be added to it. Item D has to be tried in each of the four possible positions $AD\!BEC$, $ABDEC$, $ABEDC$,

TABLE 18.3. Final scale for the five emotions.

Item	p_i	H_i	q_i^o	q_i^e
A – Calm	0.80	0.54	21	45.5
B – Happy	0.68	0.48	24	46.1
E – Nervous	0.14	0.48	21	40.2
C – Sad	0.10	0.45	22	39.9
D – Passive	0.52	0.48	24	46.3
H-Coefficient		0.49	28	54.5

and $ABEC\boldsymbol{D}$. Trying Item D in the first position, $A\boldsymbol{D}BEC$, requires that all ordered quadruples containing item D ($ADBE$, $ADBC$, $ADEC$, and $DBEC$) have positive H-values. The H-values for these quadruples are $-0.16, -0.13, 0.63$, and 0.22, respectively. Two quadruples have negative H-values, so Item D is not representable as $A\boldsymbol{D}BEC$. Quadruple $ABDE$ has a positive H-value only in the permutation $ABED$. This implies that item D can also not be represented as $AB\boldsymbol{D}EC$. Whether Item D is representable as $ABE\boldsymbol{D}C$ or $ABEC\boldsymbol{D}$, respectively, depends on ordered quadruples that contain Item C as well as Item D. Quadruple $ABDC$ has a negative H_{ABDC} value (-0.34), while H_{ABCD} is positive (0.51). Hence, Item D is also not representable as $ABE\boldsymbol{D}C$. Whether Item D is scalable in the remaining position depends now on the quadruples $AECD$ and $BECD$. Both quadruples have a positive H-value, 0.53 and 0.22, respectively. We conclude that Item D is only scalable in one position (a unique representation), namely, the fifth position ($ABEC\boldsymbol{D}$).

The next step in deciding whether Item D can be added to the scale is to evaluate the scalability of Item D, (H_D), and of the scale as a whole (H). The values of these scalability coefficients should be above the lower boundary of 0.30. To calculate H_D we have to determine q_D^o and q_D^e as the sum of the observed and expected error frequencies, respectively, of all the ordered quadruples containing Item D. The number of observed errors $q_D^o = 6(ABED) + 7(ABCD) + 4(AECD) + 7(BECD) = 24$. The number of expected error frequencies $q_D^e = 14.6 + 14.3 + 8.4 + 9.0 = 46.3$. The value of H_D is therefore $1 - 24/46.3 = 0.48$. The H-coefficient is calculated by summing the amounts of expected and observed errors over all ordered quadruples in the scale. The total observed error frequency $q^o = 4(ABEC) + 6(ABED) + 7(ABCD) + 4(AECD) + 7(BECD) = 28$. The number of expected error frequencies $q^e = 8.2 + 14.6 + 14.3 + 8.4 + 9.0 = 54.5$. The H-value for the scale is therefore $1 - 28/54.5 = 0.49$. Table 18.3 shows the final results of the procedure. All H_i values are higher than 0.30 after Item D is added to the scale. Coefficient H is also higher than 0.30. The quantities p_i in Table 18.3 are the item popularities.

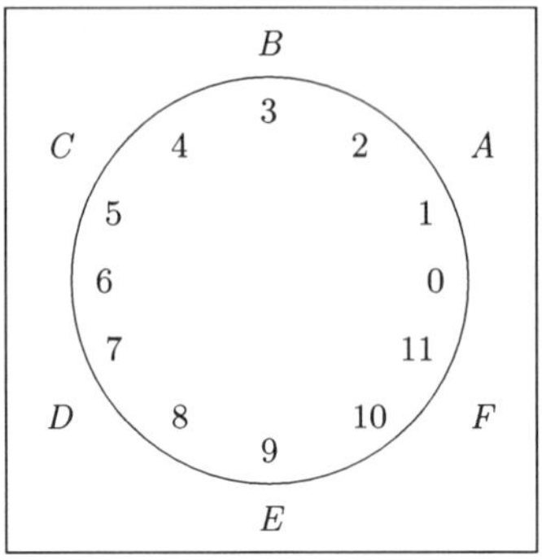

FIGURE 18.2. Scale values of the ordered circumplex scale $ABCDEF$.

We conclude that Item D can be added at the end (or, equivalently, at the start) of the ordered scale $ABEC$ to form the circumplex $ABECD$. If there are additional remaining items, then the same procedure as followed for Item D has to be repeated for these items. We also conclude that the five items are representable as a circumplex in the unique circular order $ABECD$. This order in Diener's data therefore corroborates the order first published by Russell (1980).

4 Parameter Estimation: A Scale Scoring Method

As the nonparametric circumplex is an ordinal scale, we need an ordinal method to determine scale values (positions along the circumference of a circle) for both items and subjects. The procedure sketched here is based on the procedure used to determine the scale value of items and respondents in a unidimensional unfolding scale as proposed by Van Schuur (1993).

4.1 Item Scores in a Circumplex

Given that measurement information is only ordinal, a circular ordering of the items can be represented by assigning odd rank numbers to the items, after choosing—arbitrarily or by judgment—the first item in the chain. Midpoints between pairs of adjacent items can be assigned as the corresponding even rank numbers; see Figure 18.2.

4.2 Subject Scores on a Circumplex

The assignment of CU scale scores to response patterns of subjects is based on the procedure of counting item steps as in MUDFOLD (Van Schuur,

TABLE 18.4. Scale values for six responses to the six items of the ordered circumplex $ABCDEF$.

Response Pattern						Extended Response Pattern								
A	B	C	D	E	F	A	B	C	D	E	F	A	B	Scale Value
1	1	1	0	0	0	1	1	1	0	0	0			1+1+1+0+0+0 = 3
0	1	1	0	0	0	0	1	1	0	0	0			2+1+1+0+0+0 = 4
0	0	0	1	1	1	0	0	0	1	1	1			2+2+2+1+1+1 = 9
1	1	0	0	0	1	0	0	0	0	0	1	1	1	2+2+2+2+2+1+1+1−12 = 1
0	1	1	0	1	0	0	1	1	0	1	0			2+1+1+0+1+0 = 5
1	0	0	1	0	1	0	0	0	1	0	1	1		2+2+2+1+2+1+1 = 11

1993; see also Figure 3.1 in Van Schuur & Post, 1998). Each dichotomous unfolding item, say, Item A, contains two item steps, dividing the scale in three areas. In the deterministic model only respondents in the middle area pick the item. If a respondent picks A then (moving from left to right on the scale) one item step has been passed. If the respondent does not pick A, then either zero or two item steps have been passed, depending on the other items that were picked. If the respondent picks items to the right of A and does not pick A itself, then the negative response to A is interpreted as two item steps of A being passed. If the picks are given to other items to the left of Item A, then zero item steps of A have been passed. For six items $ABCDEF$ forming an unfolding scale in this order, the response pattern 110000 means $1 + 1 + 0 + 0 + 0 + 0 = 2$, the pattern 000011 means $2 + 2 + 2 + 2 + 1 + 1 = 10$, and the pattern 001100 means $2 + 2 + 1 + 1 + 0 + 0 = 6$ item steps passed.

The assignment of CU scale scores to the response patterns of subjects follows this procedure with an adaptation to deal with circular patterns. There are a few differences, however, with respect to perfect, imperfect, and unscalable patterns. Consider an example of six ordered items, $ABCDEF$, located on a CU; see Figure 18.2. In the CU model all perfect patterns in which both "end" items A and F are picked (e.g., 110001 or 100111) are not CU errors. For CU the item step method, as used for the unfolding model for calculating the scale value, involves extending the item ordering anticlockwise in Figure 18.2.

The scoring method, which is equivalent to extended item step counting, involves assigning the (extended) rank number of the median of the picked items (for odd k, and the average rank number of the middle two ones for even k). In Table 18.4 an example is given.

Imperfect response patterns can be handled the same way, or by some estimate based on allocating the value for a "nearest" perfect response pattern. However, the two extreme patterns (no items picked, or all items picked) are not scalable, as these subjects cannot be assigned to a position on the latent trait.

5 Model Fit

So far we have discussed the circumplex model as a deterministic model with an error theory. Since the H-coefficient is based on an error criterion it can be used for both the unfolding and circumplex models. However, the coefficient of scalability H alone is not sufficient as a measure of model fit for an unfolding probabilistic model, because the H-coefficient is related to the assumption of monotone item characteristic functions, and is focused on the detection of violations of this monotonicity assumption (Mokken, 1971, 1997; Molenaar & Sijtsma, 2000). Moreover, monotonic models can be seen as special cases of unfolding, which again is a special case of the circumplex model. Hence for probabilistic versions additional criteria are necessary for investigating model fit and for discriminating circular from linear unfolding.

5.1 *Fit of the Probabilistic Circular Unfolding Model*

A probabilistic version of the circumplex model can be formulated in such a way that the item characteristic curve (ICC), that is, the probability of a positive response by a person to a dichotomous item, is highest for an item close to that person's position on the latent trait, and decreases monotonically as their distance increases.

Denote the ICC as

$$P_{\beta_i}(\theta_v) = P(X_{vi} = 1 \mid \theta_v, \beta_i) \;, \tag{18.4}$$

where θ_v, $v = 1, \ldots, n$, and β_i, $i = 1, \ldots, k$, denote the subject and item parameters, respectively.

Then a probabilistic CU model can be defined as a nonnegative, (weakly) unimodal function $C_\beta(\theta)$ on the interval $(0, 2\pi]$ which is *simply periodic* in 2π for all θ, $\beta \in (0, 2\pi]$, where θ and β are measured in radians,

$$P_\beta(\theta) = P_\beta(\theta \pm k2\pi) \;, \qquad \theta, \beta \in (0, 2\pi] \;. \tag{18.5}$$

Notice that $P_\beta(\theta)$ is maximal for $\theta = \beta_i$, and is minimal for $\theta = \beta_i \pm \pi$. Notice further that $P_\beta(0) = P_\beta(2\pi)$. Choosing $P_\beta(\theta)$ on $(0, 2\pi]$ as (weakly) unimodal within the class of functions treated by Post and Snijders (1993), considering corresponding densities for θ, and observing that, due to simple periodicity, integration over definite integrals of length 2π are equal to those over $(0, 2\pi]$, the diagnostic criteria derived by them for certain GU models may well be valid for related CU models as well. Eventually, parametric models, satisfying (18.4) and (18.5), and the Post–Snijders conditions, can be studied and estimated by conventional optimization methods.

Such approaches might be useful for the evaluation of the probabilistic circular unfolding model, because they will substantiate the adaptation of related diagnostics from Mokken (1971, 1997) and Molenaar (1997; see also Molenaar & Sijtsma, 2000) for the dominance model, and from Post (1992; see also Post & Snijders, 1993) for the linear unfolding model. Three diagnostic matrices can then be used: the correlation matrix, the conditional adjacency matrix, and the conditional score group matrix. These matrices are checked by visual inspection, but statistical evaluations, as developed by Molenaar for MSP (Molenaar & Sijtsma, 2000), are possible. The order of cell proportions in the matrices can be predicted by assumptions underlying the probabilistic model. Violations of these proportions can indicate items that contradict the model assumptions.

6 Discussion

We formulated a nonmetric circumplex model and scale procedure for dichotomous data along the lines of Mokken (1971, 1997) and Van Schuur (1984) in terms of a nonparametric IRT approach. It was hypothesized that procedures of parameter estimation and evaluation of goodness-of-fit can be further developed, analogous to those worked out for nonparametric cumulative scaling and unfolding. More elaborate testing of the model on real data is needed.

The use and applicability of the scale scoring values also need further research. Scale values on a circumplex do not have an immediate linear or ordinal interpretation. The highest subject and item scale values are next to the lowest ones, which means that the interpretation of the scale values of subjects cannot be used straightforwardly in linear models. More satisfactory methods of allocating scale scores to imperfect patterns need to be developed.

At the moment it is uncertain how well the diagnostic matrices perform as a measure of goodness-of-fit. A model including polytomous rating data would be more general than the present model for dichotomous data, as dichotomization means loss of information (Molenaar, 1997; Molenaar & Sijtsma, 2000). A more general model might also help solve the problem of unscalability, since more information would then be available about the response behavior of persons. We expect that the circumplex model, as formulated here, could be extended to polytomous data, by analogy to similar extensions of the Mokken model (e.g., Molenaar & Sijtsma, 2000) and the unfolding model (e.g., Van Schuur, 1993).

It may also be straightforward to extend the model to full or partial ranking data, as in the unfolding model (Van Schuur, 1984). Partial ranking data can be described as a *rank* m/k situation (Coombs, 1964); in the case of dichotomous data, every person ranks exactly m items from a set of k items and considers the other $m - k$ as having lower but unspecified rank.

A T-statistic can be developed for the H-values of the quadruples, items, and the scale as a whole, as developed by Mokken (1971) for the cumulative scaling model, and by Post (1992) for the unfolding model, but this will involve intricate and quite lengthy calculations. The T-statistic can be used to check whether H-values differ significantly from the null model.

The circumplex model has been applied only to specific domains in psychology and sociology, where its applicability has often been demonstrated on the basis of factor analysis, or other models that were not specifically developed for the circumplex. The procedure offered in this chapter may be helpful to researchers in their attempts to test the circumplex model for their data.

Acknowledgments: The initial work on the circumplex was started by Leeferink (1997) for his masters thesis, developing earlier ideas of Van Schuur, with Van Schuur and Mokken as his supervisors. The authors owe thanks to Professor Edward Diener (University of Illinois at Urbana-Champaign) for providing the data for Section 3.3. We also thank the editors for their efforts to improve earlier drafts of this chapter.

References

Andrich, D., & Luo, G. (1993). A hyperbolic cosine latent trait model for unfolding dichotomous single-stimulus responses. *Applied Psychological Measurement, 17*, 253–276.

Benjamin, L.S. (1974). Structural analysis of social behavior. *Psychological Review, 81*, 392–425.

Browne, M.W. (1992). Correlation matrices for circumplex models. *Psychometrika, 57*, 469–497.

Coombs, C.H. (1964). *A theory of data.* New York: Wiley.

De Raad, B., & Hofstee, W.K.B. (1993). A circumplex approach to the five factor model: A facet structure of trait adjectives supplemented by trait verbs. *Personality and Individual Differences, 15*, 493–505.

Gurtman, M.B. (1992). Trust, distrust, and interpersonal problems: A circumplex analysis. *Journal of Personality and Social Psychology, 62*, 989–1002.

Guttman, L. (1954). A new approach to factor analysis: The radex. In P.F. Lazarsfeld (Ed.), *Mathematical thinking in the social sciences* (pp. 258–348). New York: Columbia University Press.

Hoijtink, H.J.A. (1991). *PARELLA: Measurement of latent traits by proximity items.* Leiden: DSWO Press.

Kiesler, D.J. (1983). The 1982 interpersonal circle: A taxonomy for complementarity in human transactions. *Psychological Review, 90*, 185–214.

Larsen, R.J., & Diener, E. (1992). Promises and problems with the circumplex model of emotion. In M.S. Clark & J.R. Averill (Eds.), *Emotion: Review of personality and social psychology*, (Vol. 13, pp. 25–59). Newbury Park, CA: Sage.

Leary, T. (1957). *Interpersonal diagnosis of personality.* New York: Ronald Press.

Leeferink, A.J. (1997). *A nonmetric circumplex model for dichotomous data: The circumplex as a nonmetric IRT model for latent variable analysis.* Unpublished doctoral thesis, University of Groningen, Department of Sociology.

Levy, S., & Guttman, L. (1975). On the multivariate structure of wellbeing. *Social Indicators Research, 2*, 361–388.

Marcus, G.E. (1988). The structure of emotional response: 1984 Presidential candidates. *American Political Science Review, 82*, 737–761.

Mokken, R.J. (1971). *A theory and procedure of scale analysis: With applications in political research.* The Hague: Mouton.

Mokken, R.J. (1997). Nonparametric models for dichotomous responses. In W.J. van der Linden & R.K. Hambleton (Eds.), *Handbook of modern item response theory* (pp. 351–367). New York: Springer-Verlag.

Molenaar, I.W. (1983). *Item steps* (Heymans Bulletin HB–83–630–EX). Groningen: University of Groningen, Vakgroep Statistiek en Meettheorie.

Molenaar, I.W. (1997). Nonparametric models for polytomous responses. In W.J. van der Linden & R.K. Hambleton (Eds.), *Handbook of modern item response theory* (pp. 369–380). New York: Springer-Verlag.

Molenaar, I.W., & Sijtsma, K. (2000). User's manual MSP5 for Windows: A program for Mokken scale analysis for polytomous items. Version 5.0 [Software manual]. Groningen: ProGAMMA.

Post, W.J. (1992). *Nonparametric unfolding models: A latent structure approach.* Leiden: DSWO Press.

Post, W.J., & Snijders, T.A.B. (1993). A nonparametric probabilistic model for parallellogram analysis. *Methodika, 7*, 130–156.

Roberts, J.S., & Laughlin, J. E. (1996). A unidimensional item response model for unfolding responses from a graded disagree-agree response scale. *Applied Psychological Measurement, 20*, 231–255.

Ross, T.A. (1938). A statistic for circular scales. *Journal of Educational Psychology, 29*, 384–389.

Russell, J.A. (1980). A circumplex model of affect. *Journal of Personality and Social Psychology, 39*, 1161–1178.

Russell, J.A., Lewicka, M., & Niit, T. (1989). A cross-cultural study of a circumplex model of affect. *Journal of Personality and Social Psychology, 57*, 848–856.

Shye, S., Elizur, D., & Hoffman, M. (1994). *Introduction to facet theory: content design and intrinsic data analysis in behavioral research.* London: Sage.

Tracey, T.J., & Rounds, J. (1993). Evaluating Holland's and Gati's vocational interest models: A structural meta-analysis. *Psychological Bulletin, 113*, 229–246.

Van Blokland-Vogelesang, A.W. (1990). *Unfolding and group consensus ranking for individual preferences.* Leiden: DSWO Press.

Van Schuur, W.H. (1984). *Structure in political beliefs: A new model for stochastic unfolding with application to European party activists.* Amsterdam: CT Press.

Van Schuur, W.H. (1993). Nonparametric unfolding models for multicategory data. *Political Analysis, 4*, 41–74.

Van Schuur, W.H. (1994). Determining subjects' scale scores in MUDFOLD for multicategory data. *Kwantitatieve Methoden, 45*, 67–69.

Van Schuur, W.H., & Post, W.J. (1998). MUDFOLD: A program for multiple unidimensional unfolding. Version 4.0 [Software manual]. Groningen: ProGAMMA.

Wiggins, J.S. (1979). A psychological taxonomy of trait descriptive terms. *Journal of Personality and Social Psychology, 37*, 395–412.

Wiggins, J.S. (1980). Circumplex models of interpersonal behavior. In L. Wheeler (Ed.), *Review of Personality and Social Psychology* (Vol. 1, pp. 265–294). Beverly Hills, CA: Sage.

19

Using Resampling Methods to Produce an Improved DIMTEST Procedure

William Stout,[1] Amy Goodwin Froelich,[2] and Furong Gao[3]

ABSTRACT Following in the nonparametric item response theory tradition, DIMTEST (Stout, 1987) is an asymptotically justified nonparametric procedure that provides a test of hypothesis of unidimensionality of a test data set. This chapter introduces a new bias correction method for the DIMTEST procedure based on the statistical principle of resampling. A simulation study shows this new version of DIMTEST has a Type I error rate close to the nominal rate of $\alpha = 0.05$ in most cases and very high power to detect multidimensionality in a variety of realistic multidimensional models. The result with this new bias correction method is an improved DIMTEST procedure with much wider applicability and good statistical performance.

1 Introduction

Following in the nonparametric item response theory (IRT) tradition, Stout (1987) and Nandakumar and Stout (1993) developed and refined the DIMTEST procedure to provide an asymptotically justified nonparametric hypothesis test of unidimensionality of a test data set. To test for unidimensionality, the DIMTEST procedure divides the test items into two groups, the (dimensionality) Assessment Subtest (AT) and the (examinee) Partitioning Subtest (PT). In order to correct for a serious statistical bias in the DIMTEST statistic, Stout (1987) introduced a second Assessment Subtest (called AT2) to function parallel to the first Assessment Subtest (called

[1]Department of Statistics, University of Illinois at Urbana-Champaign, 725 S. Wright Street, Champaign, IL 61820, USA; e-mail: *stout@stat.uiuc.edu*

[2]Department of Statistics, University of Illinois at Urbana-Champaign, 725 S. Wright Street, Champaign, IL 61820, USA; e-mail: *amyf@stat.uiuc.edu*

[3]CTB/McGraw-Hill Research Department, 20 Ryan Ranch Road, Monterey, CA 93940, USA; e-mail: *FGao@ctb.com*

AT1). Unfortunately, this experimental design had two serious weaknesses. One, for certain applications, some statistical bias still remained in the procedure, resulting in unacceptable levels of Type I hypothesis testing error. Two, the use of DIMTEST was precluded on short tests or when the Assessment Subtest contained more than one-third of the test items.

The purpose of this chapter is to develop an improved version of DIMTEST that removes the need for the selection of the AT2 Subtest of items by introducing a bias correction method based on statistical resampling. This new version of DIMTEST should have better statistical performance and provide greater flexibility than the original DIMTEST procedure. This bias correction method can also be extended to situations where choosing an AT2 Subtest is either very difficult or not possible, such as with polytomous items or items on a Computer Adaptive Test (CAT) (Froelich, 2000).

Improving DIMTEST offers benefits for theoreticians and practitioners alike. Because many IRT-based procedures, such as BILOG (Mislevy & Bock, 1990), assume unidimensionality of test data, it is essential to have a reliable procedure to test for unidimensionality of test data. By providing a test of hypothesis for unidimensionality, DIMTEST can be used to confirm the dimensional distinctiveness of item clusters found through commonly used methods of exploratory statistical analysis, such as factor analysis with tetrachoric correlations, Mokken scaling, cluster analysis, multidimensional scaling, the item pair conditional covariance-based procedures HCA/CCPROX (Roussos, Stout, & Marden, 1998), or DETECT (Zhang & Stout, 1999b). DIMTEST can also be used to confirm the dimensional distinctiveness of substantively based item clusters. Especially with the advent of Computer Adaptive Testing, performance-based testing, and the recent emphasis on the relationship between test equity and latent multidimensionality (Roussos & Stout, 1996), it is important to reliably determine the multidimensional latent variable structure of a test.

In order to fully discuss the DIMTEST procedure, the following review of the literature and notation is necessary; that is, throughout, our notation follows conventions in DIMTEST literature to make referencing more straightforward. Let X_i be the response of a randomly sampled examinee on the ith item of a test, and let $\mathbf{X}^{(n)} = (X_1, X_2, \dots, X_n)'$ denote the response vector of a randomly sampled examinee to an n-item test. Only dichotomous items are considered here, so X_i is either 1 if an examinee answers item i correctly or 0 if an examinee answers item i incorrectly.

Examinee responses to test items depend on both the characteristics of the items themselves and the value of the random latent ability vector $\boldsymbol{\theta}$. For a particular value of $\boldsymbol{\theta}$, the probability of a randomly sampled examinee correctly answering item i is given by the conditional probability

$$P_i(\boldsymbol{\theta}) = P(X_i = 1 \mid \boldsymbol{\theta}) \ . \tag{19.1}$$

The function in Equation 19.1 is defined as the item characteristic curve or ICC for item i. Likewise, the probability of obtaining a particular response pattern $\mathbf{x}^{(n)}$ for a randomly sampled examinee is given by the conditional probability

$$P(\mathbf{X}^{(n)} = \mathbf{x}^{(n)} \mid \boldsymbol{\theta}) \ . \tag{19.2}$$

For the latent variable models defined in (19.1) and (19.2) to be psychometrically reasonable, two requirements are often considered desirable. First, the latent variable probability model should be monotone. A latent variable model is *monotone* if the ICC in Equation 19.1 increases coordinatewise in $\boldsymbol{\theta}$ for each item i. Thus, the probability of a correct response on item i increases as the value of the latent examinee ability vector $\boldsymbol{\theta}$ increases. Second, the latent variable probability model should be locally independent. Local independence requires that examinee responses to all test items be independent conditioned on examinee ability. Thus, a latent variable model is *locally independent* if for all $\boldsymbol{\theta}$ and all possible response patterns $\mathbf{x}^{(n)}$,

$$P(\mathbf{X}^{(n)} = \mathbf{x}^{(n)} \mid \boldsymbol{\theta}) = \prod_{i=1}^{n} P(X_i = x_i \mid \boldsymbol{\theta}) \ . \tag{19.3}$$

The dimensionality of a test is defined as the minimum number of dimensions of $\boldsymbol{\theta}$ required to produce a locally independent and monotone latent variable model for Equation 19.2. When the dimensionality of a test is one, the latent variable model is called *unidimensional*, and when the dimensionality of a test is greater than one, the latent variable model is called *multidimensional*.

In practice, the requirement of local independence is often replaced with the less stringent requirement of weak local independence. A latent variable model displays *weak local independence* if for all $\boldsymbol{\theta}$ and all item pairs (i, i'),

$$\text{Cov}\,(X_i, X_{i'} \mid \boldsymbol{\theta}) = 0 \ , \quad i \neq i' \ . \tag{19.4}$$

Thus, a test is unidimensional if Equation 19.4 holds for all item pairs (i, i') and for a unidimensional θ.

Investigating dimensionality through the use of conditional covariances can be developed further using a geometric representation for the items of a multidimensional test (see, e.g., Ackerman, 1996). Assume a test has two dimensions, labeled θ_1 and θ_2. The dimensions θ_1 and θ_2 can be represented by a coordinate system with θ_1 corresponding to the horizontal axis and θ_2 corresponding to the vertical axis, as shown in Figure 19.1. The use of orthogonal axes in no way implies the correlation between dimensions θ_1 and θ_2 is zero. In Figure 19.1, Item 1 represents an item that measures the θ_1 ability but not the θ_2 ability, while Item 2 represents an item that measures the θ_2 ability but not the θ_1 ability. Item 3 represents an item that measures both the θ_1 and θ_2 abilities equally. Thus, Item 3 measures the composite ability $\theta_1 + \theta_2$.

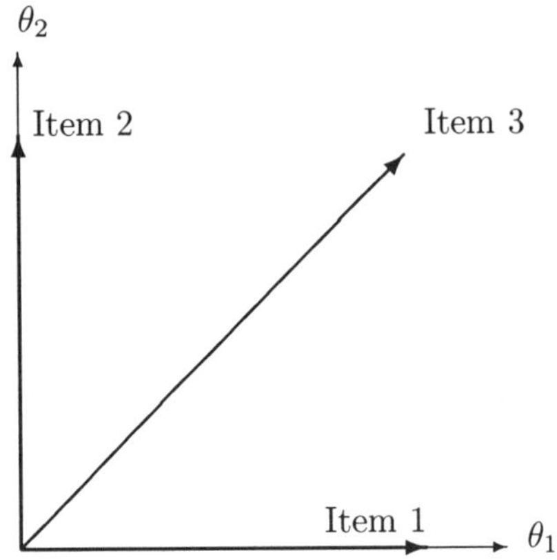

FIGURE 19.1. Two-dimensional coordinate system.

This multidimensional representation of item vectors can be used to explain the role of conditional covariances in determining the dimensionality of a test. Figure 19.2 represents a two-dimensional test with the dominant ability the test is designed to measure, also called the direction of best measurement of the test, denoted θ. (The direction of best measurement of both an item and a test is rigorously defined in Zhang and Stout, 1999a.)

The conditional covariances between two items on the same side of the latent conditioning variable θ (Items 1 and 2, and Items 3 and 4) will be positive, and the conditional covariances between two items on different sides of the latent conditioning variable θ (Items 1 and 3, Items 1 and 4,

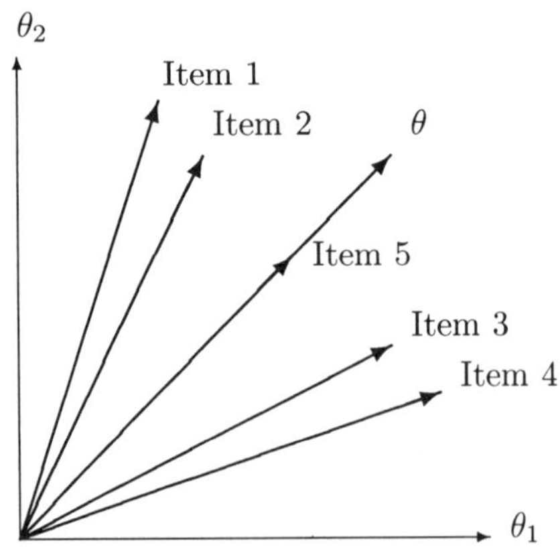

FIGURE 19.2. Graphical representation of conditional covariances.

Items 2 and 3, and Items 2 and 4) will be negative. In addition, if either item of an item pair lies near the latent conditioning variable θ (all item pairs including Item 5), the conditional covariance for that item pair will be approximately zero. (Zhang & Stout, 1999a, show similar results hold for conditional covariances when the dimensionality of a test is greater than two.) Thus, the magnitude and sign of the conditional covariance of an item pair provides information about how dimensionally similar the two items are to each other. This serves as the basis for the DIMTEST procedure.

2 Review of the DIMTEST Procedure

The DIMTEST procedure is a test of hypothesis of unidimensionality of a data set. For the new resampling version of DIMTEST to be clearly understood, we first give a step-by-step description of the original DIMTEST procedure found in Stout (1987) and Nandakumar and Stout (1993).

Step 1. Select the Assessment Subtest 1 (AT1), Assessment Subtest 2 (AT2), and the examinee Partitioning Subtest (PT). The AT1 subtest is chosen first. Then based on the items chosen for the AT1 subtest, the items for the AT2 subset are chosen. The PT subtest then consists of the remaining test items.

Step 1.1. Select a group of m items from the total n items on the test to form the Assessment Subtest 1 (AT1). The AT1 items are tested for dimensional distinctiveness relative to the remaining items. AT1 can be selected based on either a substantive analysis of item content or an exploratory statistical data analysis approach, such as factor analysis, cluster analysis, the DETECT procedure (Zhang & Stout, 1999b), multidimensional scaling, dimensionality-based cluster analysis, and so on. When multidimensionality is present in the test data, the goal is to choose AT1 so that it has the best possible chance of being dimensionally homogeneous on a dimension distinct from the dimension or dimensions measured by the remaining items on the test.

Obviously, the choice of AT1 is extremely important to successfully determining the dimensional structure of the test. For the DIMTEST procedure to provide a meaningful test of the null hypothesis of unidimensionality, the items in the AT1 Subtest should be as dimensionally distinct from the direction of best measurement of the remaining items as possible. Two different two-dimensional tests are represented by Figures 19.3 and 19.4, with the AT1 Subtest and the direction of best measurement of the remaining test items, θ, marked.

In Figure 19.3, since the AT1 items are all on the same side of the conditioning variable θ, the conditional covariances of all the AT1 item pairs

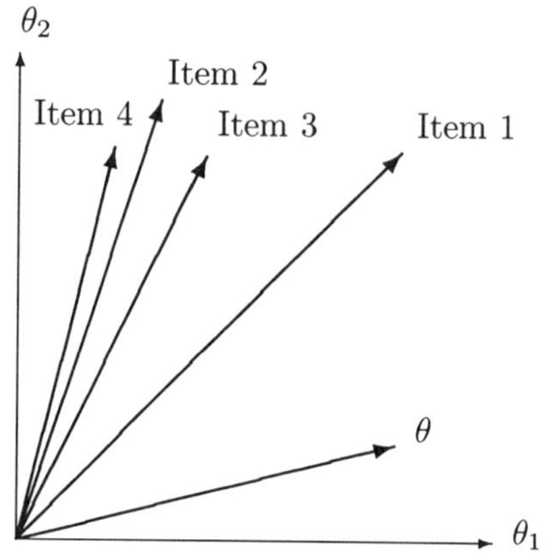

FIGURE 19.3. A good choice for the AT1 subtest.

will be positive. Thus, DIMTEST will likely reject the null hypothesis of unidimensionality and correctly conclude the AT1 items are dimensionally distinct from the remaining test items. However, in Figure 19.4, only the conditional covariances between Items 1 and 2 and between Items 3 and 4 will be positive. The rest of the conditional covariances between the four items will be negative since the items lie on different sides of the conditioning variable θ. Thus, DIMTEST will likely fail to reject the null hypothesis of unidimensionality for this test and incorrectly conclude the AT1 items are dimensionally similar to the remaining test items.

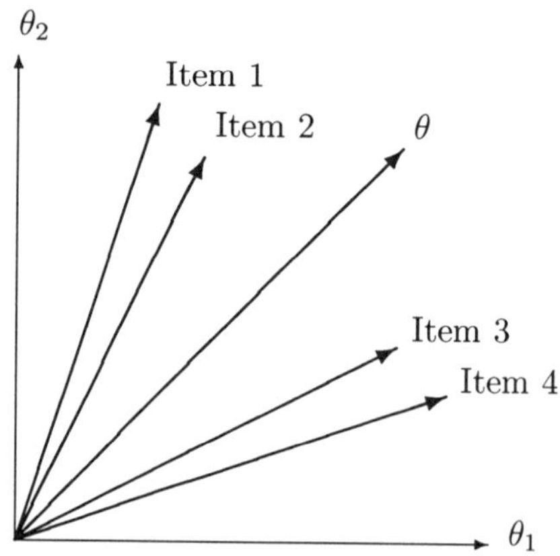

FIGURE 19.4. A poor choice for the AT1 subtest.

Step 1.2. Select a second group of m items from the remaining $n - m$ items on the test to form the Assessment Subtest 2 (AT2). The AT2 items are chosen to have a similar item difficulty distribution as the AT1 items. (See page 595 of Stout, 1987, for details of the selection of AT2 Subtest.)

Step 1.3. Form the PT Subtest using the remaining $n - 2m$ test items.

Step 2. Assign examinees to different subgroups according to their total score on the PT Subtest, denoted Z_{PT}. Let $X_{ij}^{(k)}$ denote the response of the jth examinee from subgroup k to the ith assessment item. Let J_k denote the number of examinees in subgroup k and let K denote the total number of subgroups. Denote the total correct score of examinee j in subgroup k on the AT1 Subtest,

$$Y_j^{(k)} = \sum_{i=1}^{m} X_{ij}^{(k)} , \tag{19.5}$$

and denote the average total correct score of the J_k examinees in subgroup k,

$$\overline{Y}^{(k)} = \frac{1}{J_k} \sum_{j=1}^{J_k} Y_j^{(k)} . \tag{19.6}$$

Then, an estimate of the variance of examinee scores on the AT1 Subtest for the PT subgroup k is

$$\hat{\sigma}_k^2 = \frac{1}{J_k} \sum_{j=1}^{J_k} (Y_j^{(k)} - \overline{Y}^{(k)})^2 . \tag{19.7}$$

Step 3. Let

$$\hat{p}_i^{(k)} = \frac{1}{J_k} \sum_{j=1}^{J_k} X_{ij}^{(k)} \tag{19.8}$$

be the observed difficulty for subgroup k on item i. The unidimensional estimate of the subgroup k variance of each item is defined to be $\hat{p}_i^{(k)}(1 - \hat{p}_i^{(k)})$. Then, the total unidimensional estimate for the variance of $Y_j^{(k)}$ is defined to be

$$\hat{\sigma}_{U,k}^2 = \sum_{i=1}^{m} \hat{p}_i^{(k)} (1 - \hat{p}_i^{(k)}) . \tag{19.9}$$

The difference between these two variance estimates is given by

$$T_{L,k} = \hat{\sigma}_k^2 - \hat{\sigma}_{U,k}^2 = 2 \sum_{i<i' \in \mathrm{AT1}} \widehat{\mathrm{Cov}}\,(X_i, X_{i'} \mid Z_{PT} = k) , \tag{19.10}$$

where $\widehat{\text{Cov}}\,(X_i, X_{i'} \mid Z_{PT} = k)$ is the usual estimate of the covariance between two items conditioned on the set of all examinees with PT Subtest score equal to k. If the AT1 items are dimensionally similar to the PT items, the direction of best measurement of each of the AT1 items will be near the direction of best measurement of the PT Subtest. Therefore, the value of the $T_{L,k}$ statistic will be small for each k. However, if the AT1 items measure a dimension different from that measured by the PT items, the direction of best measurement of the AT1 items will be far away from the direction of best measurement of the PT Subtest. Therefore, the value of the $T_{L,k}$ statistic will be large for each k. Thus, for each examinee subgroup k, the statistic $T_{L,k}$ is a measure of the dimensional distinctiveness between the AT1 Subtest and the PT Subtest.

Step 4. The asymptotic variance of $T_{L,k}$ is given by

$$S_k^2 = \frac{(\hat{\mu}_{4,k} - \hat{\sigma}_k^4) - \hat{\delta}_{4,k}}{J_k} \,, \tag{19.11}$$

where

$$\hat{\mu}_{4,k} = \frac{1}{J_k} \sum_{j=1}^{J_k} (Y_j^{(k)} - \bar{Y}^{(k)})^4 \tag{19.12}$$

and

$$\hat{\delta}_{4,k} = \sum_{i=1}^{m} \hat{p}_i^{(k)}(1 - \hat{p}_i^{(k)})(1 - 2\hat{p}_i^{(k)})^2 \,. \tag{19.13}$$

The DIMTEST statistic T_L is now given by

$$T_L = \frac{\sum_{k=1}^{K} T_{L,k}}{\sqrt{\sum_{k=1}^{K} S_k^2}} \,. \tag{19.14}$$

Stout (1987) showed that the DIMTEST statistic in Equation 19.14 has an asymptotic standard normal distribution under the null hypothesis of unidimensionality when the number of items and the number of examinees tends to infinity. However, for a finite length test, Stout (1987) found the DIMTEST statistic T_L tended to have its expected value shifted in the positive direction for both unidimensional and multidimensional tests. According to the theory of conditional association from Holland and Rosenbaum (1986), when the test is unidimensional, the theoretical quantity $\text{Cov}\,(X_i, X_{i'} \mid Z_{PT} = k)$ is positive for any two non-PT items i and i'. Hence, we expect its estimate, $\widehat{\text{Cov}}\,(X_i, X_{i'} \mid Z_{PT} = k)$, to be positive as well. Thus, the DIMTEST statistic T_L will generally be greater than 0, making T_L positively biased even under the null hypothesis of unidimensionality. Stout (1987) corrected for this bias in the DIMTEST statistic by calculating another DIMTEST statistic using the second Assessment Subtest (AT2) items.

Step 5. Repeat Steps 2 through 4 using the m items in AT2 as the assessment subtest to calculate another DIMTEST statistic, denoted T_B. If the AT2 and PT Subtests are dimensionally homogeneous, the DIMTEST statistic T_B will not be influenced by multidimensionality in the test data. The bias corrected DIMTEST statistic is then given by

$$T = \frac{T_L - T_B}{\sqrt{2}} . \tag{19.15}$$

Under the null hypothesis of unidimensionality, Stout (1987) showed that the DIMTEST statistic T has an asymptotic standard normal distribution as the number of items and the number of examinees tends to infinity. Thus, the null hypothesis is rejected with level α if the value of T is greater than the $100(1-\alpha)$th percentile of the standard normal distribution.

Remark. Stout (1987) found the asymptotic form of the DIMTEST statistic T had an unacceptably high level of Type I error in simulation studies. To lower the Type I error rate of the DIMTEST procedure, Stout made two changes in the computation of the DIMTEST statistic. The asymptotic variance of $T_{L,k}$ was changed to

$$S_k^{2'} = \frac{(\hat{\mu}_{4,k} - \hat{\sigma}_k^4) + \hat{\delta}_{4,k} + 2\sqrt{(\hat{\mu}_{4,k} - \hat{\sigma}_k^4)\hat{\delta}_{4,k}}}{J_k} , \tag{19.16}$$

where $\hat{\mu}_{4,k}$ and $\hat{\delta}_{4,k}$ are the same as in Step 4 above. In addition, the form of the T_L statistic was changed to

$$T_L' = \frac{1}{\sqrt{K}} \sum_{k=1}^{K} \frac{T_{L,k}}{S_k'} . \tag{19.17}$$

The final DIMTEST statistic,

$$T' = \frac{T_L' - T_B'}{\sqrt{2}} , \tag{19.18}$$

now has only an approximate asymptotic standard normal distribution as both the number of items and the number of examinees tends to infinity under the null hypothesis of unidimensionality.

3 Correcting the Bias in T_L

Generally, the AT2 method of bias correction used by Stout (1987) has been successful (shown to usually have Type I error rates near the nominal rate and good power to detect multidimensionality in test data) in simulation and real data studies (see Stout, 1987; Nandakumar & Stout, 1993; Hattie, Krakowski, Rogers, & Swaminathan, 1995). However, there are situations where the DIMTEST statistic can have an inflated Type I error rate when unidimensionality holds. Simulation studies by Nandakumar and Stout (1993) have shown when the AT1 items have relatively large discriminations and/or similar difficulties, the AT2 items often fail to correct enough of the bias in the T_L statistic to produce acceptable results. Also, these studies have shown that DIMTEST can produce unreliable results when the test is relatively short, or when the AT1 Subtest is relatively long. The following equation helps to explain these simulation results.

Let Z_{PT} be the total score on the PT subset of items and let X_i and $X_{i'}$ be the examinee responses to AT1 items i and i', respectively. The following equality holds when the data are unidimensional (see Froelich, 2000, for a proof).

$$\mathrm{Cov}\,(X_i, X_{i'} \mid Z_{PT} = k) = \mathrm{Cov}\,[P_i(\theta), P_{i'}(\theta) \mid Z_{PT} = k]\ . \qquad (19.19)$$

Clearly, from Equation 19.19, the theoretical value of the conditional covariance between two AT1 items conditioned on the PT score depends on three functions: the items' ICCs $P_i(\theta)$ and $P_{i'}(\theta)$, and the conditional distribution of θ given the total score on the PT items.

For the AT2 bias correction method, since the DIMTEST statistics from both the AT1 and AT2 items are calculated with the same conditioning PT score, the conditional distribution of θ given the total score on the PT items is the same for both T_L and T_B. Thus, to correct for the bias in the quantity $\mathrm{Cov}\,(X_i, X_{i'} \mid Z_{PT} = k)$ for items i and i' in AT1, the corresponding items in AT2 should have roughly the same ICCs throughout the range of θ.

In the current DIMTEST procedure, the AT2 items are chosen to have roughly the same item difficulty distribution as the AT1 items. Generally, this method will produce AT2 items with ICCs close enough to the ICCs of the AT1 items to give a reliable DIMTEST statistic, thereby giving good Type I error results. However, when the AT1 items have similar item difficulties, the AT2 method will often be forced to choose items with significantly smaller and/or larger item difficulties than the AT1 items. Also, when the AT1 items have large discriminations, the AT2 items will generally have lower discriminations than the AT1 items. When either of these two situations occurs, the ICCs of the AT2 items are not close enough to the ICCs of the AT1 items throughout the range of θ to sufficiently correct for the bias in the T_L statistic.

4 New Bias Correction Method

As the above discussion indicates, when the data are unidimensional, the value of the estimated conditional covariances in the numerator of the

DIMTEST statistic T_L depends on the ICCs for the AT1 items, and on the conditional distribution of θ given the total score on the PT Subtest. To correct for the bias in the DIMTEST statistic calculated from the original data, a unidimensional data set is generated using a nonparametric model estimated from the original data. If the null hypothesis of unidimensionality holds, this generated data will have approximately the same ICCs and the same conditional distribution of θ given the PT score as the original data. Thus, according to (19.19), the DIMTEST statistic calculated from this unidimensional generated data set (denoted T_G) should have the same amount of bias as the DIMTEST statistic calculated from the original data (denoted T_L). In other words, when the null hypothesis is true, $E(T_L - T_G)$ will be close to zero, giving an approximately unbiased statistic $T_L - T_G$. If the null hypothesis is false and the assessment subtest well chosen, $E(T_L - T_G)$ will be large, reflecting the multidimensional structure of the original data.

In order to generate a second unidimensional data set using a model estimated from the original data, we must estimate the ICCs for all items on the test. The nonparametric method of kernel smoothing as applied to ICC estimation by Ramsay (1991) is used to estimate these ICCs. Douglas (1997) showed that, under certain regularity conditions, the method of kernel smoothing to jointly estimate ICCs and examinee abilities is consistent as both the number of items and the number of examinees tends to infinity. Using a unidimensional estimate of examinee ability, kernel smoothing estimates the ICC for a test item at θ by essentially calculating a weighted moving average of examinee responses to the item. Examinee responses are weighted by the kernel function κ, which we take to be the quadratic function $\kappa(x) = 1 - x^2, |x| \leq 1; \kappa(x) = 0$, elsewhere. Kernel smoothing also uses a bandwidth h to control the amount of smoothing or borrowing of information from examinees judged to be close to θ in ability. The widely accepted optimal value of h that minimizes the mean squared error or the mean integrated squared error of the kernel-smoothed estimates is proportional to $J^{-0.2}$, where J is the number of examinees. Generally, kernel-smoothed estimates of the ICC value $P_i(\theta)$ are calculated for a fixed number of points equally spaced throughout the range of θ.

The kernel-smoothing estimation procedure tends to produce a good estimate of the ICC for each item when the value of θ is towards the middle of the θ range (Ramsay 1991). However, since the kernel-smoothing estimate of an ICC at θ is essentially a weighted average of examinee item responses, for a low value of θ, a disproportionate number of examinees with abilities greater than θ are used to obtain an estimate of $P_i(\theta)$, and for a high value of θ, a disproportionate number of examinees with abilities less than θ are used to obtain an estimate of $P_i(\theta)$. Thus, the kernel-smoothing procedure will produce positively biased estimates of $P_i(\theta)$ at the low end of the θ scale and negatively biased estimates of $P_i(\theta)$ at the high end of the θ scale.

To correct this bias near the endpoints of the θ scale, Rice (1984) used a second smoothing with a wider bandwidth near those endpoints. Müller (1991) corrected for this bias by using a different kernel function at the endpoints of the θ scale. Habing and Nitcheva (2000) found Müller's kernel function with a wider bandwidth minimized the mean square error of the ICC estimates in a variety of simulated data. Thus, for values of θ less than one bandwidth away from the endpoints of the θ scale, the kernel-smoothed estimate of $P_i(\theta)$ is found using this variation of Müller's method.

5 DIMTEST Without AT2

The new version of DIMTEST uses the resampling bias correction method from Section 4 to correct for the bias in the T_L statistic. Kernel smoothing is used to estimate the ICCs of each test item. Using these estimated ICCs and a specified examinee ability distribution, another test data set is generated and the DIMTEST statistic, denoted T_G, is calculated from these data. The data are resampled N times, and the average of the T_G values is calculated. If the test data are unidimensional, the expected difference between T_L and the average of the N T_G values will be approximately zero.

The seven steps in the new DIMTEST procedure are explained below. The differences between the original DIMTEST procedure and the new version are highlighted.

Step 1. Choose m items for the Assessment Subtest (AT). (Since there is no AT2 subset, we simply denote the assessment subtest AT.) The methods available for selecting the AT items are the same as in the original DIMTEST procedure. The goal is still to choose AT so that it has the best possible chance of being dimensionally homogeneous on a dimension distinct from the dimension or dimensions measured by the remaining items on the test.

Step 2. Define the PT Subtest to be the remaining $n - m$ items. Assign examinees to different subgroups according to their total score on the PT Subtest, denoted Z_{PT}. Define $X_{ij}^{(k)}$ and calculate $Y_j^{(k)}$, $\overline{Y}_j^{(k)}$, and $\hat{\sigma}_k^2$ as before (cf. (19.5), (19.6), and (19.7), respectively).

Step 3. Define $\hat{p}_i^{(k)}$ and calculate $\hat{\sigma}_{U,k}^2$ and $T_{L,k}$ as before (cf. (19.8), (19.9), and (19.10), respectively).

Step 4. Calculate S_k^2 as before, (19.11). The DIMTEST statistic T_L is now given by

$$T_L = \frac{\sum_{k=1}^{K} T_{L,k}}{\sqrt{\sum_{k=1}^{K} S_k^2}} . \tag{19.20}$$

Step 5. Estimate the unidimensional ICC of each test item using kernel smoothing. Obtaining these kernel-smoothed ICC estimates requires four steps.

Step 5.1. Calculate each examinee's total score on the PT items. Add a variable from the uniform $\mathcal{U}(0,1)$ distribution to each examinee's total score to break any ties. This random breaking of ties makes the kernel-smoothing procedure computationally easier to implement.

Step 5.2. Rank the adjusted total scores from Step 5.1 in order from the smallest to the largest. For each examinee j, divide the examinee's rank by $J+1$, where J is the total number of examinees. Denote this value $\hat{\theta}_j, j = 1, \ldots, J$.

Step 5.3. Choose a lattice of evenly spaced points, denoted $\theta^{(l)}$, on the θ scale at which to evaluate the ICC estimate $\hat{P}_i(\theta)$. Taking the θ scale as $\mathcal{U}(0,1)$, $\theta^{(l)} = l/51$, for $l = 1, \ldots, 50$.

Step 5.4. For the ith item and for $l = 1, \ldots, 50$, calculate

$$\hat{P}_i(\theta^{(l)}) = \frac{\sum_{j=1}^{J} \kappa\left(\frac{\theta^{(l)} - \hat{\theta}_j}{h}\right) X_{ij}}{\sum_{j=1}^{J} \kappa\left(\frac{\theta^{(l)} - \hat{\theta}_j}{h}\right)}, \tag{19.21}$$

where X_{ij} is the response of examinee j to item i, $\kappa(x) = 1 - x^2, |x| \le 1$, $\kappa(x) = 0$, elsewhere, is the kernel function, and $h = 0.8J^{-0.2}$ is the bandwidth. When $\theta^{(l)} < h$ or when $\theta^{(l)} > 1 - h$, the kernel-smoothing function tends to produce biased estimates. To correct for this bias at the endpoints, Müller's correction method with a wider bandwidth is used to obtain $\hat{P}_i(\theta^{(l)})$; see Habing and Nitcheva (2000) for details.

Step 6. Examinee responses to all test items are generated using the estimated ICCs calculated in Step 5. Each examinee's ability, denoted θ_j, is generated from the $\mathcal{U}(0,1)$ distribution. A response pattern for examinee j is obtained as follows. If $\theta_j < \theta^{(1)}$, set $P_i(\theta_j) = \hat{P}_i(\theta^{(1)})$. If $\theta_j > \theta^{(50)}$, set $P_i(\theta_j) = \hat{P}_i(\theta^{(50)})$. Otherwise, for $\theta^{(l-1)} < \theta_j < \theta^{(l)}$,

$$P_i(\theta_j) = \left[\frac{\theta_j - \theta^{(l-1)}}{\theta^{(l)} - \theta^{(l-1)}}\right] \hat{P}_i(\theta^{(l)}) + \left[1 - \frac{\theta_j - \theta^{(l-1)}}{\theta^{(l)} - \theta^{(l-1)}}\right] \hat{P}_i(\theta^{(l-1)}) \,.$$

Compare $P_i(\theta_j)$ to a randomly generated value from the $\mathcal{U}(0,1)$ distribution. If $P_i(\theta_j) > \mathcal{U}(0,1)$, $X_{ij} = 1$; otherwise, $X_{ij} = 0$.

Using this resampled data set and the same (AT, PT) partition as the original data, calculate another DIMTEST statistic according to Steps 2 through 4 and denote this statistic T_G.

Step 7. To reduce the random variation in the T_G statistic, the resampling procedure in Step 6 is repeated N times and the average, denoted $\overline{T}_G$, of the N T_G values is calculated. The final DIMTEST statistic,

$$T = \frac{T_L - \overline{T}_G}{\sqrt{1 + 1/N}}, \tag{19.22}$$

has an asymptotic standard normal distribution under the null hypothesis of unidimensionality (Froelich, 2000). Thus, the null hypothesis of unidimensionality is rejected at level α if T is larger than the $100(1-\alpha)$th percentile of the standard normal distribution.

Remark. Using the bias correction method from Steps 5 and 6 above, the asymptotic form of the DIMTEST statistic T had a slightly inflated Type I error rate in preliminary simulations. To improve the Type I error rate of the new DIMTEST procedure, the asymptotic variance of $T_{L,k}$ was changed to

$$S_k^{2''} = \frac{(\hat{\mu}_{4,k} - \hat{\sigma}_k^4) + \hat{\delta}_{4,k}}{J_k}, \tag{19.23}$$

where $\hat{\mu}_{4,k}$ and $\hat{\delta}_{4,k}$ are defined by (19.12) and (19.13), respectively. The DIMTEST statistic T_L is then given by

$$T_L'' = \frac{\sum_{k=1}^K T_{L,k}}{\sqrt{\sum_{k=1}^K S_k^{2''}}}. \tag{19.24}$$

The final DIMTEST statistic,

$$T'' = \frac{T_L'' - \overline{T}_G''}{\sqrt{1 + 1/N}}, \tag{19.25}$$

now has only an approximate asymptotic standard normal distribution as both the number of items and the number of examinees tend to infinity under the null hypothesis of unidimensionality.

6 Monte Carlo Simulation Study

The purpose of this simulation study is to assess the performance of the new resampling version of DIMTEST. The simulation study is split into a Type I error study and a power study. For the Type I error study, data were simulated using the unidimensional three-parameter logistic model (3PLM) for the ICCs. DIMTEST was then run on the simulated data and the number of rejections of the null hypothesis of unidimensionality was recorded. For the power simulations, one-third of the items were randomly chosen to be generated using the unidimensional 3PLM with examinee ability θ_2; the rest of the items were generated using the unidimensional 3PLM with examinee ability θ_1. DIMTEST was then run on the simulated data and the total number of rejections of the null hypothesis of unidimensionality was recorded. For all DIMTEST runs, the number of resamplings used for $\overline{T}_G$ was $N = 30$ and the items in the AT Subtest were chosen using an exploratory factor analysis program based on tetrachoric correlations; see Stout (1987) for details.

6.1 *Type I Error Study*

The items in the Type I error study were simulated using the unidimensional three-parameter logistic model,

$$P(X_i = 1 \mid \theta) = \gamma_i + \frac{1 - \gamma_i}{1 + \exp[-1.7\alpha_i(\theta - \beta_i)]}, \tag{19.26}$$

with examinee abilities θ generated from the standard normal distribution. The α and β item parameters were estimated from three real tests; an Armed Services Vocational Aptitude Battery Test (ASVAB) Auto Shop test with 25 items (from Mislevy & Bock 1984), an ACT Math (ACTM) test with 40 items (from Drasgow, 1987), and a SAT Verbal (SATV) test with 80 items (from Lord, 1968). For all tests, the γ parameter for each item was fixed at 0.15. The three tests were simulated with three different examinee sample sizes (750, 1500, 2000). For each DIMTEST run, the factor analysis program was run with (250, 500, 750) examinees and the DIMTEST program was run with the remaining (500, 1000, 1250) examinees, respectively. The number of items in the AT subset was fixed at roughly $n/4$ or $n/3$, or was allowed to vary based on the size of the item's second factor loadings (see Nandakumar & Stout, 1993, for details). The number of rejections out of 100 DIMTEST runs for each test is given in Table 19.1. The nominal rate of rejection is $\alpha = 0.05$.

The average Type I error rate is 0.053 for the 80-item SAT Verbal test, 0.050 for the 40-item ACT Math test, and 0.083 for the 25-item ASVAB Auto Shop test, with a total average Type I error rate of 0.062. Clearly, in

TABLE 19.1. Number of rejections out of 100 trials.

Test	ASVAB			ACTM			SATV		
AT Size	5	8	Vary	8	12	Vary	15	25	Vary
$J = 750$	6	7	1	5	4	3	8	3	7
$J = 1500$	6	10	4	5	8	3	2	5	6
$J = 2000$	10	22	9	4	6	7	4	7	6

most cases, the new DIMTEST procedure performs very near the nominal rate of $\alpha = 0.05$.

There is some potentially serious Type I error inflation present in the 25-item ASVAB Auto Shop test, especially when there are 8 items in the AT Subtest. In this situation, the ability estimates used to obtain the estimated ICCs are based on the examinee's total score on only the 17 PT items. Based on such a small number of items, these ability estimates will usually not be very close to their true values. This ability estimation error will likely cause poor kernel-smoothed estimates of the test items' ICCs. As discussed in Section 3, poor estimation of the ICCs can lead to Type I error inflation by not sufficiently correcting enough of the bias present in the T_L statistic. Possible remedies for these types of situations are under study and are discussed in Section 7.

6.2 Power Study

The intent of the power study was to design a realistic set of multidimensional models containing typical departures from unidimensionality that researchers would wish to detect. For this power study, each of the items was generated to measure either θ_1 or θ_2 alone using (19.26), thus producing a two-dimensional simple structure. For each test, roughly one-third of the items were randomly selected to be generated to measure θ_2 and the rest of the items were generated to measure θ_1. The same three standardized tests and estimated parameters were used in the power study. The same examinee sample sizes, factor analysis examinee sample sizes, and AT sizes used in the Type I error study were also used in the power study. Examinee abilities θ_1 and θ_2 were generated using the bivariate normal distribution with both means equal to 0, both variances equal to 1, and the correlation between dimensions equal to either $\rho = 0.4$ or 0.7. The number of rejections out of 100 DIMTEST runs for each of the nine simulated models with $\rho = 0.4$ was 100. The number of rejections out of 100 DIMTEST runs for each test with $\rho = 0.7$ is given in Table 19.2.

The simulation results show the power of the DIMTEST procedure is very high, reaching a 100% rejection rate in almost every simulated case.

TABLE 19.2. Number of rejections out of 100 trials.

Test	ASVAB			ACTM			SATV		
AT Size	5	8	Vary	8	12	Vary	15	25	Vary
$J = 750$	99	100	100	100	100	98	100	100	99
$J = 1500$	100	100	100	100	100	100	100	100	100
$J = 2000$	100	100	100	100	100	100	100	100	100

7 Discussion and Conclusions

The nonparametric procedure DIMTEST was developed by Stout (1987) and refined by Nandakumar and Stout (1993) to test the null hypothesis of unidimensionality of a test data set. To test for unidimensionality, the DIMTEST procedure divides the test items into two groups, the (dimensionality) Assessment Subtest (AT) and the (examinee) Partitioning Subtest (PT). To correct for the statistical bias in the DIMTEST statistic, Stout (1987) introduced a second set of assessment items, called AT2, to function parallel to the AT1 Subtest. This AT2 bias correction method has been shown to work fairly well in most cases. However, the Type I error rates of the DIMTEST procedure can be inflated when the AT1 Subtest items have similar difficulties or large discriminations. In addition, the required selection of an additional subtest of items can lead to problems in the procedure, especially when the test is relatively short or the AT1 Subtest is relatively long.

The new version of DIMTEST presented in this chapter attempts to eliminate the problems associated with AT2 by replacing it with a resampling method. By resampling the original data under the null hypothesis of unidimensionality, the T_G DIMTEST statistic calculated from this generated data will be approximately equal to the T_L DIMTEST statistic calculated from the original data when the null hypothesis of unidimensionality is true. If the null hypothesis is false, the difference between the two statistics will reflect the amount of multidimensionality in the original data.

A simulation study showed the new version of DIMTEST using resampling has a Type I error rate close to the nominal rate of $\alpha = 0.05$ in most cases. However, when the test was relatively short with a relatively long AT Subtest, the Type I error rate of the new DIMTEST was significantly larger than the nominal level of $\alpha = 0.05$. In this case, the kernel-smoothed estimates of the ICCs were likely not close enough to the ICCs of the original data to sufficiently correct for the bias in the T_L statistic.

In order to improve the estimation of the test items' ICCs, several changes in the kernel-smoothing procedure are being studied. First, the kernel-smoothing procedure only uses the total score on the PT items to estimate examinee ability. For short tests with a relatively long AT Subtest, this can lead to poor estimation of examinee ability and, in turn, poor estimation of the ICCs. To correct this problem, more effective examinee ability estimates are being studied, such as weighting examinee responses by an estimate of item difficulty or using the total score on all test items as an estimate of examinee ability. Preliminary results of this research show major improvement in the Type I error rate if examinee ability is estimated using the total score on all test items (Froelich, 2000). Second, the kernel-smoothing estimates are calculated for 50 fixed points throughout the θ scale. For a test with a small number of items, better kernel-smoothed estimates of the test's ICCs might be obtained with a smaller number of evaluation points.

A simulation study also showed the new version of DIMTEST using resampling has very high power to detect multidimensionality present in realistic simulated test data. A more extensive power study, including three- and four-dimensional data models, is currently being developed, in part to determine realistic situations where the new DIMTEST procedure might have difficulty detecting the multidimensional structure of the test.

The new version of DIMTEST using resampling removes a number of the problems associated with the AT2 Subtest. By closely approximating the ICCs of the original data, the new DIMTEST statistic T has better theoretical and statistical properties than the original DIMTEST statistic. This resampling method can also be applied in situations where choosing a reasonable AT2 Subtest is either not possible or very difficult, such as with polytomous or CAT items (Froelich, 2000).

Acknowledgments: The research presented in the chapter is part of the thesis research of the second and third authors under the direction of the first author.

References

Ackerman, T. (1996). Graphical representation of multidimensional item response theory. *Applied Psychological Measurement, 20*, 311–329.

Douglas, J. (1997). Joint consistency of nonparametric item characteristic curve and ability estimation. *Psychometrika, 62,* 7–28.

Drasgow, F. (1987). A study of the measurement bias of two standardized psychological tests. *Journal of Applied Psychology, 72* 19–30.

Froelich, A.G. (2000). *Assessing unidimensionality of test items and some asymptotics of parametric item response theory.* Unpublished doctoral dissertation, University of Illinois at Urbana-Champaign, Department of Statistics.

Gao, F. (1997). *DIMTEST enhancements and some parametric IRT asymptotics.* Unpublished doctoral dissertation, University of Illinois at Urbana-Champaign, Department of Statistics.

Habing, B., & Nitcheva, D. (2000). *Boundary correction for kernel smoothing IRF estimation.* Manuscript submitted for publication.

Hattie, J., Krakowski, K., Rogers, H.J., & Swaminathan, H. (1995). An assessment of Stout's index of essential unidimensionality. *Applied Psychological Measurement, 20,* 1–14.

Holland, P.W., & Rosenbaum, P.R. (1986). Conditional association and unidimensionality in monotone latent variable models. *The Annals of Statistics, 14,* 1523–1543.

Lord, F.M. (1968). An analysis of the verbal scholastic aptitude test using Birnbaum's three parameter logistic model. *Education and Psychological Measurement, 28*, 989–1020.

Mislevy, R.J., & Bock, R.D. (1984). *Item operating characteristics of the Armed Services Aptitude Battery (ASVAB), Form 8A*, (Technical Report N00014-83-C-0283). Washington, DC: Office of Naval Research.

Mislevy, R.J., & Bock, R.D. (1990). BILOG [Computer software]. Mooresville, IN: Scientific Software.

Müller, H. (1991). Smooth optimum kernel estimators near endpoints. *Biometrika, 78,* 521–530.

Nandakumar, R., & Stout, W. (1993). Refinements of Stout's procedure for assessing unidimensionality. *Journal of Educational Statistics,18,* 41–68.

Ramsay, J. (1991). Kernel smoothing approaches to nonparametric item characteristic curve estimation. *Psychometrika, 56,* 611–630.

Rice, J. (1984). Boundary modification for kernel regression. *Communications in Statistical Theory and Methods, 13,* 893–900.

Roussos, L., & Stout, W. (1996). A multidimensionality-based DIF analysis paradigm. *Applied Psychological Measurement, 20*, 335–371.

Roussos, L., Stout, W., & Marden, J. (1998). Using new proximity measures with hierarchical cluster analysis to detect multidimensionality. *Journal of Educational Measurement, 35,* 1–30.

Stout, W. (1987). A nonparametric approach for assessing latent trait unidimensionality. *Psychometrika, 52,* 589–617.

Stout, W. (1990). A new item response theory modeling approach with applications to unidimensionality assessment and ability estimation. *Psychometrika, 55,* 293–325.

Zhang, J., & Stout, W. (1999a). Conditional covariance structure of generalized compensatory multidimensional items. *Psychometrika, 64*, 129–152.

Zhang, J., & Stout, W. (1999b). The theoretical DETECT index of dimensionality and its application to approximate simple structure. *Psychometrika, 64*, 213–249.

20

Person Fit Across Subgroups: An Achievement Testing Example

Rob R. Meijer and
Edith M.L.A. van Krimpen-Stoop[1]

ABSTRACT Item response theory (IRT) models are used to describe answering behavior on tests and examinations. Although items may fit an IRT model, some persons may produce misfitting item score patterns, for example, as a result of cheating or lack of motivation. Several statistics have been proposed to detect deviant item score patterns. Misfitting item score patterns may be related to group characteristics such as gender or race. Investigating misfitting item score patterns across different groups is strongly related to differential item functioning (DIF). In this study the usefulness of person fit to compare item score patterns for different groups was investigated. In particular, the effect of misspecification of a model due to DIF on person fit was explored. Empirical data of a math test were analyzed with respect to misfitting item score patterns and DIF for men and women and blacks and whites. Results indicated that there were small differences between subgroups with respect to the number of misfitting item score patterns. Also, the influence of DIF on the fit of a score pattern was small for both gender and ethnic groups. The results imply that person-fit analysis is not very sensitive to model misspecification on the item level.

1 Introduction

Model fit is important when item response theory (IRT) models are used in practice. Let n persons answer k items; then fit of items can be investigated across persons or fit of persons can be investigated across items. Item fit is important because in psychological and educational measurement, instruments are developed that are used in a population of persons; item fit then can help the test constructor to develop an instrument that fits an IRT model in a particular population. Item fit statistics have been proposed by, for example, Molenaar (1983) and Glas (1999). As a next step, the fit of an

[1]Both authors are at the Department of Educational Measurement and Data Analysis, University of Twente, P.O. Box 217, 7500 AE Enschede, The Netherlands; e-mail: *meijer@edte.utwente.nl* and *krimpen@edte.utwente.nl*

individual item score pattern can be investigated. Although a test may fit an IRT model, persons may generate patterns that are unlikely given the model, for example, because they have foreknowledge of the correct answers to some of the most difficult items. Investigating person fit may help the researcher to obtain additional information to the number-correct score of a person. Given that the IRT model holds, the fit of a score pattern can be determined by means of a person-fit statistic. Some statistics can be used to obtain information at a subtest level and a more diagnostic approach can be followed.

Meijer and Sijtsma (in press) give an overview of person-fit statistics proposed for various IRT models. They also discuss studies in which person-fit statistics are applied to empirical data. From the few empirical studies that exist it can be concluded that there is some evidence that persons, such as test takers lacking motivation, may produce item scores that are unlikely given the model. However, as Rudner, Bracey, and Skaggs (1996), and Reise and Flannery (1996) emphasized, more research is needed using empirical data.

A number of authors have suggested that aberrant response behavior may be related to group membership (e.g., Van der Flier, 1982; Meijer, 1998; Schmitt, Chan, Sacco, McFarland, & Jennings, 1999). Some examples are given in Section 3. Person fit in the context of group membership is strongly related to differential item functioning (DIF). DIF analysis is concerned with difference in item responses between two persons with the same proficiency level belonging to different groups. An item shows DIF if, conditionally on the proficiency level, the probability of a correct response differs between groups. For example, assume that boys and girls with the same proficiency level have to answer items on a math test that are also related to verbal proficiency. Then, these items are subject to DIF when boys and girls with the same proficiency level in math differ in their verbal proficiency. In DIF research, items are flagged that perform differently for different subgroups of persons.

When person-fit analysis is used to investigate group differences, the complete item score pattern is taken into account, whereas in a DIF analysis group differences on the item level are investigated. Little research has been conducted in which both DIF analysis and person-fit analysis have been taken into account. Reise and Flannery (1996) suggested that person-fit analysis may help to decide when significant DIF results should be taken into consideration. Suppose that a DIF analysis results in a number of items that show DIF. Then, Reise and Flannery (1996) suggested that a person-fit analysis that results in a large difference between the percentage of misfitting item score patterns for different groups would indicate that model misspecification due to DIF has also practical importance.

One possibility for relating DIF and person fit is to compare person-fit results when DIF is incorporated into the model and when it is not incorporated. When DIF is incorporated into the model, different item parameters are used for different groups; otherwise the same item parameters are assumed to be equal for each group.

The aim of this study is twofold: to investigate the effect of DIF on the classification of item score patterns as fitting or misfitting, and to investigate if there are systematic differences between item score patterns of different groups of persons as has been suggested in the literature. In the present study we first discuss some existing person-fit statistics and discuss results of recent empirical research. Second, we conduct a person-fit analysis on an empirical data set in which we compare results in which DIF is taken into account with results in which DIF is not taken into account for different ethnic and gender groups. Finally, we discuss the implications of the results for further DIF and person-fit research.

2 Person Fit in IRT Models

Person-fit statistics have been proposed mostly in the context of IRT. However, analyzing item score patterns is almost as old as standardized testing. Already in the 1920s, patterns were analyzed to detect cheating behavior (Bird, 1929). In the 1960s group-based person-fit statistics were proposed to determine the consistency of an individual item score pattern with the other score patterns in the sample, where item scores were compared with the item proportion correct score. Many correct scores on items with low proportion correct scores and incorrect scores on items with high proportion correct scores indicated aberrant response behavior. With the increasing interest in IRT modeling, person fit has been investigated more and more within the IRT context.

In IRT the probability of obtaining a correct answer on item I_i, $i = 1, \ldots, k$, is a function of the latent trait value (θ) and characteristics of the item such as the location β (Hambleton & Swaminathan, 1985; Van der Linden & Hambleton, 1997).

Let X_i denote the binary (0,1) response to item I_i, α_i the item discrimination parameter, β_i the item difficulty parameter, and γ_i the item guessing parameter. Then the conditional probability $P_i(\theta)$ comprises the item response function which for the three-parameter logistic model (3PLM) can be written as

$$P_i(\theta) = \gamma_i + \frac{(1 - \gamma_i)\exp[\alpha_i(\theta - \beta_i)]}{1 + \exp[\alpha_i(\theta - \beta_i)]} . \tag{20.1}$$

The two-parameter logistic model (2PLM) can be obtained by fixing $\gamma_i = 0$ for all items; and the 1PLM or Rasch model (RM) can be obtained by additionally fixing $\alpha_i = 1$ for all items.

A major advantage of IRT models is that the goodness of fit of a model to empirical data can be investigated. To investigate the goodness of fit of item score patterns, several IRT-based person-fit statistics have been proposed.

Let $w_i(\theta)$ and $w_0(\theta)$ be suitable functions. Following Snijders (in press), a general form in which most person-fit statistics can be expressed is

$$\sum_{i=1}^{k} X_i w_i(\theta) - w_0(\theta) \ . \tag{20.2}$$

To have a person-fit statistic with expectation 0, many person-fit statistics are expressed in the centered form

$$V = \sum_{i=1}^{k} [X_i - P_i(\theta)]\, w_i(\theta) \ . \tag{20.3}$$

Most studies have been conducted using the loglikelihood function

$$l = \sum_{i=1}^{k} \{X_i \ln P_i(\theta) + (1 - X_i) \ln[1 - P_i(\theta)]\} \ . \tag{20.4}$$

This statistic, first proposed by Levine and Rubin (1979), was further developed by Drasgow, Levine, and Williams (1985) and Drasgow, Levine, and McLaughlin (1991). Two problems arise when l is used as a fit statistic. The first problem is that l is not standardized, implying that the classification of an item score pattern as normal or aberrant depends on θ. The second problem is that for classifying an item score pattern as aberrant, a distribution of the statistic under the null hypothesis of fitting response behavior, the null distribution, is needed. Solutions proposed for these two problems are the following.

To overcome the problem of dependence on the trait level and the problem of the unknown sampling distribution, Drasgow et al. (1985) proposed a standardized version of l, denoted l_z, which was less confounded with θ and which was purported to be asymptotically standard normally distributed. l_z was defined as

$$l_z = \frac{l - E(l)}{\sqrt{\text{Var}(l)}} \ , \tag{20.5}$$

where $E(l)$ and $\text{Var}(l)$ denote the expectation and the variance of l, respectively. These quantities are given by Drasgow et al. (1985):

$$E(l) = \sum_{i=1}^{k} \{P_i(\theta) \ln[P_i(\theta)] + [1 - P_i(\theta)] \ln[1 - P_i(\theta)]\} \ , \tag{20.6}$$

$$\text{Var}(l) = \sum_{i=1}^{k} P_i(\theta)[1 - P_i(\theta)] \left\{ \ln\left[\frac{P_i(\theta)}{1 - P_i(\theta)}\right] \right\}^2 \ . \tag{20.7}$$

Large negative values of l_z indicate aberrant response behavior. It can easily be shown that $l - E(l)$ can be written in the form of (20.2) choosing

$$w_i = \ln\left[\frac{P_i(\theta)}{1 - P_i(\theta)}\right].$$

Snijders (in press) derived the asymptotic sampling distribution for a family of person-fit statistics that have the form given in (20.3) and for which θ was replaced by $\hat{\theta}$, the estimated latent trait value for the responding person. Snijders (in press) also derived expressions for the first two moments of the distribution, $E[V(\hat{\theta})]$ and $\text{Var}[V(\hat{\theta})]$. He performed a simulation study for relatively small tests consisting of 8 and 15 items and for large tests consisting of 50 and 100 items, fitting the 2PLM, and using maximum likelihood estimation for θ. The results showed that the approximation was satisfactory for nominal Type I errors $\alpha = .05$ and $\alpha = .10$, but that the empirical Type I error was smaller than the nominal Type I error for smaller values of α. In fact, both the distribution of l_z and the version of l_z corrected for $\hat{\theta}$, denoted l_z^*, are negatively skewed (Snijders, in press; Van Krimpen-Stoop & Meijer, 1999). This skewness influences the difference between nominal and empirical Type I error rates for small αs. For example, Snijders (in press) found that for a 50-item test at $\alpha = .05$ the discrepancy between the nominal and the empirical Type I error for l_z and l_z^* at $\theta = 0$ was small (.001), whereas at $\alpha = .001$ for both statistics it was approximately .005. Van Krimpen-Stoop and Meijer (1999) found that increasing the item discrimination resulted in a distribution that was more negatively skewed. An alternative may be to use a χ^2 distribution for l_z; statistical theory that incorporates the skewness of the distribution is not yet available for the 2PLM or 3PLM. Note, however, that Molenaar and Hoijtink (1990) and Klauer and Rettig (1990) use χ^2 distributions for person-fit statistics in the Rasch model. In the empirical example given below, we use the standard normal distribution with $\alpha = .05$ and $\alpha = .025$ as critical values.

Most person-fit statistics only give information at the level of the complete item score pattern. When additional information can be obtained indicating the parts of the item score pattern where deviant behavior is present, this may help the researcher to interpret the type of aberrant behavior. For example, if the items in a test are ordered from easy to difficult, then many unexpected item scores at the beginning of the test may indicate careless behavior (Trabin & Weiss, 1983, p. 85). Sijtsma and Meijer (in press) discussed a statistical method in which an item score pattern is divided into G subsets of items. Let k increasingly more difficult items be collected in disjoint and mutually exclusive vectors $\mathbf{X}_g$ such that $\mathbf{X} = (\mathbf{X}_1, \mathbf{X}_2, \ldots, \mathbf{X}_G)$. Consider newly defined vectors $\mathbf{X}_{(g)}$ each of which contains two adjacent subsets $\mathbf{X}_g$ and $\mathbf{X}_{g+1}$: $\mathbf{X}_{(1)} = (\mathbf{X}_1, \mathbf{X}_2)$, $\mathbf{X}_{(2)} = (\mathbf{X}_2, \mathbf{X}_3), \ldots, \mathbf{X}_{(G-1)} = (\mathbf{X}_{(G-1)}, \mathbf{X}_{(G)})$. Let X_{+e} denote the total score on the easiest subset. A useful question in person-fit research is whether the number of correct answers on the more difficult items is exceptionally low given the total score and the subset scores. To test this, Sijtsma and Meijer (in press; see also Rosenbaum, 1987) showed that for each pair of subsets a conservative bound based on the hypergeometric distribution can be calculated for the probability that a person has at most $X_{+e} = x_{+e}$ correct scores on the easiest item subset. If for a particular pair this probability is lower than, say, .05, then the conclusion is that the total score on the first subset in this pair is unlikely, given that all items in the first subset are easier than the items in the second subset. Although this method is based on item response functions that are nonintersecting, Sijtsma and Meijer (in press) showed that the results are robust against violations of this assumption.

An example of a score pattern is

$$[0^5\ 1^5\ 0^2\ 1^2\ 0\ 1\ 0^3\ 1^6\ 0\ 1\ 0\ 1^4\ 0\ 1^5\ 0\ 1^2\ 0\ 1^3\ 0\ 1^2\ 0]\ ,$$

where a superscript denotes the number of consecutive 0 or 1 scores; thus $0^2\ 1^2$ denotes 0011. Inspecting this item score pattern, it can be noticed that this examinee with a total score of 31 has 14 correct answers on the 25 easiest items and 17 correct answers on the 25 most difficult items. This is especially clear by inspecting the Items 1 through 5 (mean proportion correct score .84) all answered incorrectly, and the Items 6 through 10 (mean proportion correct score .71) that were all correctly answered. Calculating the hypergeometric probability for these 10 items we find a probability of .004. Sijtsma and Meijer (in press) discussed that the real probability is lower; therefore, at a .01 level, we conclude that on the first 10 items of this score pattern deviant response behavior is present.

3 Empirical Research with Person-Fit Statistics

As several authors have emphasized, few studies analyze empirical data with respect to misfitting item score patterns. Exceptions are, for example, Birenbaum (1985) who investigated the relation between irregular response behavior and the scores on an anxiety scale and a lie scale of a personality test; Rudner et al. (1996) who investigated the usefulness of a person-fit statistic on a large-scale assessment program; and Schmitt, Cortina, and Whitney (1993) who investigated the influence of the presence of misfitting item score patterns on criterion-related validity estimates. Meijer and Sijtsma (in press) discussed the use of empirical person-fit research. They speculated that one of the reasons for the lack of empirical person-fit studies may be that person-fit statistics only inform the researcher that a score pattern does not fit the model without giving extra information about the type of aberrance. Furthermore, there is the difficulty of distinguishing between examinees with item score patterns for which the wrong IRT model is used and those with item score patterns that can be explained using additional information. They concluded that studies are needed that analyze empirical data together with background variables to obtain extra information about the type of aberrance.

Background variables were used by Van der Flier (1982). He explored the use of a nonparametric person-fit statistic in a cross-cultural setting, and found that using person-fit scores in addition to the number-correct scores on a predictor test improved prediction of the scores on a criterion test compared to using only the number-correct score on a predictor test.

Schmitt et al. (1999) investigated the relationship between testtaking motivation and conscientiousness with a multitest extension of l_z, denoted l_{zm}, for personality tests and a cognitive test. Also, the relationship between gender and race (blacks and whites) and l_{zm} was investigated. They suggested that misfit may be related to gender and race. Based on literature on visual inspection of item score patterns, men and blacks were expected to produce more unexpected item scores than women and whites, respectively. These differences may be explained by different testtaking motivation across race and gender subgroups. It was interesting that men had lower mean l_{zm} values (indicating aberrance) than women on both cognitive and personality tests, although controlling for conscientiousness reduced or eliminated this association; this was not the case for testtaking motivation. For cognitive tests, blacks had lower mean person-fit scores than whites. These scores were not affected by controlling for conscientiousness or testtaking motivation. For personality tests there was no difference in person-fit scores. Schmitt et al. (1999) concluded that carelessness may explain the misfit by men, whereas a more general traitlike measure may influence test reactions of blacks. If this conclusion is correct, then also in other test situations and with other tests, blacks should produce more irregular patterns than whites.

To explore the usefulness of person-fit statistics in empirical research and to investigate the difference between consistency of test behavior for different ethnic groups, we conducted an empirical study.

4 An Empirical Example

4.1 Test and Examinees

We analyzed data from a Math for Technology Test used by CTB/McGraw-Hill for a high school assessment program. The test consists of 50 items; each item has four options; one option contains the correct answer. The test measures the performance of high school students near completion of instruction in a math course. The respondents were between 14 and 18 years of age and came from different schools. Note that because this is an achievement test, lack of motivation is not a plausible reason for inconsistent test behavior. A data file of 5161 examinees was available; for 261 examinees, the values on the background variables were unknown; these patterns were removed from the data set. The data set contained 364 examinees with an ethnic background that was not registered. We also removed these patterns from the data set. So a data set consisting of 4536 examinees was analyzed. This data set consisted of 3811 whites, 725 blacks, 1843 women, and 2693 men.

4.2 Procedure and Data Analysis

At CTB/McGraw-Hill the 3PLM is used to estimate the item parameters. We reanalyzed the data set using the program BILOG-MG (Mislevy & Bock, 1990). Furthermore, we used l_z^* to determine the fit of the item score pattern.

Glas (2001) investigated the same CTB/McGraw-Hill data set with respect to DIF. To model DIF we used the same procedure as Glas (2001), who proposed a DIF procedure based on Lagrange multiplier (LM) tests in which DIF is considered as a special case of misfit of an IRT model. This procedure can be used in the context of the 3PLM, and DIF can be systematically related to background variables. This is done by entering an additional parameter into the model denoted β_{i1} for gender and β_{i2} for race. We used the results in which gender and race were entered as background variables. Glas found for 21 items significant main DIF effects using LM tests when gender was incorporated into the 3PLM, and 14 items with significant main effects when race was modeled. However, because of the large sample size the power of the statistics was high. Therefore, effect sizes were taken into account and only those items were considered as items subjected to DIF, for which the difference between the item difficulties for the different groups (men and women or blacks and whites) was larger than .50. Using this additional criterion, 9 items were classified with gender DIF and 6 items were classified with race-related DIF.

We conducted several analyses. First, we analyzed the data set with respect to person fit using the same item parameters for the whole group (no DIF modeling). Second, we conducted a person-fit analysis using different item estimates for men and women for the 9 items with gender DIF, and third we conducted a person-fit analysis with different item estimates for blacks and whites for the 6 items with race-related DIF. It is now interesting to investigate if a person-fit analysis is influenced by taking DIF into account.

4.3 Results

In Table 20.1 the mean and variance of $\hat{\theta}$ using maximum likelihood estimation is given for the total group and the different ethnic and gender groups

TABLE 20.1. Mean and variance of $\hat{\theta}$ for different groups.

DIF Model	Group	Mean	Variance
No Model	Total	.0325	1.332
	Men	.0344	1.573
	Women	−.1303	.964
	Whites	.0888	1.285
	Blacks	−.6698	1.093
Gender	Total	−.0697	1.164
	Men	−.0006	1.369
	Women	−.1635	.850
Race	Total	−.1156	1.263
	Whites	−.0010	1.210
	Blacks	−.7283	1.094

with and without correcting for DIF. It can be seen that without correcting for DIF (denoted No Model in Table 20.1) blacks had lower mean $\hat{\theta}$ values than whites and men scored somewhat higher than women. Taking gender-related DIF into account, the difference between the mean $\hat{\theta}$ values for men and women was similar compared to when DIF was not taken into account. Also, taking race-related DIF into account had only a small influence on the difference between the mean $\hat{\theta}$ values for blacks and whites.

In Table 20.2 the mean and variance of l_z^* and the proportion of item score patterns classified as misfitting at a .05 and .025 level are given both without correcting for DIF and when DIF is incorporated into the model. Note that the mean l_z^* values for most groups are somewhat larger than zero and the variance somewhat larger than one. The proportion of item score patterns classified as misfitting is larger than the nominal Type I errors for all groups, although not much larger. Snijders (in press) found using the 2PLM and simulated item scores for a 50-item test at a .05 nominal level empirical Type I errors between .052 and .056 across realistic θ values (between 0 and 2). At a nominal .01 Type I error rate Snijders (in press) found empirical Type I errors between .015 and .017 across different θ values. Thus, the empirical Type I errors may be somewhat larger even using item score patterns generated under an IRT model. Note, however, that because we do not know the number of *true* misfitting item score patterns we cannot report the detection rate. It is clear, however, that the number of misfitting item score patterns is modest.

From the results obtained by Schmitt et al. (1999), we expected that blacks would produce more inconsistent item score patterns than whites. In Table 20.2 it can be seen that without modeling DIF, the mean l_z^* was

TABLE 20.2. Mean and variance of l_z^* values, and proportion of item score patterns classified as misfitting at a nominal Type I error of .05 and .025.

DIF Model	Group	Mean	Variance	$l_z^* \leq -1.65$	$l_z^* \leq -1.96$
No Model	Total	.038	1.086	.063	.037
	Men	−.019	1.123	.070	.044
	Women	.120	1.021	.054	.027
	Whites	.045	1.083	.065	.038
	Blacks	.002	1.101	.054	.036
Gender	Total	.038	1.077	.062	.038
	Men	.007	1.114	.066	.042
	Women	.084	1.021	.055	.031
Race	Total	.042	1.077	.064	.036
	Whites	.043	1.076	.065	.036
	Blacks	.034	1.083	.058	.034

somewhat larger than zero for the total group, and that the mean l_z^* for blacks was smaller than for the whites. However, at a .05 level the proportion of patterns classified as misfitting was smaller for blacks than for whites. At a .025 level the proportions were almost the same. When correcting for race-related DIF at a .05 level, the proportion of item score patterns classified as misfitting for whites stayed the same, whereas the proportion of blacks increased by .004. With respect to gender it can be seen that, without correcting for DIF, women were on average more consistent than men (higher mean l_z^* value for women than for men). The number of men classified as misfitting was also higher than the number of women both at a .05 and .025 level. Correcting for gender-related DIF had a very small effect on the proportion of item score patterns classified as misfitting both for men and women. At a .05 level for men this proportion decreased by .004, whereas for women this proportion increased by .001. Similar differences can be observed at a .025 level.

To investigate whether the same examinees were classified as misfitting with and without correcting for DIF, we determined the overlap of item score patterns classified as misfitting for the different conditions. In Table 20.3 it can be seen that for both significance levels the overlap was between .931 and .878.

To illustrate how we can investigate local violations from model fit, we used the method proposed by Sijtsma and Meijer (in press); by inspecting local violations in an item score pattern, information may be obtained about the type of misfit. They suggested two strategies that can be followed when inspecting item score patterns on a subtest level: to inspect particular

TABLE 20.3. Proportion overlap in different analyses.

Analysis	$l_z^* \leq -1.65$	$l_z^* \leq -1.96$
No Model & Gender	.906	.913
No Model & Race	.931	.923
Gender & Race	.899	.878

subsets of items for which the researcher has a priori expectations that aberrant behavior may be present; and, when the researcher has no idea about the type of misfit expected, local violations can be investigated by subsequently investigating different subsets of item scores in a pattern. Although we did not have specific ideas about the type of misfit for this test, it has been suggested in the literature (Trabin & Weiss, 1983) that for an achievement test, some examinees may be careless in answering the easiest items, resulting in unexpected item scores at the easiest items. To investigate this, we first ordered the items in decreasing item popularity and then divided the first 10 items (the easiest items in the test) into two subsets of five items each and calculated the hypergeometric probability using the scores on both subtests. We found that .07% of the item score patterns were classified as aberrant using this method.

5 Discussion

In this study, we investigated the effect of DIF on the classification of an item score pattern as fitting or misfitting. The first conclusion is that person-fit results were similar with and without taking DIF into account. Our analysis with respect to gender-related DIF showed that there existed a large overlap between the percentage of testees classified as misfitting when DIF was modeled and when DIF was not taken into account. This implies that, in this study, misspecification of a model due to DIF had little effect on the fit of an item score pattern. Although a substantial number of items showed DIF, the person-fit study showed only small differences between the number of item score patterns classified as misfitting in the different analyses. This also implies that person-fit statistics are not informative as an indicator when DIF results have practical importance. For example, we found that when we did not model DIF, the proportion of item score patterns that was classified at $\alpha = .05$ as misfitting for the whole group was .063 (Table 20.2), whereas this was .062 when corrected for gender DIF and .064 when corrected for race-related DIF. These numbers suggest that there are no differences in the fit of item scores under the different conditions and that on the item level the different item parameters had no effect on the fit of the item score patterns. This was confirmed by the fact that the overlap between the item score patterns classified as aberrant under the different conditions was high. In the literature it has been suggested (Reise & Flannery, 1996) that person fit may help to decide when significant DIF results should to be taken into consideration. Results of this study indicate that it may be very difficult to evaluate the practical importance of DIF by means of a person-fit statistic. An alternative may be to formulate specific models in which DIF is incorporated into the alternative model. It would be interesting to formulate such extensions for the 2PLM and the 3PLM.

With respect to the number of item score patterns classified as misfitting, the results of this study are in agreement with the results obtained by Rudner et al. (1996), who did not find many misfitting item score patterns in a large-scale assessment program. We also found no differences in the number of misfitting item score patterns for blacks and whites. This may be due to the type of test: both in the Rudner et al. (1996) study and the present study achievement tests were used. For these types of tests examinees are highly motivated to obtain high scores and misfit due to carelessness or motivation problems is unlikely. In the situation described by Schmitt et al. (1993) examinees took a test for validating purposes and in these situations the presence of aberrant response behavior is more likely.

In the present study we used significance levels of .05 and .025. In empirical person-fit research and also in research to detect cheating (Holland, 1996) often very small Type I errors are used because of the possible consequences of classifying a fitting score pattern or an honest person as misfitting or cheater, respectively. Holland (1996), for example, reports that at Educational Testing Service Type I errors of .0001 are used. However, for such small values empirical Type I error rates for l_z^* are not in agreement with the nominal Type I errors (Snijders, in press).

References

Bird, C. (1929). An improved method of detecting cheating on objective examinations. *Journal of Educational Research*, *19*, 341–348.

Birenbaum, M. (1985). Comparing the effectiveness of several IRT based appropriateness measures in detecting unusual response patterns. *Educational and Psychological Measurement*, *45*, 523–534.

Drasgow, F., Levine M.V., & McLaughlin, M.E. (1991). Appropriateness for some multidimensional test batteries. *Applied Psychological Measurement*, *15*, 171–191.

Drasgow, F., Levine, M.V., & Williams, E.A. (1985). Appropriateness measurement with polychotomous item response models and standardized indices. *British Journal of Mathematical and Statistical Psychology*, *38*, 67–86.

Glas, C.A.W. (1999). Modification indices for the 2-PL and nominal response model. *Psychometrika, 64*, 273–294.

Glas, C.A.W. (2001). Differential item functioning depending on general covariates. In A. Boomsma, M.A.J. van Duijn, & T.A.B. Snijders (Eds.), *Essays on item response theory* (pp. 131–148). New York: Springer-Verlag.

Hambleton, R.K., & Swaminathan, H. (1985). *Item response theory: Principles and applications.* Boston: Kluwer-Nijhoff.

Holland, P.W. (1996). *Assessing unusual agreement between the incorrect answers of two examinees using the K-index: Statistical theory and empirical support* (ETS Technical Report No. 96–4). Princeton, NJ: Educational Testing Service.

Klauer, K.C., & Rettig, K. (1990). An approximately standardized person test for assessing consistency with a latent trait model. *British Journal of Mathematical and Statistical Psychology, 43*, 193–206.

Levine, M.V., & Rubin, D.B. (1979). Measuring the appropriateness of multiple-choice test scores. *Journal of Educational Statistics, 4*, 269–290.

Meijer, R.R. (1998). Consistency of test behavior and individual difference in precision of prediction. *Journal of Occupational and Organizational Psychology, 71*, 147–160.

Meijer, R.R., & Sijtsma, K. (in press). A review of methods for evaluating the fit of item score patterns on a test. *Applied Psychological Measurement.*

Mislevy R.J., & Bock, R.D. (1990). BILOG user's guide [Software manual]. Chicago: Scientific Software.

Molenaar, I.W. (1983). Some improved diagnostics for failure of the Rasch model. *Psychometrika, 48*, 49–73.

Molenaar, I.W., & Hoijtink, H. (1990). The many null distributions of person fit indices. *Psychometrika, 55*, 75–106.

Reise, S.P., & Flannery, W.P. (1996). Assessing person-fit measurement of typical performance applications. *Applied Measurement in Education, 9*, 9–26.

Rosenbaum, P.R. (1987). Probability inequalities for latent scales. *British Journal of Mathematical and Statistical Psychology, 40*, 157–168.

Rudner, L.M., Bracey, G., & Skaggs, G. (1996). The use of a person-fit statistic with one high quality achievement test. *Applied Measurement in Education, 9*, 91–109.

Schmitt, N., Chan, D., Sacco, J.M., McFarland, L.A., & Jennings, D. (1999). Correlates of person fit and effect of person fit on test validity. *Applied Psychological Measurement, 23*, 41–53.

Schmitt, N., Cortina, J.M., & Whitney, D.J. (1993). Appropriateness fit and criterion-related validity. *Applied Psychological Measurement, 17*, 143–150.

Sijtsma, K., & Meijer, R.R. (in press). The person response function as a tool in person-fit research. *Psychometrika.*

Snijders, T.A.B. (in press). Asymptotic null distributions of person-fit statistics with estimated person parameter. *Psychometrika.*

Trabin, T.E., & Weiss, D.J. (1983). The person response curve: Fit of individuals to item response theory models. In D.J. Weiss (Ed.), *New horizons in testing* (pp. 83–108). New York: Academic Press.

Van der Flier, H. (1982). Deviant response patterns and comparability of test scores. *Journal of Cross-Cultural Psychology, 13*, 267–298.

Van der Linden, W.J., & Hambleton, R.K. (Eds.). (1997). *Handbook of modern item response theory.* New York: Springer-Verlag.

Van Krimpen-Stoop, E.M.L.A., & Meijer, R.R. (1999). Simulating the null distribution of person-fit statistics for conventional and adaptive tests. *Applied Psychological Measurement, 23*, 327–345.

21

Single-Peaked or Monotone Tracelines? On the Choice of an IRT Model for Scaling Data

Wendy J. Post,[1] Marijtje A.J. van Duijn,[2] and Berna van Baarsen[3]

ABSTRACT It is investigated how a motivated choice can be made for the analysis of an item set using either a cumulative item response model with monotone tracelines, modeling dominance relations, or a unimodal item response model with single-peaked tracelines, modeling proximity relations. The focus is on item sets consisting of positively and negatively formulated items (with respect to the latent trait to be measured), where the common practice is to reverse one type of item. The differences between the cumulative and unimodal model are studied theoretically, in terms of item location and item order, and empirically, in a reanalysis of a sample of De Jong Gierveld loneliness scale data. Item locations, and, in the case of the unimodal model, also subject locations are shown to be important determinants of the differences. For the loneliness scale data the analysis with the unimodal model is preferred over the cumulative model. An outline of a recommended strategy for an IRT analysis of scaling data is given.

1 Introduction

In item response theory (IRT) a distinction is made between models with monotone tracelines and models with single-peaked or unimodal tracelines. In both types of IRT models it is assumed that there is an underlying unidimensional latent trait that governs the response behavior to items

[1]Department of Medical Statistics, Leiden University Medical Center, P.O. Box 9604, 2300 RC Leiden, The Netherlands; e-mail: *w.j.post@mta.azg.nl*

[2]Department of Statistics and Measurement Theory, University of Groningen, Grote Kruisstraat 2/1, 9712 TS Groningen, The Netherlands; e-mail: *m.a.j.van.duijn@ppsw.rug.nl*

[3]Department of Philosophy and Medical Ethics, Free University, Van der Boechorststraat 7, 1081 BT Amsterdam, The Netherlands; email: *b.vbaarsen.metamedica@med.vu.nl*

chosen to measure that latent trait, and that both subjects and items can be represented on the same scale. The traceline, or item characteristic curve (ICC), is a function of the latent trait that represents the probability of a certain response to an item, given the subject's value on the latent trait.

In cumulative IRT models, the probability of a positive response to an item by a subject is an increasing function of the difference between the subject's location on the latent trait, and the item location. Thus, *dominance relations* are modeled. The latent trait is usually interpreted as ability and the item location as item difficulty.

In unimodal IRT models, also called unfolding models, *proximity relations* are modeled: the probability of subjects giving a positive response to an item is a decreasing function of the distance (i.e., the absolute difference) between a subject's location and the item location, yielding a unimodal or single-peaked traceline. In unfolding models, the latent trait often represents a certain attitude or preference.

In this chapter, we investigate the theoretical implications of using a cumulative IRT model when the true model has single-peaked ICCs and vice versa. Our interest in this topic arose from an application of the De Jong Gierveld loneliness scale, a well-established 11-item cumulative scale with positively and negatively formulated items (De Jong Gierveld & Kamphuis, 1985). In a large sample of elderly people from the NESTOR study *Living Arrangements and Social Networks of Older Adults* (Knipscheer, De Jong Gierveld, Van Tilburg, & Dykstra, 1995), hereafter called the loneliness scale data, an analysis with a cumulative IRT model revealed two scales, one with only positively formulated items, the other with only negatively formulated items (Van Baarsen, Snijders, Smit, & Van Duijn, in press). The applied research question is whether an analysis of these data with a unimodal IRT model results in one scale instead of two scales.

In the next section the background of the research question is further explored. In Section 3 the differences between analyses with cumulative and unimodal IRT models under different conditions are studied. In Section 4 first a short description is given of the De Jong Gierveld loneliness scale and the problems researchers face when using this item set. Next, the loneliness scale data are reanalyzed using MUDFOLD, a nonparametric scaling approach for items with single-peaked tracelines (Van Schuur, 1984; Van Schuur & Post, 1998). The chapter is completed with a summary of the conclusions and a discussion.

2 Choosing an IRT Model

The research question focuses on scale analysis of item sets containing both positively and negatively formulated items. To avoid response biases such as agreement response tendency, many questionnaires are constructed in that way (Rorer, 1965; see also Mellenbergh, 2001, Sect. 6.1).

Theoretically, the choice for a unimodal or a cumulative IRT model is determined by the type of items selected to measure a certain latent trait (cf. Mellenbergh, 2001). In practice, however, it may be unclear whether an item represents a dominance or a proximity relation. The cumulative model is applied more often than the unfolding model, also for the analysis of item sets that cannot be assumed a priori to represent dominance relations. That is, for example, the case with item sets containing both positively and negatively formulated items.

To see why, suppose that the positively formulated items are indicative of a low amount of the latent trait, and that the negatively formulated items are indicative of a high amount of the latent trait. Then, the probability of an affirmative response to a negatively formulated item increases with the latent trait, whereas the probability of an affirmative answer to a positively formulated item decreases with the latent trait. This implies that the tracelines of the negatively formulated items are increasing, and those of the positively formulated items are decreasing. Hence the assumption of increasing tracelines made in cumulative IRT models is violated. The common practice in the analysis of such an item set is to reverse the items with decreasing tracelines (here, the positively formulated items) so that all items can be assumed to be indicative of the same pole (here, a high amount) of the latent trait.

This reversal strategy leading to a scale with regular and reverse-scored items is not necessary in an analysis with a unimodal IRT model, where it is assumed that items are answered affirmatively by subjects close to these items. For the case presented above, the positively formulated items are answered affirmatively by subjects sufficiently close to these items, that is, with a low amount of the latent trait, and the negatively formulated items are answered affirmatively by subjects who possess a high amount of the latent trait. Monotone increasing or decreasing tracelines can be viewed as a limiting case of single-peaked tracelines, and therefore, the cumulative IRT model (possibly with reverse-scored items) is a special limiting case of the unimodal IRT model.

In addition, when no a priori assumptions are made about the item type—dominance or proximity—a unimodal IRT analysis will reveal whether the tracelines are in fact monotone increasing and decreasing or unimodal, as shown theoretically in Section 3 and empirically in Section 4. Thus, it is possible to choose between a cumulative and a unimodal IRT model.

2.1 Previous Research

Recent IRT research dealt with some of the differences between unimodal and cumulative IRT models. Klinkenberg (2001) showed that the use of a cumulative model with item reversal for the analysis of items with single-peaked tracelines is justified under special conditions. These conditions include that the tracelines are known to be either increasing or decreasing. Klinkenberg's case is discussed in more detail in Section 3.2.

For the opposite case, Van Schuur (1998) showed that sometimes two cumulative scales are mistaken for one unimodal scale. This may occur if the items of the first cumulative scale have increasing tracelines and are located to the left of the items of the second cumulative scale that have decreasing tracelines. Earlier, Van Schuur and Kiers (1994) demonstrated that factor-analyzing unidimensional scaling data with single-peaked item tracelines resulted in two factors, indicating two different dimensions (or two different scales), which are in fact the two poles of an unfolding scale.

2.2 Item Reversal and Item Content

The reversal strategy described above should not be performed uncritically. Certain implicit assumptions may be violated. First, questions can be raised about item content in relation to reverse-scored items. In general, researchers implicitly assume that regular and reverse-scored items measure the same latent trait (Schriesheim & Eisenbach, 1995). There is empirical evidence, however, that this is not always the case (Ahlawat, 1985; Chang, 1995). Responding affirmatively to a positively formulated item does not necessarily imply not responding affirmatively to the associated negatively formulated item, and vice versa (Rorer, 1965; Holleman, 2000). Using reverse-scored items can also have consequences for the reliability and/or validity of the item set (Schriesheim, Eisenbach, & Hill, 1991; Schriesheim & Hill, 1981).

Second, if an item set containing both regular and reverse-scored proximity items is analyzed with a cumulative IRT model, it is (wittingly or unwittingly) assumed that Klinkenberg's case holds: only decreasing and increasing tracelines are present in the data. This assumption may not hold in a particular item set. Hemker (2001) defines a condition under which item reversal is allowed by requiring that the traceline of the reversed item have the same form as the traceline of the unreversed item. Klinkenberg's case conforms to this requirement: the tracelines of regular and reverse-scored items are all monotone. Reversing items with single-peaked tracelines, however, yields items with differently shaped tracelines ("single-dipped" instead of single-peaked).

In the next section the theoretical consequences of ignoring violations of the assumptions entailed by item reversal are investigated.

3 Monotone versus Single-Peaked Tracelines

For the theoretical considerations in this section it is assumed that item reversal can be done without consequences for the item content. Under different conditions the differences between cumulative IRT models and unfolding models with respect to item tracelines, item locations, and item order are studied. In the first condition the ICCs are monotone (i.e., increasing or decreasing). In that case both cumulative IRT models (after reversal of the decreasing tracelines) and unimodal IRT models hold, because the monotone tracelines are boundary versions of single-peaked tracelines. In the second condition the ICCs are unimodal so that cumulative IRT models do not hold. For both conditions it is investigated whether the choice of the IRT model has consequences for the item order and, if so, in which cases choosing a "wrong" IRT model has serious consequences.

In the following, for the benefit of Figures 21.1 and 21.2, monotone tracelines are postulated conformable with the requirements of the Rasch model—hereafter called the cumulative model. The single-peaked tracelines are assumed to be a unimodal equivalent of the Rasch model, conformable with the unfolding requirements of tracelines made by Post and Snijders (1993)—hereafter called the unfolding model. These unfolding requirements involve unimodality of the tracelines and two regularity conditions necessary to derive testable consequences in diagnostic statistics.

3.1 Monotone Tracelines

Assume that there are two items, A and B, with decreasing tracelines, and two items, C and D, with increasing tracelines. For four cases with different item locations, depicted in Figure 21.1, the item orders in the cumulative and unfolding model are compared.

In the cumulative model, the location of an item is defined as the value of the latent trait where the probabilities of a positive and a negative answer to the item are equal, that is, 0.5. In this model reversing an item does not alter its location. The definition of the location of an item in the unfolding model is based on the proximity character of the items: the smaller the distance between subject and item, the larger the probability of a positive response to this item. An unfolding item is therefore represented by the mode of its traceline. Reversing an unfolding item alters its location, but in an analysis with the unfolding model, items do not need to be reversed.

The item locations of the unfolding model are given at the top of the panel, and those of the cumulative model at the bottom.

Case a. First consider the case depicted in Figure 21.1a where the ICCs of Items A and B are situated left of those of Items C and D. In both the unfolding and the cumulative model item order $ABCD$ results. The item locations of the two models, however, are different.

The difference between the item locations of decreasing and increasing tracelines is larger in the unfolding model than in the cumulative model. The difference in location for items with either decreasing or increasing tracelines is the same in both IRT models, as long as the tracelines are

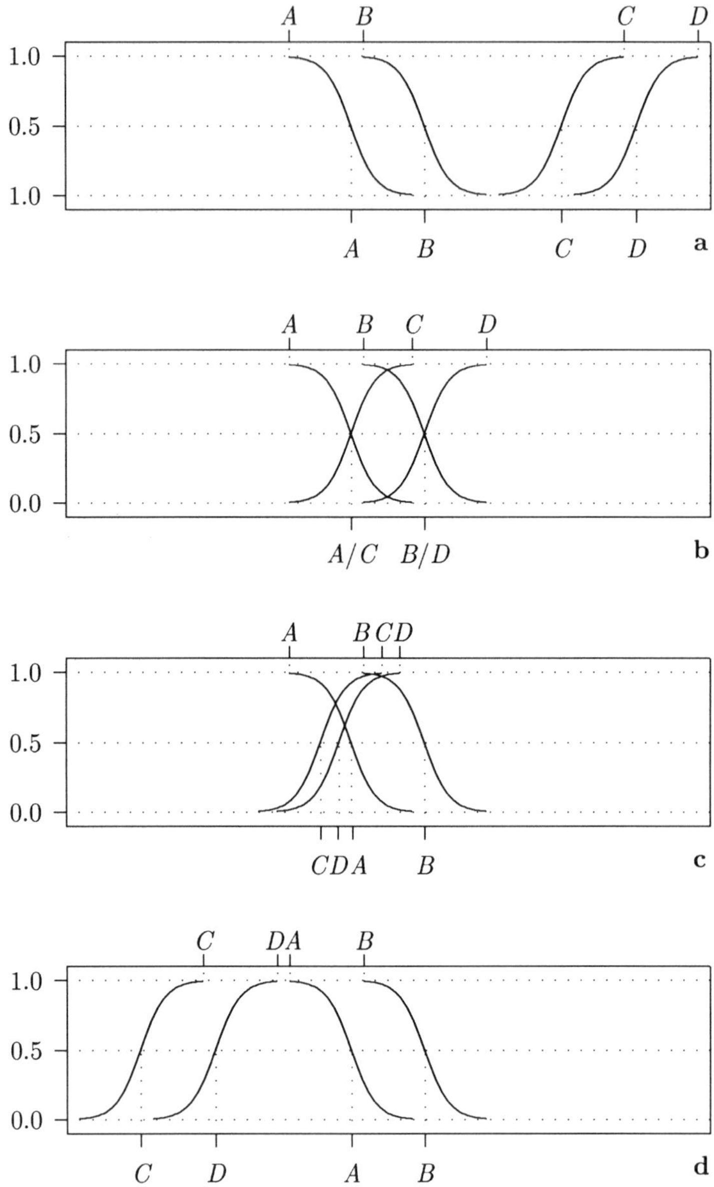

FIGURE 21.1. Four cases with different item locations of four items having monotone tracelines; item locations of the unfolding model are given at the top of the panels, those of the cumulative model at the bottom.

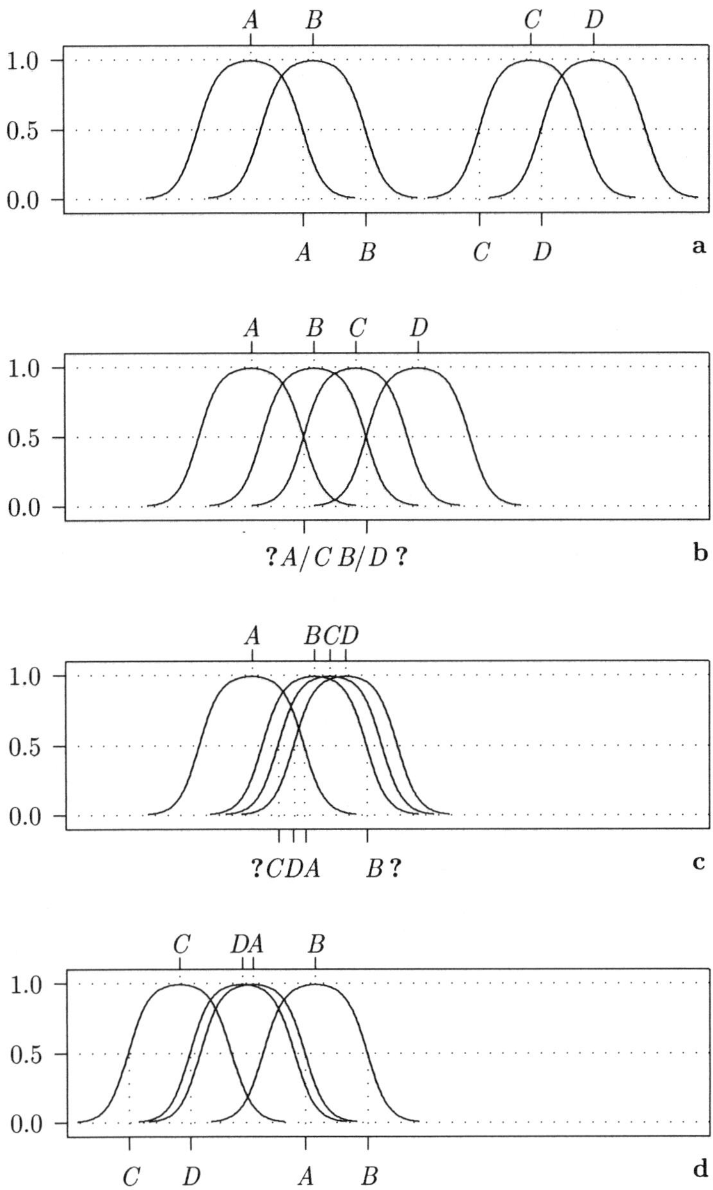

FIGURE 21.2. Four cases with different item locations of four items having single-peaked tracelines; item locations of the unfolding model are given at the top of the panels, those of the cumulative model (after reversal of Items A and B) at the bottom.

assumed to be equally shaped. Allowing the tracelines to have different slopes (not shown) causes different shifts in item locations (the steeper the slope, the smaller the shift), which may also alter the item order among items with the same type of monotone tracelines.

Case b. In Figure 21.1b the ICCs of Items C and D are moved to the left in such a way that they are exactly equal to the reversed ICCs of Items A and B. The unfolding model distinguishes four different items with item order $ABCD$. In the cumulative model, however, Item A (after reversal) and Item C are indistinguishable (i.e., equivalent), as are Items B and D. The item order is therefore $(A/C)(B/D)$. This case represents a boundary case in the sense that the outer Items A and D can still be considered to be the "extreme" items, but the order of the middle Items B and C is already reversed.

Case c. In Figure 21.1c the tracelines of Items C and D are moved still farther to the left. The result is that the unfolding order remains $ABCD$, whereas the cumulative order becomes $CDAB$. So here the order is changed even more: the extreme Items A and D and middle Items B and C have traded places. With unfolding order $ABCD$ all other cumulative orders are possible, if the items are close together and different traceline slopes are allowed (not shown).

Case d. In Figure 21.1d the case is depicted where both models result in the same item order $CDAB$. The ICCs of Items C and D are moved so far to the left that their modes are now to the left of those of the ICCs of Items A and B. Note, however, that in the cumulative model the item locations are more extreme than in the unimodal model: Items C and D are located farther to the left, and Items A and B are farther to the right. This case corresponds to the situation described by Van Schuur (1998).

From the four situations depicted in Figure 21.1 it is clear that, although both the cumulative and the unfolding model hold, the results for both models in terms of item order are not always the same. This is caused by the difference in the definition of the item locations. When the items are relatively close to each other, as in cases b and c, the item order resulting from using a cumulative IRT model is different from its unimodal counterpart.

3.2 Single-Peaked Tracelines

Assume that there are four items, A, B, C, and D, with single-peaked tracelines, so that only the unimodal model is valid. Four cases, with the same item locations under the unfolding model as in Figure 21.1 are depicted in Figure 21.2. The item orders in the cumulative and unimodal model are compared for these cases.

Case a. In Figure 21.2a Items A and B represent the left side of the latent trait, and Items C and D the right side. No items are located in the middle of the scale. If Items A and B are reversed, and the cumulative model is used, item order $ABCD$ results in item locations as indicated at the bottom of Figure 21.2a, identical to those in Figure 21.1a. Klinkenberg (2001) described this case with the additional assumption that the increasing parts of the tracelines of Items A and B and the decreasing parts of the tracelines of Items C and D are not "visible" in the data. This means that the tracelines show no "peak," that is, do not change from increasing to decreasing. Then Figure 21.2a can be approximated by Figure 21.1a.

The visibility of the tracelines depends on the location of both items and subjects. In Klinkenberg's case the items were all extreme (i.e., located either on the left or right side of the scale), with a gap between them, where most of the subjects were located. If subjects are located on both sides of an item, however, the peak of a unimodal traceline is visible, because subjects may not choose that item for two reasons: they are either too far to the left or too far to the right of the item. Empirically, the visibility of an ICC also depends on the numbers of subjects located to the left and to the right of the item.

In Figure 21.2a Klinkenberg's case implies that the subjects are located between unimodal item locations B and C. When, instead, the subjects are distributed over the complete range of the latent trait, it is unclear where the items are located in a cumulative model and Figure 21.2a cannot be approximated by Figure 21.1a.

Cases b and c. In Figures 21.2b and 21.2c, where the tracelines of Items C and D are moved farther to the left, the unfolding order is still $ABCD$. The order given at the bottom of these panels is the order that would be obtained with a cumulative model after reversing Items A and B, if the subjects were located between (unfolding locations) B and C, that is, if the subject distribution were over an interval in which no changing from increasing to decreasing of the single-peaked traceline occurs. The smaller the distance between the items, the less likely that situation. If the subjects are instead distributed over a larger part of the latent trait, the cumulative order cannot be deduced.

Case d. Figure 21.2d represents the case where the single-peaked items have the same locations as in Figure 21.1d. If it is assumed that only the increasing parts of the ICCs of Items C and D are present in the data, and only the decreasing parts of Items A and B, then it is implied that subjects are located either at the right-hand side of the scale (more to the right than Item B), or at the left-hand side (farther to the left of Item C). This model may be viewed as a boundary case of the circumplex model proposed by Mokken, Van Schuur, and Leeferink (2001). Theoretically, both the unfolding and the cumulative model will give the same item order:

CDAB. Algorithms for finding an unfolding scale, however, will mostly transform the scale to the case depicted in Figure 21.2a, by putting the modes of the tracelines in the center of the scale, resulting in item order *ABCD*.

From the four situations shown in Figure 21.2, where the unimodal IRT model is the true model, the consequences of analyzing these items with a cumulative model depend on the subject locations. In general, an analysis with the cumulative model will not give valid results in terms of item order. The only exception is the case where the subjects are located in an interval without traceline peaks, which is most likely to occur if the items are located on either side of the latent trait as depicted in Figure 21.2a.

Summary. On the basis of Figures 21.1 and 21.2, it can be concluded that analyses with cumulative and unimodal models give different results. If the cumulative model is the true model, the unimodal model is not wrong, although it defines item locations differently which may result in a different item order (Figures 21.1b and 21.1c). If the unfolding model is the true model, the cumulative model is wrong in general, except for one case (Figure 21.2a) with the additional assumption that the subjects are located between unimodal item locations *B* and *C*.

Therefore, when it is not clear a priori what the form of the tracelines is, an analysis with an unfolding model is to be preferred. If the tracelines are found to be monotone, an analysis with a cumulative model can be performed. In that case, the item order defined by the cumulative item locations may be considered to be more meaningful than its unfolding counterpart (cf. Klinkenberg, 2001).

In the next section, this analysis strategy is applied to empirical data of the De Jong Gierveld loneliness scale.

4 Reanalysis of the Loneliness Scale Data

First the De Jong Gierveld Loneliness scale and some previous research on the scale is introduced in Section 4.1. After describing the loneliness scale data in Section 4.2, a unimodal IRT analysis is performed with MUDFOLD (Van Schuur & Post, 1998) in Section 4.3 and compared in Section 4.4 with the results of a cumulative IRT analysis with the Mokken model using MSP (Molenaar & Sijtsma, 2000). In addition, sample invariance is examined for some subgroups (men/women, with/without a partner), and the subject distributions and empirical tracelines are estimated in Section 4.5.

4.1 The De Jong Gierveld Loneliness Scale

The De Jong Gierveld loneliness scale (De Jong Gierveld & Kamphuis, 1985), also known as the loneliness scale, was devised to measure high as

TABLE 21.1. The 11 items of the De Jong Gierveld loneliness scale.

A	There is always someone I can talk to about my day to day problems	+
B	I miss having a really close friend	–
C	I experience a general sense of emptiness	–
D	There are plenty of people I can lean on in case of trouble	+
E	I miss the pleasure of company of others	–
F	I find my circle of friends and acquaintances too limited	–
G	There are many people that I can count on completely	+
H	There are enough people that I feel close to	+
I	I miss having people around	–
J	Often I feel rejected	–
K	I can call on my friends whenever I need them	+

well as low levels of loneliness and has been used in various research projects (Van Baarsen, Smit, Snijders, & Knipscheer, 1999; Dykstra, 1995; De Jong Gierveld & Van Tilburg, 1995; Stevens, 1989; for an overview see De Jong Gierveld & Van Tilburg, 1999). In Table 21.1 the five positively formulated items (indicated by +) *A*, *D*, *G*, *H*, and *K*, and the six negatively formulated items (indicated by –) *B*, *C*, *E*, *F*, *I*, and *J*, of the loneliness scale are given.

Considering their content, the items of the loneliness scale may be viewed as proximity items. For instance a respondent may answer "no" to a negatively formulated item such as Item *I*, "I miss having people around," because of not being lonely at all and having enough people around, or because of being indeed lonely but used to not having people around and not missing the company of others at all.

The items of the loneliness scale are commonly analyzed, however, as dominance items. Using three possible response categories, no, more or less, and yes, the items are dichotomized by assigning a one to yes and a zero to the no and more or less answer options, and recoding the positively formulated items (cf. De Jong Gierveld & Van Tilburg, 1999). This means that the positively and negatively formulated items are scored differently, the more or less response coded as 1 in both types of items. This asymmetric way of coding presumes that respondents are unwilling to admit that they are lonely.

De Jong Gierveld and Kamphuis (1985) showed that the loneliness scale meets the criteria of the dichotomous Rasch model reasonably well. The homogeneity of the scale is not very strong, however, since Loevinger's H-coefficient typically ranges from 0.30 to 0.50 (De Jong Gierveld & Van Tilburg, 1999, p. 4).

In the literature, the unidimensionality of the loneliness scale was a topic of research and debate. Moorer and Suurmeijer (1993) questioned the coding instructions of De Jong Gierveld and Van Tilburg to dichotomize the scores as described above. On the basis of a Mokken scale analysis for polytomous items they found one weak scale, consisting of 10 of the 11 items, with an H-coefficient of 0.37. Item J was excluded because its H-coefficient was lower than the critical value of 0.30. Using both factor analysis and Mokken scale analysis, Van Baarsen et al. (1999) detected two subscales, one consisting of the six negatively formulated items, interpreted as social loneliness, and the second consisting of the five positively formulated items, interpreted as emotional loneliness. Van Baarsen et al. (in press) found evidence for the same two-dimensional scale using the Rasch model. The difficulty of interpreting two-dimensionality due to the confounding of item formulation and item content was approached by testing the construct validity of the emotional and social loneliness scale (Zeller & Carmines, 1980). Significantly different correlations of the two subscales with theoretically relevant external variables (such as marital status and a measure of self-reported loneliness) were interpreted as support for the presence of two latent loneliness traits.

Both subscales found by Van Baarsen et al. (in press) showed gender bias, that is, a different item order was found for men and women. This implies that the scales were not sample invariant and therefore, men and women could not be compared on the same scale. On both scales some of the items were located very close to each other.

4.2 The Loneliness Scale Data

Data from the NESTOR study *Living Arrangements and Social Networks of Older Adults* of Knipscheer et al. (1995) are reanalyzed. The data consist of a random sample of almost 4500 respondents stratified according to sex, age (average 73 years, range 55 to 89 years), and residence (11 municipalities, distributed over the western, northeastern, and southern part of the Netherlands). The loneliness questionnaire was administered to the participants in face-to-face interviews. Complete data on the loneliness scale items were available for 3987 respondents consisting of 2040 women (51%) and 1947 men (49%). At the time of the interview, 2394 elderly (60%) cohabited (married or unmarried), and 1593 (40%) did not. To all these respondents the items of the loneliness scale were presented. Respondents were asked whether each statement applied to their situation and could answer yes, more or less, or no.

In the MUDFOLD analysis where items are not reversed, only the response yes is coded as one, and the more or less and no responses as zero. In Table 21.2 the item popularities (i.e., the proportion of respondents who answered yes to the statements) are given. Although the negatively formulated items (indicative of loneliness) are far less frequently answered with yes than the positively formulated items, their popularities range from 0.06 to 0.15, which is sufficient for a MUDFOLD analysis (cf. Post, 1992, p. 124).

TABLE 21.2. Item popularities for the total sample and different subgroups.

Item	Total	Men	Women	Partner	No Partner
A	0.80	0.80	0.80	0.85	0.72
B	0.14	0.12	0.16	0.09	0.22
C	0.15	0.11	0.18	0.06	0.29
D	0.82	0.81	0.84	0.85	0.79
E	0.13	0.10	0.15	0.06	0.24
F	0.14	0.14	0.14	0.09	0.20
G	0.72	0.69	0.75	0.74	0.69
H	0.77	0.74	0.80	0.80	0.73
I	0.13	0.11	0.15	0.07	0.22
J	0.06	0.06	0.07	0.04	0.10
K	0.79	0.77	0.81	0.80	0.78
n	3987	2040	1947	2394	1593

The same holds for the different subsamples of men and women, and of respondents with and without a partner. Item J, "Often I feel rejected," is the least popular item, and Item D, "There are plenty of people I can lean on in case of trouble," the most popular.

4.3 The MUDFOLD Analysis

The analysis strategy as recommended in the MUDFOLD manual (Van Schuur & Post, 1998) is followed. As a first step the program searches a scale using default options. In the search procedure, the H-coefficient is an important criterion for scalability. The H-coefficient is defined per triple of items (H_{ijk}), per item (H_i), and for the total scale (H). In the default options, H_{ijk} has to be larger than 0 for all triples, and H_i larger than 0.3. Then H is also larger than 0.3, which, analogous to the Mokken model, is considered to indicate sufficient scalability. Scales with $H > 0.4$ are regarded as medium strong scales, and those with $H > 0.5$ as strong.

The subset of items obtained by the default search procedure is used as a start set in the second step. The default options are weakened to include all items in the scale. In the third step, the scale found in the second step is inspected with the conditional adjacency matrix (CAM). The CAM, developed by Post and Snijders (1993), is used as a diagnostic to check the shape of the tracelines and the order of the items and is independent of the subject distribution. The ijth entry of the CAM is calculated as the number of subjects choosing items I_i and I_j divided by the number of subjects choosing item I_j, and is thus an estimate of the conditional

TABLE 21.3. Scales resulting from two MUDFOLD analyses ($n = 3885$).

Item		Formulation	*H*-Coefficient First Step	*H*-Coefficient Second Step
J	$-$	Feel rejected	Rejected	0.55
F	$-$	Circle too limited	0.52	0.48
B	$-$	Miss close friend	Rejected	0.42
I	$-$	Miss people around	0.55	0.50
E	$-$	Miss pleasure company	0.57	0.51
C	$+$	Experience emptiness	0.55	0.51
A	$+$	Someone to talk to	Rejected	0.43
K	$+$	Friends when needed	0.51	0.47
D	$+$	People to lean on	0.51	0.48
H	$+$	Feel close to	0.52	0.50
G	$+$	Count on completely	0.54	0.53
Total Scale			0.54	0.49

probability of choosing (row) item I_i given that (column) item I_j is chosen. The rows of the CAM correspond to the ordered items in the scale, and can be considered to represent rough estimates of the ICCs.

If the items conform to an IRT model with single-peaked tracelines, each row must exhibit a unimodal shape, or at least show no violations against unimodality. The latter requirement implies that the rows must exhibit a nondecreasing, nonincreasing, or constant series of numbers, that is, one of the patterns shown by items with monotone tracelines. A unimodal row pattern clearly suggests a model with single-peaked tracelines, whereas with a nondecreasing, nonincreasing, or constant row pattern one cannot differentiate between IRT models with single-peaked or monotone tracelines.

If the item order is correct, the maxima of the different rows—indicating the top of the traceline and thus estimating the item location—move from the top left downwards to the right. This "moving maxima" property may show inversions around the diagonal, because maxima just left or right of the diagonal position are indistinguishable. As was shown by Post (1992), violation of the moving maxima property is a clear sign that the items are not ordered in accordance with the unfolding model.

In the first step of the MUDFOLD analysis, a rather strong scale of eight items was found: all items have H_i-coefficients larger than 0.5 and the H-coefficient for the total scale is 0.54. Items A, B, and J are not scalable. The H-coefficients and the item order are presented in Table 21.3.

The results of the second step, also reported in Table 21.3, show a moderately strong scale in terms of the H-coefficient: somewhat lower H_i-

TABLE 21.4. The conditional adjacency matrix for all loneliness items in the unfolding order ($n = 3885$).

	J	*F*	*B*	*I*	*E*	*C*	*A*	*K*	*D*	*H*	*G*
J	–	24	24	**30**	29	27	4	4	4	3	3
F	*52*	–	*46*	**53**	52	40	10	*9*	*10*	8	8
B	**53**	48	–	*47*	*48*	45	10	10	10	*9*	10
I	*60*	*50*	*43*	–	**62**	52	10	10	10	*9*	*10*
E	*60*	*48*	*43*	**62**	–	58	10	10	10	*9*	*10*
C	*62*	*43*	47	60	**67**	–	12	12	12	*11*	*12*
A	57	*60*	*58*	*62*	60	65	–	85	86	86	**87**
K	45	54	59	*62*	61	65	85	–	87	87	**89**
D	49	57	63	*65*	63	67	88	91	–	90	**92**
H	40	46	54	*55*	54	58	83	85	85	–	**89**
G	37	42	49	54	54	58	78	*81*	*80*	83	–

coefficients for items included in the first scale, whose order is maintained, and a lower H-coefficient of 0.49 for the total scale. The now included Items A and B have the lowest H_i-coefficients compared to the other items, nevertheless they are both larger than 0.4, which indicates that they are medium strong unfolding items according to this criterion.

In the third step the conditional adjacency matrix, presented in Table 21.4, is inspected. Recall that this matrix should display two properties if the model holds: unimodality and moving maxima. The italic numbers indicate violations against unimodality, and the bold numbers are the row maxima. The standard errors of the CAM entries (not given in the table) are approximately 2.5 for the negatively formulated items, and 1.5 for the positively formulated items. Using 95% confidence intervals for these entries based on the normal approximation, this implies that differences between adjacent row numbers of roughly 10 for negatively and of roughly 6 for positively formulated items can be considered to be significant.

Departures from unimodality are observed in the columns for Items J and B, causing serious violations, especially in the rows of Items I, E, and C. The item pairs (I, E), (K, D), and (H, G) exhibit minor unimodality violations suggesting that the order of these pairs is reversed. Except Item B, the negatively formulated items (J, F, I, E, and C) all display unimodality, implying a unimodal IRT model. The positively formulated items (A, K, D, H, and G) all show increasing rows. The rows of these items are quite similar, especially in the columns J through C, indicating that the item locations are almost identical.

The property of the moving maxima is violated once, for Item B, which can be caused either by this item or by Item J.

TABLE 21.5. The conditional adjacency matrices of the eight-item loneliness scale for the men ($n = 1882$) and for the women ($n = 2003$).

	Men								Women							
	F	*I*	*E*	*C*	*K*	*D*	*H*	*G*	*F*	*I*	*E*	*C*	*K*	*D*	*H*	*G*
F	–	53	**55**	41	10	10	8	8	–	**54**	50	40	*9*	10	8	8
I	43	–	**61**	49	10	9	*7*	*9*	56	–	**63**	53	11	11	11	11
E	41	57	–	**58**	8	8	*7*	*8*	55	**66**	–	58	*11*	12	10	*11*
C	33	50	**63**	–	9	9	*8*	*9*	52	66	**69**	–	15	15	14	14
K	58	*67*	64	65	–	85	85	**88**	50	59	59	64	–	89	89	**91**
D	57	*65*	*63*	68	89	–	89	**91**	58	*64*	*63*	67	92	–	92	**93**
H	45	50	53	57	82	82	–	**88**	47	*58*	*54*	59	88	88	–	**90**
G	43	53	53	58	*78*	*77*	**81**	–	42	54	54	57	*84*	*83*	**84**	–

Summarizing the evidence gathered from inspection of the CAM, we conclude that one of the Items *B* or *J* is certainly not a good unfolding item, because of violation of the moving maxima and the unimodality property. There is also some doubt about the scalability of Item *A*, showing some violations of the unimodality property, and suggesting a rather flat traceline. Therefore, we prefer the strong unfolding scale *FIECKDHG* consisting of the eight items obtained in the first step. The CAM belonging to this eight-item scale can be deduced directly from Table 21.4 by deleting the rows and columns of the rejected Items *A*, *B*, and *J*. (A MUDFOLD analysis with the asymmetric coding described in Section 4.1 gave similar results.)

To check whether the eight-item scale satisfies the psychometric property of sample invariance, that is, the same item order can be assumed for different subgroups in the sample, the scale diagnostics (H-coefficient and CAM) of this scale are compared for the sample of men and for the sample of women. For men the H-coefficient for the eight-item scale is 0.51 and for women 0.56; the H_i-coefficients for all items are 0.48 or larger in both groups. The CAMs for men and women are presented in Table 21.5. In both matrices the moving maxima property is satisfied. Only a few small violations of the unimodal pattern per row are observed. For the negatively formulated items unimodality is found and the row numbers of the positively formulated items are monotone increasing. Equally satisfactory results with respect to the maxima and unimodality are obtained for the split of the sample into two groups of respondents with partner and respondents without a partner (not further reported here). The results imply that the eight-item scale satisfies the sample invariance property with respect to both gender and marital status, making a comparison between the subject scores of the subgroups valid. The means and standard deviations of the average scores of men, women, and respondents with and without a partner, are 5.92 (1.13), 5.88 (1.12), 6.12 (0.97), and 5.57 (1.26), respectively, implying that respondents without a partner are significantly lonelier than respondents with a partner, and that the loneliness scores of men and women do not differ significantly.

4.4 Interpretation of the MUDFOLD Scale and Comparison with a Mokken Analysis

Looking at the item order *FIECKDHG*, the unfolding scale can be described as ranging from lonely to not lonely. Following Van Baarsen et al. (in press), this description can be refined by taking the item content into account. The right pole of the scale corresponds to the quantitative aspect of loneliness (here, absence of loneliness), which can be interpreted as the degree of "social integration," or the amount of people around. Even though the CAM indicates that the items are quite similar, the absence of loneliness increases with the type of support given, from practical ("needed" or "in case of trouble") to affectionate ("count on completely" and "close"). The left pole of the scale corresponds to the qualitative aspect of loneliness, signifying feelings of loneliness, which can be labeled "emotional loneliness." Emotional loneliness increases with the specificity of the lacking support, from general ("emptiness") to personal ("circle of friends and acquaintances").

The eight-item unfolding scale was tested with a Mokken scale analysis after recoding the negatively formulated Items (C, E, F, and I) to obtain a latent trait roughly indicating "lack-of-loneliness," resulting in a weak scale with item order *EIFCDKHG* ($H = 0.36$). In both the MUDFOLD and the Mokken scales the negatively formulated items are located on the left side of the scale and the positively formulated items on the right side of the scale. Note that the order among the four items on the left and right sides of the scale is different in both scale analyses.

The differences between the Mokken scale and the MUDFOLD scale can be analyzed in terms of the cases treated in Section 3, with four instead of two items with decreasing and increasing tracelines, respectively. Because of the unimodal pattern in the CAM for Items C, E, and I found in the MUDFOLD analysis (see Tables 21.4 and 21.5) it is clear that there are unimodal tracelines in the data and that Figure 21.1 cannot represent the true model nor an approximation of Figure 21.2. Klinkenberg's (2001) situation does not hold, as is confirmed by a histogram of the subject locations on the MUDFOLD scale; see Figure 21.3.

4.5 Further Analyses to Establish Unimodality of the ICCs

The CAM establishes unimodality of Items C, E, and I. Since the histogram of subject locations (see Figure 21.3) shows that relatively few respondents

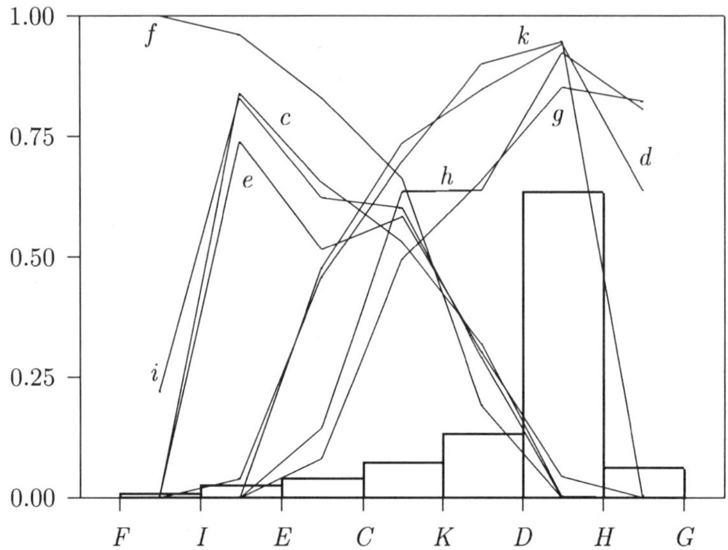

FIGURE 21.3. Histogram of the subject scores on the MUDFOLD loneliness scale and the empirical tracelines of its eight items (denoted by lowercase letters) based on $n = 3885$ respondents.

are located in the neighborhood of these items, two further analyses are performed to gain more information about the unimodality of the items.

Empirical tracelines. To estimate the tracelines of the items in the loneliness scale, empirical tracelines are calculated. Given the item order, empirical tracelines are defined as the observed item popularities given the subject location, that is, the proportion of subjects with the same location choosing an item. In the loneliness scale data, the discretized subject distribution has seven categories. The empirical tracelines are given in Table 21.3, together with the histogram of the subject distribution. Excluded from this analysis are 102 respondents because they are unscalable on an unfolding scale: they answered no or more or less to all items.

The tracelines of all items except Item F show unimodality. A small disturbance of unimodality can be observed in traceline e. The similarity in the tracelines of Items C, E, and I is striking. The tracelines of the positively formulated Items D, G, H, and K all have their peaks between D and H, that is, at the mode of the subject distribution. This implies that these items are in fact very close together at the location of the majority of the respondents. It can thus be inferred that the loneliness scale only discriminates well between the two types of items (and two types of respondents), that is, emotionally lonely or socially integrated, but not within the two groups of items. Returning to the cases sketched in Section 3, this means that Figure 21.2a depicts the loneliness scale data best.

Analysis of response patterns. Because of the quite skewed subject distribution, it is worthwhile to investigate the response patterns in some detail, especially of those respondents who answered three or more items with yes since the MUDFOLD algorithm is based on triples of items. These 3334 respondents (84%) are located over the complete item range, with a similar skew distribution. Almost all possible response patterns occur in the loneliness scale data. Hence, a certain subject location on the unfolding scale may be implied by many response patterns. One important exception, however, is the large number of respondents (46%) who answered yes to all positively formulated items and no or more or less to all negatively formulated items. This is the group located between Items *D* and *H*, causing the peak in the subject distribution. It is clear that this group of socially integrated people greatly influences the analysis.

5 Summary and Discussion

The results for the MUDFOLD analysis of the loneliness scale data from the NESTOR study are summarized. The implications of these findings for scales with positively and negatively formulated items are discussed. Finally, a recommended strategy for the analysis of such scales is given.

5.1 Summary of the Results for the Loneliness Scale Data

On the basis of the MUDFOLD analysis it can be concluded that the De Jong Gierveld loneliness scale can be used as a unidimensional scale for the NESTOR study. For other data sets, the unidimensionality of the loneliness scale would need to be investigated.

The MUDFOLD analysis revealed that there is doubt about the scalability of Items *A*, *B*, and *J*. In line with findings of Moorer and Suurmeijer (1993) using the (cumulative) Mokken model, and Van Baarsen et al. (in press) using the Rasch model, Item *A*, "There is always someone I can talk to about my day-to-day problems," was identified as a weak item in the scale. In the emotional loneliness scale *JBFIEC* consisting of the negatively formulated items proposed by Van Baarsen et al. (in press), the scalability of Items *B* and *J* also raised some questions, because their item locations were much farther apart than the other items in the scale. The least-chosen Item *J*, "Often I feel rejected," could be rejected on the basis of its content: it may measure a personality trait rather than loneliness experienced in a particular time frame. Item *B*, "I miss having a really close friend," may be confounded due to different response options regarding (1) missing a close confidant(e), and/or (2) missing a (recently deceased) partner. Following Collins (1996), who reasoned that the relevance of items may be different over time, they argued that inclusion of these items might be valuable for the assessment of individual change and development over time.

The eight-item loneliness unfolding scale *FIECKDHG* has satisfactory properties. The H-coefficient indicates a strong scale and the conditional adjacency matrix diagnoses no serious violations of the assumptions of an unfolding model. The unimodality of all items (except Item F) is contra-indicative of two cumulative scales as found by Van Baarsen et al. (in press) and the case described by Van Schuur (1998). Probably, the case described by Van Schuur and Kiers (1994) applies: the two (cumulative) subscales found in other research (Van Baarsen et al., 1999) are in fact one unimodal scale. In the MUDFOLD analysis the eight-item loneliness scale was found to be sample invariant, at least with respect to the variables gender and marital status.

The analysis also revealed that the positively and negatively formulated items are located very closely together in two item clumps, as was also found by Van Baarsen et al. (in press). This implies that the scale can distinguish well between emotionally lonely and socially integrated but is less well suited for measuring the degree of loneliness, at least in this sample where 46% of the respondents answered yes to all positively formulated items and no or more or less to the negatively formulated items. To distinguish better between degrees of emotional loneliness for instance, more items reflecting aspects of this type of loneliness are needed.

The MUDFOLD analysis for a dichotomous unimodal IRT model did not solve the problem of the scoring of the three answer categories (no, more or less, yes). Possibly the application of a polytomous unimodal IRT model (Van Schuur, 1993) could help to find an answer to this issue, although it does not solve the problem of a "socially desirable" response tendency with respect to sensitive topics such as loneliness, where respondents may be reluctant to admit that they are lonely, as is reflected in the coding instructions (De Jong Gierveld & Van Tilburg, 1999; see also Section 4.1).

5.2 Strategy for the Analysis of Scaling Data

The practice of item reversal in scale construction with both positively and negatively formulated items is debatable, for several methodological reasons.

1. Under the assumption that item reversal does not affect item content, it was shown in Section 3 that the item locations and the form of the ICCs are crucial for the consequences (in terms of item order) of the choice of an IRT model. The subject locations are important if the unfolding model is the true model.

2. A positive answer to an item cannot be considered to be the same as a negative answer to the opposite of that item, as was shown by Rorer (1965) and supported by Holleman (1999, 2000).

3. The use of only one type of item to avoid item reversal as proposed by Schriesheim and Eisenbach (1995) does not avoid the problem of choosing an appropriate IRT model, because (a) this approach neglects the fact that "Our language is not random. Certain questions are characteristically asked in certain ways." (Rorer, 1965, p. 143) The negation of items may be ambiguous, as was illustrated for Item *I* in Section 4.1; (b) one might encounter the problem of response style; and (c) items that seem to characterize dominance relations may in fact represent proximity relations.

Using a unimodal IRT model for the analysis of positively and negatively formulated items provides a solution to the aforementioned difficulties since it does not require item reversal. Moreover, the analysis will show whether all items are in fact monotone, so that—when appropriate—a cumulative analysis can be performed as a second step. In theory, the type of items (dominance or proximity) should be decisive for the choice of a particular model. In practical applications it may be unclear whether the scale, perhaps after reversing some items, consists of items with only cumulative tracelines and it may be difficult to determine the consequences of item reversing in terms of item content.

The strategy we recommend, illustrated in Section 4, is to start with an unfolding analysis that requires no item reversal and to determine the scalability of the items. If the items are found to be scalable, then the unimodality of the tracelines should be investigated using the CAM and the empirical tracelines based on the item order and the subject distribution, and the subject range inspected. If the tracelines are found to be (approximately) monotone, a cumulative model can be used for the analysis.

We conclude that, especially for the analysis of scales containing positively and negatively formulated items, the use of a cumulative IRT model where either the positively or the negatively formulated items are reversed, should be reconsidered in favor of a unimodal IRT model using unreversed item responses.

Acknowledgments: The NESTOR study *Living Arrangements and Social Networks of Older Adults* is supported by a program grant from the Netherlands Program for Research on Aging (NESTOR), funded by the Ministry of Education and Sciences and the Ministry of Health, Welfare, and Sport. The authors thank Anne Boomsma and Tom Snijders for their valuable comments on this chapter, and Edwin Klinkenberg for his comments and help with the figures.

References

Ahlawat, K.S. (1985). On the negative valence items in self-report measures. *Journal of General Psychology, 112*, 89–99.

Chang, L. (1995). Connotatively consistent and reversed connotatively inconsistent items are not fully equivalent: Generalizability study. *Educational and Psychological Measurement, 55*, 991–997.

Collins, L.M. (1996). Measurement of change in research on ageing: Old and new issues from an individual growth perspective. In J.E. Birren & K.W. Schaie (Eds.), *Handbook of the psychology of ageing* (pp. 38–56). London: Academic Press.

De Jong Gierveld, J., & Kamphuis, F.H. (1985). The development of a Rasch-type loneliness-scale. *Applied Psychological Measurement, 9*, 289–299.

De Jong Gierveld, J., & Van Tilburg, T.G. (1995). Social relationships, integration, and loneliness. In C.P.M. Knipscheer, J. de Jong Gierveld, T.G. van Tilburg, & P.A. Dykstra (Eds.), *Living arrangements and social networks of older adults* (pp. 155–172). Amsterdam: Free University Press.

De Jong Gierveld, J., & Van Tilburg, T.G. (1999). *Manual of the loneliness scale.* Amsterdam: Free University, Department of Social Research Methodology.

Dykstra, P.A. (1995). Loneliness among the never and formerly married: The importance of supportive friendships and a desire for independence. *Journal of Gerontology, 50B*, 321–329.

Hemker, B. (2001). Reversibility revisited and other comparisons of three types of polytomous IRT models. In A. Boomsma, M.A.J. van Duijn, & T.A.B. Snijders (Eds.), *Essays on item response theory* (pp. 277–296). New York: Springer-Verlag.

Holleman, B.C. (1999). The nature of the forbid/allow asymmetry: Two correlational studies. *Sociological Methods and Research, 28*, 209–244.

Holleman, B.C. (2000). *The forbid/allow asymmetry: On the cognitive mechanisms underlying wording effects in surveys.* Amsterdam: Rodopi.

Klinkenberg, E.L. (2001). A logistic IRT model for decreasing and increasing item characteristic curves. In A. Boomsma, M.A.J. van Duijn, & T.A.B. Snijders (Eds.), *Essays on item response theory* (pp. 173–192). New York: Springer-Verlag.

Knipscheer, C.P.M., De Jong Gierveld, J., Van Tilburg, T.G., & Dykstra, P.A. (Eds.). (1995). *Living arrangements and social networks of older adults.* Amsterdam: Free University Press.

Mellenbergh, G.J. (2001). Outline of a faceted theory of item response data. In A. Boomsma, M.A.J. van Duijn, & T.A.B. Snijders (Eds.), *Essays on item response theory* (pp. 415–432). New York: Springer-Verlag.

Mokken, R.J., Van Schuur, W.H., & Leeferink, A.J. (2001). The circles of our minds: A nonparametric IRT model for the circumplex. In A. Boomsma, M.A.J. van Duijn, & T.A.B. Snijders (Eds.), *Essays on item response theory* (pp. 339–356). New York: Springer-Verlag.

Molenaar, I.W., & Sijtsma, K. (2000). User's manual MSP5 for Windows: A program for Mokken scale analysis for polytomous items. Version 5.0 [Software manual]. Groningen: ProGAMMA.

Moorer, P., & Suurmeijer, T.P.B.M. (1993). Unidimensionality and cumulativeness of the loneliness scale using Mokken scale analysis for polychotomous items. *Psychological Reports, 73*, 1324–1326.

Post, W.J. (1992). *Nonparametric unfolding models: A latent structure approach.* Leiden: DSWO Press.

Post, W.J., & Snijders, T.A.B. (1993). Nonparametric unfolding models for dichotomous scaling data. *Methodika, 7*, 130–156.

Rorer, L.G. (1965). The great response-style myth. *Psychological Bulletin, 63*, 129–156.

Schriesheim, C.A., & Eisenbach, R.J. (1995). An exploratory and confirmatory factor-analytic investigation of item wording effects on the obtained factor structures of survey questionnaire measures. *Journal of Management, 21*, 1177–1193.

Schriesheim, C.A., Eisenbach, R.J., & Hill, K.D. (1991). The effect of negation and polar opposite item reversals on questionnaire reliability and validity: An experimental investigation. *Educational and Psychological Measurement, 51*, 67-78.

Schriesheim, C.A., & Hill, K.D. (1981). Controlling acquiescence response bias by item reversals: The effect on questionnaire validity. *Educational and Psychological Measurement, 41*, 1101-1114.

Stevens, N.L. (1989). *Well-being in widowhood: A question of balance.* Unpublished doctoral thesis, University of Nijmegen.

Van Baarsen, B., Smit, J.H., Snijders, T.A.B., & Knipscheer, C.P.M. (1999). Do personal conditions and circumstances surrounding partner loss explain loneliness in newly bereaved older adults? *Ageing and Society, 19*, 441–469.

Van Baarsen, B., Snijders, T.A.B., Smit, J.H., & Van Duijn, M.A.J. (in press). Lonely but not alone: Emotional isolation and social isolation as two distinct dimensions of loneliness in older people. *Educational and Psychological Measurement.*

Van Schuur, W.H. (1984). *Structure in political beliefs: A new model for stochastic unfolding with application to European party activists.* Amsterdam: CT Press.

Van Schuur, W.H. (1993). Nonparametric unfolding models for multicategory scaling data. *Political Analysis, 4*, 41–74.

Van Schuur, W.H. (1998). From Mokken to MUDFOLD and back. In C. van der Eijk, M. Fennema, & H. Schijf (Eds.), *In search of structure: Essays in social science and methodology* (pp. 45–62). Amsterdam: Het Spinhuis.

Van Schuur, W.H., & Kiers, H.A.L. (1994). Why factor analysis often is the incorrect model for analyzing bipolar concepts and what model to use instead. *Applied Psychological Measurement, 18*, 97–110.

Van Schuur, W.H., & Post, W.J. (1998). MUDFOLD. A program for multiple unidimensional unfolding. Version 4.0 [Software manual]. Groningen: ProGAMMA.

Van Tilburg, T.G., & De Leeuw, E.D. (1991). Stability of scale quality under different collection procedures: A mode of comparison on the "De Jong Gierveld loneliness scale." *International Journal of Public Opinion Research, 3*, 69–85.

Zeller, R.A., & Carmines, E.G. (1980). *Measurement in the social sciences: The link between theory and data.* Cambridge: Cambridge University Press.

22

Outline of a Faceted Theory of Item Response Data

Gideon J. Mellenbergh[1]

ABSTRACT Coombs' (1964) *Theory of Data* is a classification system of behavioral data. In his system, item response data are individual–comparison or individual–stimulus differences data. In this contribution, item response data are further classified using the following facets: Intended Behavior Type, Task Type, Intermediating Variable Type, Construct Level Type, Construct Dimension Type, Recording Type, Scaling Type, Person × Stimulus Interaction Type, Construct Relation Type, and Step Task Type. These facets cannot be completely crossed because some of the facets are nested within certain (combinations of) cells of crossed facets. The facets constitute a theory of item responses (TIR). The TIR and item response theory (IRT) complete each other: the TIR structures the item response data, while IRT explains these data. The TIR can be used to facilitate the choice of an appropriate IRT model.

1 Introduction

In psychometrics, latent variable models are used for the description and explanation of a person's response data from behavioral and social science measurement instruments. The first latent variable models were presented between 1940 and 1950 by Lawley (Lord & Novick, 1968, p. 369), Guttman (1950), and Lazarsfeld (1950). Their work was followed by the publications of Lord (1952), Rasch (1960), Birnbaum (1968), Lazarsfeld and Henry (1968), Samejima (1969), and Mokken (1971), who presented models which are the basis of modern item response theory (IRT); an overview of a large number of IRT models is given in a volume edited by Van der Linden and Hambleton (1997). In this chapter, the term IRT is used in a broad sense: it includes models having different types of latent variables.

In general terms, the usual measurement situation of the behavioral and social sciences consists of the following phases: (1) a set of stimuli is administered to a person under standardized conditions, (2) the person is asked to

[1]Department of Psychology, University of Amsterdam, Roetersstraat 15, 1018 WB Amsterdam, The Netherlands; e-mail: *ml_Mellenbergh@macmail.psy.uva.nl*

respond to the stimuli, (3) a selection of the person's responses is recorded, and (4) the person's recorded responses are converted into response data. In the third phase a selection of the person's responses is made and only the selected responses are recorded, for example, an examinee's choice of one out of four options of an achievement item is recorded, but his moaning or galvanic skin response is usually not recorded. Moreover, the third and fourth phase must be distinguished because the recorded responses of the third phase are evaluated in the fourth phase, for example, the examinee's recorded option choice of a four-choice item is evaluated as a correct or incorrect answer in the fourth phase; see Coombs (1964, Chap. 1).

An IRT model is a mini-theory to explain a person's item response data obtained from behavioral and social science measurement instruments. In practice, different types of item response data occur, and different IRT models are applied to different item response data. Therefore, a systematic classification of item response data can be a foundation for the choice of an IRT model.

Coombs (1964) presented a general structure of behavioral data, which he called a *Theory of Data*. In his structure, item response data are individual–stimulus comparison (Quadrant II) or individual–stimulus differences (Quadrant I) data. In this contribution, a further classification of item response data is proposed. A number of facets are described which can be used to structure item response data. This faceted classification system structures item response data, while IRT explains these data. Consistent terms for these two theoretical levels would be theory of item response data (TIRD) and item response data theory (IRDT), respectively. Because of the convention of using the abbreviation IRT, and not IRDT, the term "theory of item responses" (TIR) is used instead of TIRD.

2 Stimulus Facets

Measurement stimuli differ widely among each other. In this section, general stimulus facets are discussed which seem to be relevant for the compatibility of a TIR and existing or conceivable IRT models. The following stimulus facets are discussed: Intended Behavior Type, Task Type, and Intermediating Variable Type.

2.1 Intended Behavior Type

Cronbach (1960, Chap. 2) distinguished two major behavior types which can be elicited by stimuli: maximum and typical performance. Maximum performance stimuli challenge the person to give the best response to the stimulus. Examples are intelligence and achievement items, where the subject is instructed to give the correct answer to the item. In contrast, typical performance stimuli elicit a response from the person that is typical for his or her behavior. Examples are attitude and personality items, where the respondent is asked to give a typical (personal) response.

It is remarked that the behavior, which the stimulus tries to elicit, does not need to coincide with the person's real behavior. An examinee, for example, can give a correct answer by copying from another examinee, or a respondent can mask his real attitude by faking socially acceptable responses. Cronbach's distinction applies to intended behavior and does not necessarily apply to real behavior; and, therefore, the intended behavior type is a stimulus property and not a person property.

2.2 Task Type

Two main types of tasks which stimuli present to a person can be distinguished: free-response and choice. Using a free-response task, the person must create his or her own response to the stimulus; for example, an examinee must formulate an answer to a mathematical problem. In a choice task, however, a person must choose one or more options from a given number of options. For example, an examinee must choose one out of four options of an arithmetic item, or a respondent must choose one out of five categories of an attitude item.

2.3 Intermediating Variable Type

A basic assumption of IRT is that a person's recorded item response is related to a latent variable (unobserved construct). A person is characterized by a value on a latent variable, such as verbal intelligence or political attitude, and this value is linked to the person's recorded item response. It is, however, not necessary to assume that the person's latent variable value is directly linked to the recorded item response. It is always possible to assume an unobserved variable, which mediates between the latent variable and the recorded item response: the latent variable is linked to the intermediating variable and the intermediating variable is linked to the recorded item response. The intermediating variable is item-specific; that is, each of the test items has its own intermediating variable.

The intermediating variable can be discrete or continuous. Therefore, the intermediating variable type facet has two elements: discrete and continuous intermediating variables.

An example of an item where a discrete intermediating variable can be assumed is:

The capital of Italy is . . . .

It is assumed that the item measures the continuous latent variable (construct) of geographical knowledge. Moreover, it is assumed that two groups of examinees can be distinguished: (1) examinees who know that Rome is the capital of Italy, and (2) examinees who do not know this fact. Therefore, the unobserved item-specific intermediating variable is discrete, and has two values: does and does not know that Rome is the capital of Italy. Examinees with high values on the latent geographical knowledge trait have a high probability of knowing the Rome-answer, whereas examinees with low values on the latent trait have a low probability of knowing the Rome-answer. Examinees who know the answer will almost certainly give the response "Rome," whereas examinees who do not know the answer will omit the item or will make a guess (they might give the Rome-response by guess). The latent trait is a continuous variable, which cannot coincide with the dichotomous intermediating variable. The intermediating variable has two values (does/does not know that Rome is the capital of Italy), which does not coincide with the recorded item response (Rome, omit, Paris, Madrid, etc.).

An example of an item, where both a discrete and a continuous intermediating variable can be assumed, is the following item of Likert's (1932) Survey of Opinions:

Our country should never declare war again under any circumstances.

Strongly Approve	Approve	Undecided	Disapprove	Strongly Disapprove

The item is from an Internationalism Attitude scale. Suppose that the internationalism construct is a continuous latent variable, that is, a latent trait. It is assumed that the latent internationalism construct is linked to a variable, which mediates between the construct and the recorded item response. The intermediating variable can be discrete or continuous. A discrete intermediating variable has five ordered categories: persons from the first category choose Strongly Approve, persons from the second category choose Approve, and so on. The five-category intermediating variable completely coincides with the five-option recorded response variable; that is, the intermediating variable is superfluous. Therefore, the latent construct can be linked directly to the recorded item response, without an intermediating variable. On the other hand, a continuous intermediating variable can be assumed. The real line is assumed, which is divided into five segments: persons from the utmost left segment choose Strongly Approve, persons from the next segment choose Approve, and so on. Under this assumption, the intermediating variable does not coincide with the construct or the recorded item response. The intermediating variable does not coincide with the construct because a separate intermediating variable is assumed for each of the test items, and the division into segments may differ between items. The intermediating variable does not coincide with the recorded response variable because it is continuous, whereas the recorded item response is discrete.

3 Person Facets

In this section, general person facets are discussed, which are relevant for the compatibility of TIR and existing or conceivable IRT models. The facets Construct Level Type and Construct Dimension Type are described.

3.1 Construct Level Type

A construct, which a stimulus tries to measure, can be conceptualized as a discrete variable (latent classes, or a continuous variable (latent trait). Furthermore, a discrete latent variable can consist of unordered or ordered latent classes. Therefore, a distinction is made between (1) a nominal latent class variable, (2) an ordinal latent class variable, and (3) a continuous latent trait.

In many areas of the behavioral and social sciences, theorists differ in opinion on the nature of the construct levels. In, for instance, educational and personality research, both discrete and continuous constructs are assumed. The construct of arithmetic ability, for example, can be conceptualized as either an ordinal latent class variable (masters and nonmasters) or a continuous latent arithmetic trait. In psychological research on personality, the dominant view is that five continuous traits (the "Big-Five") are needed to describe a person's personality (Goldberg, 1990). In psychiatric research, however, the dominant view is that discrete nominal classes are needed to describe mental disorders. This view is, for example, clearly expressed in the Introduction of the *Diagnostic and Statistical Manual of Mental Disorders* (American Psychiatric Association, 1994, p. XXII): "DSM-IV is a categorical classification that divides mental disorders into types based on criteria set with defining features."

3.2 Construct Dimension Type

It seems plausible that always more than one construct is used by a person responding to a stimulus; one construct is intended to be measured, but usually other constructs are involved. For example, an examinee must be able to read in order to respond to the content of an arithmetic item; and, therefore, at least two constructs are needed, that is, reading and numerical ability. But sometimes, these other constructs can be neglected. For example, if all examinees of a population sufficiently understand the content of an arithmetic item, their reading ability can be neglected. Therefore, a distinction is made between two main construct dimension types: unidimensional, where predominantly one intended construct is measured, versus multidimensional, where, besides the intended construct, other constructs are measured as well.

The other constructs can be further subdivided into auxiliary and nuisance constructs. An auxiliary construct is a construct that is necessary for or facilitates the person's responding. An example is reading skill, which is necessary for answering an arithmetic item. A nuisance construct is a construct that violates the measurement of the intended construct; for a discussion of the violator concept, the reader is referred to Oort (1996, Chap. 1). Examples of nuisance constructs are response sets and demand characteristic constructs.

Rorer (1965, p. 133) defined a response set as "a conscious or unconscious desire on the part of a respondent to answer in such a way as to produce a certain picture of himself. An individual may have a set to dissimulate, to malinger, to appear aggressive, to fake good, to get the job etc." Response sets are considered to be rather stable personality traits and are not intended to be measured by the stimuli.

The measurement situation is mostly standardized to prevent the influence of unintended constructs. But the measurement situation, standardized or not, has, in Orne's (1962) terms, "demand characteristics," that is, cues to which the person reacts and that influence his or her responses. The demand characteristics of the measurement situation can induce unintended constructs, such as test anxiety in examination and job selection testing situations.

4 Recording Type Facet

In the third phase of the measurement situation, the person's responses are recorded. As remarked before, only a selection of a person's responses is recorded, which means that relevant information might be lost in this phase. The responses are recorded in a number of discrete categories.

A convenient distinction is that between two (dichotomous) and more than two (polytomous) categories which are used for recording the item responses. Examples of dichotomous recorded responses are the person's option choice of true/false or yes/no items. Examples of polytomous recorded responses are the person's answers to open-ended achievement questions and option choices of four-choice achievement or five-point attitude items.

5 Scaling Type Facet

In the fourth phase of the measurement situation, the recorded responses are converted into item response data. This phase is denoted the scaling of the recorded responses. The scaling phase is distinguished from the recording phase because scaling implies an evaluation of the recorded responses. For example, in the recording phase an examinee's option choice of a multiple choice item is recorded, and in the scaling phase the chosen option is evaluated as a correct or incorrect answer.

The elements of the scaling type facet are the conventional scaling levels: (1) nominal, (2) ordinal, and (3) metric (interval and ratio). An example of nominal item response data is a respondent's choice of a presidential candidate, which is scaled as Democrat or Republican. An example of ordinal item response data is an examinee's answer to an open-ended achievement question, which is evaluated as a correct, half-correct, or incorrect answer. An example of metric item response data is a subject's time in seconds to react to a stimulus of a psychological experiment.

6 Nested Facets

Stimulus, person, recording, and scaling facets were distinguished. The crossing of the various facets yields a 2 (maximum/typical behavior) × 2 (free-response/choice) × 2 (discrete/continuous intermediating variable) × 3 (nominal latent class/ordinal latent class/latent trait) × 2 (uni-/multidimensional construct) × 2 (dichotomous/polytomous recorded response variable) × 3 (nominal/ordinal/metric scale) facet design. Most of the cells of this design contain existing or conceivable item response data. Exceptions are cells, where a continuous intermediating variable is combined with a nominal scale. These combinations are logically impossible because a continuous intermediating variable cannot yield nominal item response data. For example, suppose that a continuous variable intermediates between a latent anxiety construct and a yes/no statement of an anxiety questionnaire. The yes-response is scaled as anxiety-indicative and the no-response as anxiety-nonindicative. The dichotomous item response data must be ordinal because the indicative yes-response indicates a higher anxiety level than the nonindicative no-response.

A number of other facets are distinguished. These facets cannot be combined with each of the cells of the above-mentioned facet design. They are nested within certain (combinations of) cells of the design. The following facets are discussed: Person × Stimulus Interaction, Construct Relation, and Step Task facets.

6.1 Person × Stimulus Interaction Type

In the measurement situation, a person interacts with a stimulus. The person's characteristic of main interest is his or her position on the intended construct, while the stimulus characteristic of main interest is its difficulty (for maximum performance behavior) or attractiveness (for typical performance behavior). The person's construct position interacts with the stimulus difficulty or attractiveness. Therefore, a person × stimulus interaction facet is used with four elements: ignoring, absolute dominance, relative dominance, and proximity.

The ignoring interaction means that the person responds to stimulus properties other than difficulty or attractiveness. Response styles belong to this type of interaction. Rorer (1965, p. 134) described a response style as "a tendency to select some response category a disproportionate amount of the time independently of item content." Examples of response styles are tendencies of yea-saying, nay-saying, and extreme response category avoiding. Response styles must be distinguished from the previously mentioned response sets: a response set is a stable person construct, such as the tendency to fake socially acceptable answers to attitude items, whereas a response style, such as yea-saying is a person $\times$ item interaction because it occurs only in certain tests and not in others.

The absolute dominance interaction applies if it can be thought that person and stimulus play a game. The only two players are the person and the stimulus, and no other parties are involved. The person wins the game if his or her construct position dominates the stimulus difficulty or attractiveness; otherwise, the stimulus wins the game. Examples are achievement items: the examinee masters the item if his or her ability level exceeds the item difficulty; otherwise, the examinee does not master the item.

The only two players of the absolute dominance game are the person and the stimulus. In the relative dominance interaction, another (third) party is involved. The outcome of the game is not determined by the dominance of the person's absolute construct position over the item attractiveness, but by the dominance of the person's relative construct position over the item attractiveness, where the person's construct position is relative to a reference population of persons. Borsboom, Mellenbergh, and Van Heerden (2000), for example, mention the following social science type of item to measure a person's height: "I think I would do good in a basketball team." A woman of 1.80 meters in height probably gives a positive answer, whereas a man of the same height probably gives a negative answer. The reason is that a reference group is involved in answering this item: a woman of 1.80 meters is rather tall compared to the population of females, whereas a man of 1.80 meters is rather small compared to the population of males.

In a proximity interaction, the person does not compete with a stimulus, but tries to give the response that is nearest to his or her construct position (ideal point). An example is a political attitude item, where the respondent is asked to choose the preferred political party out of six parties. It is thought that the respondent chooses the party that is nearest to his or her "ideal" party. Note that the proximity interaction can only apply to typical performance behavior because the person must give a response typical for his or her behavior, and the person does not try to beat the stimulus in a game.

The Person $\times$ Stimulus Interaction facet has the following elements: the person tries to (1) ignore the stimulus difficulty or attractiveness, (2) dominate the stimulus difficulty or attractiveness in an absolute sense, (3) dominate the stimulus difficulty or attractiveness in a relative sense, and (4) approximate the stimulus attractiveness. This interaction facet is nested within certain cells of the above-mentioned facet design: the first three interaction types (ignoring and the two types of dominance) can be combined with both maximum and typical performance behavior, but the last interaction type (proximity) can only be combined with typical performance behavior. Some types of ignoring, such as yea-saying, exclude the other interaction types. However, other types of ignoring can be compatible with other interaction types; for example, extreme response category avoiding can occur simultaneously with each of the other three interaction types.

6.2 Construct Relation Type

A distinction was made between unidimensional and multidimensional construct measurement. Multidimensional construct measurement means that, in addition to the intended construct, other constructs are measured by the stimulus. These constructs were further subdivided into auxiliary and nuisance constructs.

Constructs can be related to each other in three different ways: compensatory, conjunctive, and disjunctive. First, the constructs are compensatory related to each other. This means that a low level at one of the constructs is compensated by a high level at another construct. An example of where a compensatory relation might apply is the measurement of reading comprehension. A reading comprehension item consists of reading a passage on a given topic, such as the solar eclipse, and one or more questions on comprehension of passage content. The intended construct is reading comprehension, while astronomical knowledge is an auxiliary construct. Reading comprehension and astronomical knowledge might be compensatory in answering the question. Second, the constructs can be conjunctively related to each other: a minimum level at another construct is a prerequisite to being able to respond to the item content. An example is the previously mentioned arithmetic item, where a minimum reading ability level is a prerequisite to the understanding of the item. Finally, constructs can be disjunctively related: a minimum level on any one of the constructs is needed for an adequate response. Coombs (1964, p. 247) gives the following example. Suppose that an item can be solved by prolonged trial and error or by reasoning. The subject can give the correct answer to the item by enough effort or by enough reasoning ability; the subject needs only to exceed a minimum value on one of the two traits and not on both.

The Construct Relation facet has three elements: compensatory, conjunctive, and disjunctive relation. This facet is nested into cells of the above-mentioned crossed facet design because it applies exclusively to multidimensional construct measurement.

6.3 Step Task Type

Crossing the three levels of the scaling facet (nominal/ordinal/metric) with the two levels of the recording facet (dichotomous/polytomous) yields six combinations. The Step Task facet applies to the ordinal-polytomous combination.

Molenaar (1983) noted that an ordinal-polytomous item presents a task to the person, which can be considered to consist of a number of "item steps." Van Engelenburg (1997, Chap. 2) generalized the item step task concept to his step approach.

The following elements of the Step Task facet are distinguished by Van Engelenburg: (1) simultaneous, (2) separate try-all, and (3) separate stop-if-failed step tasks. A step task is called simultaneous if a person is instructed to try all item steps in one action and the number of completed steps is counted. An example from athletics is long jumping: the person jumps (one action) and the number of centimeters of the jump (steps) is counted. Van Engelenburg gives the following example from achievement testing: "Explain why...," where the examinee's open answer is evaluated as "poor," "fair," "well," or "excellent." A step task is of the separate and try-all type if the person is instructed to try each of the item steps in separate actions and the number of completed steps is counted. Van Engelenburg gives the following example: "Name the three Baltic States," where the number of correctly answered states is counted. The task consists of three separate steps (the three Baltic States), and each of these steps is tried. Finally, a step task is of the separate and stop-if-failed type if the person is instructed to take separate ordered steps until he or she fails. An example from athletics is high jumping: if the athlete fails at 200 centimeters he stops; otherwise he continues at 201 centimeters, and so on. Van Engelenburg gives the following example from achievement testing: "(a) What is the difference between A and B? (b) What is the usual explanation for this difference? (c) What is problematic about this explanation?"

7 TIR and IRT

The TIR is a taxonomy of item response data, while IRT tries to explain these data by using at least one latent variable. Taxonomies of IRT models were presented by Heinen (1996, Chap. 1) and Thissen and Steinberg (1986). Heinen classified latent structure models into a 3×3 facet design, which can be subsumed under the TIR. His two facets are identical to the TIR facets Construct Level Type (nominal latent classes/ordered latent classes/latent trait) and Scaling Type (nominal/ordinal/metric). However, Thissen and Steinberg's taxonomy cannot be subsumed under the TIR. They distinguished between binary models and more complex models, and the more complex models were subdivided into difference, divide-by-total, left-side added, and left-side added divide-by-total models. The distinction between binary and more complex models partly corresponds to the TIR distinction between dichotomously and polytomously recorded item responses: binary models only apply to dichotomous item responses; however, the more complex left-side added models not only apply to polytomous items, but also to dichotomously scored multiple choice performance items. Moreover, Thissen and Steinberg's subdivision of more complex models looks closest at the TIR Step Task facet. However, these two distinctions are based on different principles: Thissen and Steinberg's distinction is based on the item trace lines, whereas the Step Task elements are based on the steps a person must take in responding to the item.

The existing IRT taxonomies only partly complement the TIR, or do not match with the TIR at all. However, the TIR can be used to specify properties of IRT models for item response data of TIR cells. In the remainder of this section, examples of such properties are discussed.

Dominance and proximity models. A maximum performance item challenges a person to give the best answer to the item; that is, a maximum performance instruction induces an absolute dominance relation between person and item. Therefore, the IRT model needs to specify this dominance relation. Some well-known examples of IRT models for absolute dominance are the one-parameter Rasch (1960), two-parameter Birnbaum (1968), and the nonparametric Mokken (1971) models for dichotomous item response data, and the parametric Samejima (1969) and the nonparametric Molenaar (1997) models for ordinal-polytomous item response data. A typical performance item asks for a response that is typical of the person's behavior. Usually, it is not obvious from the item content which type of person × stimulus interaction will be induced by the item: absolute dominance, relative dominance, or proximity. A proximity relation needs IRT models other than dominance relations. Typical performance items that induce a dominance relation can be modeled with the same type of IRT models as maximum performance items, but typical performance items that induce a proximity relation need other types of models. Examples of proximity relation IRT models are Van Schuur's (1984) nonparametric MUDFOLD and Hoijtink's (1991) parametric PARELLA models.

Guessing and nonguessing models. IRT models for free-response items do not need to account for guessing. Neither do IRT models for typical performance choice items, such as Likert-scale attitude items, need to account for guessing. However, IRT models for maximum performance choice items must model guessing behavior. Examples of IRT guessing models are Birnbaum's (1968) three-parameter, and Thissen and Steinberg's (1984) multiple choice item models.

Discrete and continuous intermediating variable models. The TIR distinguishes discrete and continuous intermediating variable types. For some items such as the previously mentioned capital of Italy item, the assumption of a discrete intermediating variable is rather plausible. On the other hand, for items such as the previously mentioned internationalism item of Likert's (1932) Survey of Opinions, it is not obvious whether a discrete or continuous intermediating variable must be assumed. However, different IRT models correspond to different assumptions on the intermediating variable. Under the assumption of a discrete intermediating variable, IRT models for ordinal-polytomous item response data, for example, Samejima's (1969) parametric graded response model or Molenaar's (1997) nonparametric model, can be applied to the data. Under the assumption of a continuous intermediating variable, however, polychoric correlation coefficients can be computed between the items, and factor-analytic IRT models can be applied to the interitem polychoric correlation matrix (Bollen, 1989, Chap. 9). A well-argued choice between the two assumptions and corresponding IRT models is hard to make because theoretical or empirical results on this issue are not available. Moreover, it is hard to imagine what kind of studies would give conclusive evidence to make this choice.

Latent trait and latent class models. The choice between latent trait and latent class IRT models largely seems to be a matter of taste. Usually, psychologists prefer latent trait IRT models because their cognitive and personality theories mostly assume continuous latent constructs, such as verbal intelligence and dominance. On the other hand, sociologists prefer latent class IRT models. Fortunately, this choice can be based on empirical research: latent trait and latent class models are applied to the same item response data, and the most tenable model is preferred above the other. Finally, it is noted that, conceptually, latent classes approximate a latent trait if a large number of ordered latent classes is used, which means that the distinction between ordered latent classes and a latent trait is gradual.

Unidimensional and multidimensional models. The unidimensionality of a measurement instrument can be violated by different kinds of nuisance constructs (Oort, 1996, Chap. 1). The discussion of constructs, which are measured by a test besides the intended construct, has a long history; for example, response sets were defined in the '60s (Rorer, 1965). IRT can detect the influence of nuisance constructs by using differential item functioning (DIF) methods. A number of DIF detection methods are described in the literature and their properties were extensively studied; for an overview, the reader is referred to Millsap and Everson (1993). IRT methods can detect DIF between groups, such as males and females, or blacks and whites. A group-membership variable, however, does not give a satisfactory explanation of DIF causes. DIF explanations must be based on the nuisance constructs, which differ between groups. The TIR distinguishes types of nuisance constructs within the Construct Dimension Type facet. In this respect, the TIR encourages researchers to study causal explanations of DIF.

Nominal, ordinal, and metric item response data models. The recorded item responses are nominally, ordinally, or metrically scaled. The most common item response data are of a nominal or ordinal nature, but metric item response data do exist, for example, the time an examinee needs to give the correct answer to the mental arithmetic item $7 \times 13 = \ldots$. The covariance matrix can be computed between metrically scaled test items, and factor-analytic models can be applied to this matrix; see, for example, Mellenbergh (1994).

Cumulative probability, continuation-ratio, and adjacent-category models. The Step Task Type facet applies to ordinal-polytomous item response data. Molenaar (1983) and Mellenbergh (1995) showed that three different types of ordinal-polytomous IRT models must be distinguished: cumulative probability, continuation-ratio, and adjacent-category models. These models correspond to different conceptual splittings of the item response variable with $m + 1$ categories into m dichotomous variables; therefore, IRT models for dichotomous item responses can be generalized to polytomous responses (Mellenbergh, 1995). Molenaar (1983) proposed his step approach to characterize different types of ordinal-polytomous item response data and this approach was generalized by Van Engelenburg (1997). The choice among these three types of IRT models can be greatly facilitated by an analysis of ordinal-polytomous items in terms of item steps.

Compensatory, conjunctive, and disjunctive models. IRT models for multidimensional item response data specify more than one latent variable. These latent variables can be combined compensatorily, conjunctively, or disjunctively. Most of the existing multidimensional IRT models assume a compensatory relation between latent traits, but a conjunctive or disjunctive combination might be more appropriate for some types of item response data. Maris (1995) gives an example of a verbal intelligence test. The items are words, for example, "woods," and the subject is instructed to mention a synonym. He assumes that the person needs two latent traits in responding to the item: a trait to generate synonyms of the word and a trait to evaluate the correctness of each of the generated synonyms. These two traits cannot compensate for each other, but they have a conjunctive relation. Maris analyzed this test using the conjunctive Rasch model, a description of which is given by Jannarone (1997).

8 Example

In this section, an example of an analysis of an item in TIR terms is given. The item is from the School Life Satisfaction subscale of Vorst's (1990) School Questionnaire.

The teacher sometimes treats us as small children.

A	B	C
That is true.	I do not know.	That is not true.

The TIR specifications of this item are as follows.

- Intended Behavior Type: Typical performance.
- Task Type: Choice.
- Intermediating Variable Type: It is not clear whether it is a discrete three-category variable or a continuous variable.
- Construct Level Type: In attitude measurement, it is usually assumed that a continuous latent trait is measured, but it is conceivable that the School Life Satisfaction construct is a discrete ordered latent class variable.
- Construct Dimension Type: The item tries to measure one construct, that is school life satisfaction, but the measurement might be violated by a nuisance construct, such as socially desirable answering.
- Recording Type: Polytomous.
- Scaling Type: The scaling type is questionable because it is not obvious that the "I do not know" category is scalable between the two other categories.
- Person × Stimulus Interaction Type: Most plausible is an absolute dominance relation, but it is possible that some students ignore the item content and use response styles.
- Construct Relation Type: The relation between constructs is not obvious when the item is multidimensional.
- Step Task Type: The student is instructed to consider the three item categories at once; therefore, a simultaneous step task type is most plausible.

The choice of an IRT model must be based on the TIR, but some of the TIR facet specifications are not yet made. The levels of the Intermediating Variable Type, Construct Level Type, Construct Dimension Type, Scaling Type, Person × Stimulus Interaction Type, and Construct Relation Type facets must be determined to make an argued choice of an IRT model. Some empirical information is available on the Construct Dimension Type and

Scaling Type. Oort (1996, Chap. 3) showed that it is socially undesirable for students to agree with the item. Therefore, the Construct Dimension Type is multidimensional because the measurement of School Life Satisfaction is violated by Social Desirability. Moreover, Oort applied a HOMALS analysis to empirical test data and showed that the three item categories are on the same scale.

Theoretical or empirical evidence on the other facets is not available, and, therefore, assumptions on these facets must be made. For example, under the assumption of a discrete intermediating variable, two compensatory latent traits (Satisfaction and Social Desirability), and the absence of response styles, a two-dimensional latent trait model for graded responses is a good choice; under the assumption of a continuous intermediating variable, two compensatory latent traits, and the absence of response styles, a two-factor model can be chosen; under the assumption of a discrete intermediating variable, a latent Satisfaction trait and two latent classes (socially desirable and socially undesirable answering persons), and the absence of response styles, a mixed latent trait/latent class model for graded responses can be used; and so on.

The example demonstrates that the TIR structures item response data and guides the choice of an IRT model. The example also shows that the specification of most of the facets is not clearcut, and must be based on either empirical research or on assumptions. The preferred strategy for TIR specification is, of course, empirical research. However, in practical applications, research on some of the facets will not be available, and assumptions on these facets must be made. The TIR forces the researcher to make explicit assumptions and to argue their choice.

9 Conclusion

A theory of item response data was proposed to characterize item response data. The TIR consists of a number of facets, and each of the facets consists of a number of elements. The proposed TIR is preliminary in the sense that other facets and other facet elements might be needed to give a complete description of item response data.

The TIR and IRT complete each other: the TIR structures item response data, and IRT explains item response data. These two theoretical levels are closely related, and the TIR, and TIR research, facilitate the choice of appropriate IRT models.

Acknowledgments: I thank a reviewer, Sanneke Schouwstra, and the editors for their valuable comments, and Norman D. Verhelst for his excellent review.

References

American Psychiatric Association. (1994). *Diagnostic and statistical manual of mental disorders* (4th ed.). Washington, DC: Author.

Birnbaum, A. (1968). Some latent trait models and their use in inferring an examinee's ability. In F.M. Lord & M.R. Novick, *Statistical theories of mental test scores* (pp. 397–424). Reading, MA: Addison-Wesley.

Bollen, K.A. (1989). *Structural equations with latent variables.* New York: Wiley.

Borsboom, D., Mellenbergh, G.J., & Van Heerden, J. (2000). *Different kinds of DIF: A distinction between relative and absolute forms of measurement invariance and bias.* Manuscript submitted for publication.

Coombs, C.H. (1964). *A theory of data.* New York: Wiley.

Cronbach, L.J. (1960). *Essentials of psychological testing* (2nd ed.). New York: Harper.

Goldberg, L.R. (1990). An alternative "Description of personality." The Big-Five factor structure. *Journal of Personality and Social Psychology, 59,* 1216–1229.

Guttman, L. (1950). The basis for scalogram analysis. In S.A. Stouffer, L. Guttman, E.A. Suchman, P.F. Lazarsfeld, S.A. Star, & J.A. Clausen (Eds.), *Measurement and prediction* (Vol. IV, pp. 60–90). Princeton, NY: Princeton University Press.

Heinen, T. (1996). *Latent class and discrete latent trait models: Similarities and differences.* Thousands Oaks, CA: Sage.

Hoijtink, H. (1991). The measurement of latent traits by proximity items. *Applied Psychological Measurement, 15,* 153–169.

Jannarone, R.J. (1997). Models for locally dependent responses: Conjunctive item response theory. In W.J. van der Linden & R.K. Hambleton (Eds.), *Handbook of modern item response theory* (pp. 465–479). New York: Springer-Verlag.

Lazarsfeld, P.F. (1950). The logical and mathematical foundation of latent structure analysis. In S.A. Stouffer, L. Guttman, E.A. Suchman, P.F. Lazarsfeld, S.A. Star, & J.A. Clausen (Eds.), *Measurement and prediction* (Vol. IV, pp. 362–412). Princeton, NY: Princeton University Press.

Lazarsfeld, P.F., & Henry, N.W. (1968). *Latent structure analysis.* Boston: Houghton Mifflin.

Likert, R. (1932). A technique for the measurement of attitudes. *Archives of Psychology, 140,* 5–55.

Lord, F.M. (1952). A theory of test scores. *Psychometric Monographs, 7.*

Lord, F.M., & Novick, M.R. (1968). *Statistical theories of mental test scores.* Reading, MA: Addison-Wesley.

Maris, E. (1995). Psychometric latent response models. *Psychometrika, 60,* 523–547.

Mellenbergh, G.J. (1994). A unidimensional latent trait model for continuous item responses. *Multivariate Behavioral Research, 29,* 223–236.

Mellenbergh, G.J. (1995). Conceptual notes on models for discrete polytomous item responses. *Applied Psychological Measurement, 19,* 91–100.

Millsap, R.E., & Everson, H.T. (1993). Methodology review: Statistical approaches for assessing measurement bias. *Applied Psychological Measurement, 17,* 297–334.

Mokken, R.J. (1971). *A theory and procedure of scale analysis with applications in political research.* The Hague: Mouton.

Molenaar, I.W. (1983). *Item steps* (Heymans Bulletin HB–83–630–EX). Groningen: University of Groningen, Vakgroep Statistiek en Meettheorie.

Molenaar, I.W. (1997). Nonparametric models for polytomous responses. In W.J. van der Linden & R.K. Hambleton (Eds.), *Handbook of modern item response theory* (pp. 369–380). New York: Springer-Verlag.

Oort, F.J. (1996). *Using restricted factor analysis in test construction.* Unpublished doctoral dissertation, University of Amsterdam.

Orne, M.T. (1962). On the social psychology of the psychological experiment: With particular reference to demand characteristics and their implications. *American Psychologist, 17,* 776–783.

Rasch, G. (1960). *Probabilistic models for some intelligence and attainment tests.* Copenhagen: The Danish Institute of Educational Research. (Expanded edition, 1980. Chicago: The University of Chicago Press.)

Rorer, L.G. (1965). The great response-style myth. *Psychological Bulletin, 63,* 129–156.

Samejima, F. (1969). Estimation of latent ability using a response pattern of graded scores. *Psychometrika Monograph, 17.*

Thissen, D., & Steinberg, L. (1984). A response model for multiple choice items. *Psychometrika, 49,* 501–519.

Thissen, D., & Steinberg, L. (1986). A taxonomy of item response models. *Psychometrika, 51,* 567–577.

Van der Linden, W.J., & Hambleton, R.K. (1997). *Handbook of modern item response theory.* New York: Springer-Verlag.

Van Engelenburg, G. (1997). *On psychometric models for polytomous items with ordered categories within the framework of item response theory.* Unpublished doctoral dissertation, University of Amsterdam.

Van Schuur, W.H. (1984). *Structure in political beliefs: A new model for stochastic unfolding with application to European party activists.* Amsterdam: CT Press.

Vorst, H.C.M. (1990). *Schoolvragenlijst SVL: Normen per leerling, per klas en per vraag* [School questionnaire: Norms per student, per class, and per item]. Nijmegen: Berkhout.

Index

H

I

K

L

Abbreviations

2PLM	Two-Parameter Logistic Model
3PLM	Three-Parameter Logistic Model
CAM	Conditional Adjacency Matrix
CCC	Category Characteristic Curve
CML	Conditional Maximum Likelihood
DIF	Differential Item Functioning
DM	Double Monotonicity
DTF	Differential Test Functioning
EAP	Expected A Posteriori Estimate
EM	Expectation Maximization
ERF	Expected Response Function
GRM	Graded Response Model
ICC	Item Characteristic Curve
IIO	Invariant Item Ordering
IRF	Item Response Function
IRT	Item Response Theory
ISRF	Item Step Response Function
LLRA	Linear Logistic model with Relaxed Assumptions
LLTM	Linear Logistic Test Model
LM	Lagrange Multiplier
LR	Likelihood Ratio
MCMC	Markov Chain Monte Carlo
MH	Monotone Homogeneity
MHD	Mantel–Haenszel DIF index
MLR	Monotone Likelihood Ratio
MML	Marginal Maximum Likelihood
NIRT	Nonparametric Item Response Theory
OEL	Ordering of the Expected Latent trait
OPLM	One-Parameter Logistic Model
PCM	Partial Credit Model
RM	Rasch Model
RSM	Rating Scale Model
SOL	Stochastic Ordering of the Latent trait
SOM	Stochastic Ordering of the Manifest score
SRM	Sequential Rating scale Model
UMA	Uniformly Most Accurate
UMP	Uniformly Most Powerful
WLE	Warm's Weighted Likelihood Estimator

Lecture Notes in Statistics

For information about Volumes 1 to 105, please contact Springer-Verlag

106: Harald Niederreiter, Peter Jau-Shyong Shiue (Editors), Monte Carlo and Quasi-Monte Carlo Methods in Scientific Computing. xiv, 372 pp., 1995.

107: Masafumi Akahira, Kei Takeuchi, Non-Regular Statistical Estimation. vii, 183 pp., 1995.

108: Wesley L. Schaible (Editor), Indirect Estimators in U.S. Federal Programs. viii, 195 pp., 1995.

109: Helmut Rieder (Editor), Robust Statistics, Data Analysis, and Computer Intensive Methods. xiv, 427 pp., 1996.

110: D. Bosq, Nonparametric Statistics for Stochastic Processes. xii, 169 pp., 1996.

111: Leon Willenborg, Ton de Waal, Statistical Disclosure Control in Practice. xiv, 152 pp., 1996.

112: Doug Fischer, Hans-J. Lenz (Editors), Learning from Data. xii, 450 pp., 1996.

113: Rainer Schwabe, Optimum Designs for Multi-Factor Models. viii, 124 pp., 1996.

114: C.C. Heyde, Yu. V. Prohorov, R. Pyke, and S. T. Rachev (Editors), Athens Conference on Applied Probability and Time Series Analysis Volume I: Applied Probability In Honor of J.M. Gani. viii, 424 pp., 1996.

115: P.M. Robinson, M. Rosenblatt (Editors), Athens Conference on Applied Probability and Time Series Analysis Volume II: Time Series Analysis In Memory of E.J. Hannan. viii, 448 pp., 1996.

116: Genshiro Kitagawa and Will Gersch, Smoothness Priors Analysis of Time Series. x, 261 pp., 1996.

117: Paul Glasserman, Karl Sigman, David D. Yao (Editors), Stochastic Networks. xii, 298, 1996.

118: Radford M. Neal, Bayesian Learning for Neural Networks. xv, 183, 1996.

119: Masanao Aoki, Arthur M. Havenner, Applications of Computer Aided Time Series Modeling. ix, 329 pp., 1997.

120: Maia Berkane, Latent Variable Modeling and Applications to Causality. vi, 288 pp., 1997.

121: Constantine Gatsonis, James S. Hodges, Robert E. Kass, Robert McCulloch, Peter Rossi, Nozer D. Singpurwalla (Editors), Case Studies in Bayesian Statistics, Volume III. xvi, 487 pp., 1997.

122: Timothy G. Gregoire, David R. Brillinger, Peter J. Diggle, Estelle Russek-Cohen, William G. Warren, Russell D. Wolfinger (Editors), Modeling Longitudinal and Spatially Correlated Data. x, 402 pp., 1997.

123: D. Y. Lin and T. R. Fleming (Editors), Proceedings of the First Seattle Symposium in Biostatistics: Survival Analysis. xiii, 308 pp., 1997.

124: Christine H. Müller, Robust Planning and Analysis of Experiments. x, 234 pp., 1997.

125: Valerii V. Fedorov and Peter Hackl, Model-oriented Design of Experiments. viii, 117 pp., 1997.

126: Geert Verbeke and Geert Molenberghs, Linear Mixed Models in Practice: A SAS-Oriented Approach. xiii, 306 pp., 1997.

127: Harald Niederreiter, Peter Hellekalek, Gerhard Larcher, and Peter Zinterhof (Editors), Monte Carlo and Quasi-Monte Carlo Methods 1996, xii, 448 pp., 1997.

128: L. Accardi and C.C. Heyde (Editors), Probability Towards 2000, x, 356 pp., 1998.

129: Wolfgang Härdle, Gerard Kerkyacharian, Dominique Picard, and Alexander Tsybakov, Wavelets, Approximation, and Statistical Applications, xvi, 265 pp., 1998.

130: Bo-Cheng Wei, Exponential Family Nonlinear Models, ix, 240 pp., 1998.

131: Joel L. Horowitz, Semiparametric Methods in Econometrics, ix, 204 pp., 1998.

132: Douglas Nychka, Walter W. Piegorsch, and Lawrence H. Cox (Editors), Case Studies in Environmental Statistics, viii, 200 pp., 1998.

133: Dipak Dey, Peter Müller, and Debajyoti Sinha (Editors), Practical Nonparametric and Semiparametric Bayesian Statistics, xv, 408 pp., 1998.

134: Yu. A. Kutoyants, Statistical Inference For Spatial Poisson Processes, vii, 284 pp., 1998.

135: Christian P. Robert, Discretization and MCMC Convergence Assessment, x, 192 pp., 1998.

136: Gregory C. Reinsel, Raja P. Velu, Multivariate Reduced-Rank Regression, xiii, 272 pp., 1998.

137: V. Seshadri, The Inverse Gaussian Distribution: Statistical Theory and Applications, xi, 360 pp., 1998.

138: Peter Hellekalek, Gerhard Larcher (Editors), Random and Quasi-Random Point Sets, xi, 352 pp., 1998.

139: Roger B. Nelsen, An Introduction to Copulas, xi, 232 pp., 1999.

140: Constantine Gatsonis, Robert E. Kass, Bradley Carlin, Alicia Carriquiry, Andrew Gelman, Isabella Verdinelli, Mike West (Editors), Case Studies in Bayesian Statistics, Volume IV, xvi, 456 pp., 1999.

141: Peter Müller, Brani Vidakovic (Editors), Bayesian Inference in Wavelet Based Models, xi, 394 pp., 1999.

142: György Terdik, Bilinear Stochastic Models and Related Problems of Nonlinear Time Series Analysis: A Frequency Domain Approach, xi, 258 pp., 1999.

143: Russell Barton, Graphical Methods for the Design of Experiments, x, 208 pp., 1999.

144: L. Mark Berliner, Douglas Nychka, and Timothy Hoar (Editors), Case Studies in Statistics and the Atmospheric Sciences, x, 208 pp., 2000.

145: James H. Matis and Thomas R. Kiffe, Stochastic Population Models, viii, 220 pp., 2000.

146: Wim Schoutens, Stochastic Processes and Orthogonal Polynomials, xiv, 163 pp., 2000.

147: Jürgen Franke, Wolfgang Härdle, and Gerhard Stahl, Measuring Risk in Complex Stochastic Systems, xvi, 272 pp., 2000.

148: S.E. Ahmed and Nancy Reid, Empirical Bayes and Likelihood Inference, x, 200 pp., 2000.

149: D. Bosq, Linear Processes in Function Spaces: Theory and Applications, xv, 296 pp., 2000.

150: Tadeusz Caliński and Sanpei Kageyama, Block Designs: A Randomization Approach, Volume I: Analysis, ix, 313 pp., 2000.

151: Håkan Andersson and Tom Britton, Stochastic Epidemic Models and Their Statistical Analysis: ix, 152 pp., 2000.

152: David Ríos Insua and Fabrizio Ruggeri, Robust Bayesian Analysis: xiii, 435 pp., 2000.

153: Parimal Mukhopadhyay, Topics in Survey Sampling, x, 303 pp., 2000.

154: Regina Kaiser and Agustín Maravall, Measuring Business Cycles in Economic Time Series, vi, 190 pp., 2000.

155: Leon Willenborg and Ton de Waal, Elements of Statistical Disclosure Control, xvii, 289 pp., 2000.

156: Gordon Willmot and X. Sheldon Lin, Lundberg Approximations for Compound Distributions with Insurance Applications, xi, 272 pp., 2000.

157: Anne Boomsma, Marijtje A.J. van Duijn, and Tom A.B. Snijders (Editors), Essays on Item Response Theory, xv, 448 pp., 2000.

158: Dominique Ladiray and Benoît Quenneville, Seasonal Adjustment with the X-11 Method, xxii, 220 pp., 2001.